AF891304

Production Economics

Evaluating Costs of Operations in Manufacturing and Service Industries

Industrial Engineering: Challenges in Production and Service Industries

Series Editor
Anil Mitel

Production Economics: Evaluating Costs of Operations in Manufacturing and Service Industries

Anoop Desai and Aashi Mital

Production Economics

Evaluating Costs of Operations in Manufacturing and Service Industries

Anoop Desai
Aashi Mital

CRC Press
Taylor & Francis Group
Boca Raton London New York

CRC Press is an imprint of the
Taylor & Francis Group, an **informa** business

CRC Press
Taylor & Francis Group
6000 Broken Sound Parkway NW, Suite 300
Boca Raton, FL 33487-2742

© 2018 by Taylor & Francis Group, LLC
CRC Press is an imprint of Taylor & Francis Group, an Informa business

No claim to original U.S. Government works

Printed on acid-free paper

International Standard Book Number-13: 978-1-138-03326-9 (Hardback)

This book contains information obtained from authentic and highly regarded sources. Reasonable efforts have been made to publish reliable data and information, but the author and publisher cannot assume responsibility for the validity of all materials or the consequences of their use. The authors and publishers have attempted to trace the copyright holders of all material reproduced in this publication and apologize to copyright holders if permission to publish in this form has not been obtained. If any copyright material has not been acknowledged please write and let us know so we may rectify in any future reprint.

Except as permitted under U.S. Copyright Law, no part of this book may be reprinted, reproduced, transmitted, or utilized in any form by any electronic, mechanical, or other means, now known or hereafter invented, including photocopying, microfilming, and recording, or in any information storage or retrieval system, without written permission from the publishers.

For permission to photocopy or use material electronically from this work, please access www.copyright.com (http://www.copyright.com/) or contact the Copyright Clearance Center, Inc. (CCC), 222 Rosewood Drive, Danvers, MA 01923, 978-750-8400. CCC is a not-for-profit organization that provides licenses and registration for a variety of users. For organizations that have been granted a photocopy license by the CCC, a separate system of payment has been arranged.

Trademark Notice: Product or corporate names may be trademarks or registered trademarks, and are used only for identification and explanation without intent to infringe.

Library of Congress Cataloging-in-Publication Data

Names: Desai, Anoop, author. | Mital, Aashi, author.
Title: Production economics : evaluating costs of operations in manufacturing and service industries / Anoop Desai and Aashi Mital.
Description: Boca Raton : Taylor & Francis, CRC Press, 2018. | Series: Industrial engineering : challenges in production and service industries | Includes bibliographical references and index.
Identifiers: LCCN 2018011968| ISBN 9781138033269 (hardback : alk. paper) | ISBN 9780429487040 (ebook)
Subjects: LCSH: Manufacturing industries–Costs. | Service industries–Costs. | Production engineering–Estimates.
Classification: LCC TS165 .D47 2018 | DDC 657/.867–dc23
LC record available at https://lccn.loc.gov/2018011968

Visit the Taylor & Francis Web site at
http://www.taylorandfrancis.com

and the CRC Press Web site at
http://www.crcpress.com

Single Payment

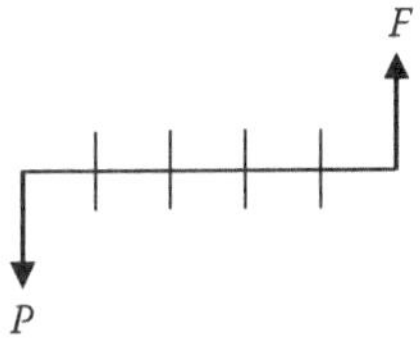

Compound Amount:

To Find F
Given P

$(F/P, i, n)$

$$F = P(1+i)^n$$

Present Worth:

To Find P
Given F

$(P/F, i, n)$

$$P = F(1+i)^{-n}$$

Uniform Series

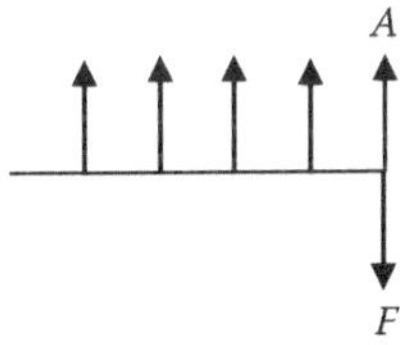

Series Compound Amount:

To Find F
Given A

$(F/A, i, n)$

$$F = A\left[\frac{(1+i)^n - 1}{i}\right]$$

Sinking Fund:

To Find A
Given F

$(A/F, i, n)$

$$A = F\left[\frac{i}{(1+i)^n - 1}\right]$$

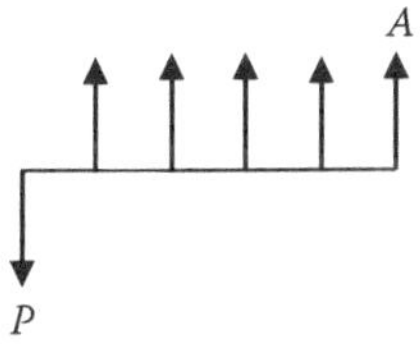

Capital Recovery:

To Find A
Given P

$(A/P, i, n)$

$$A = P\left[\frac{i(1+i)^n}{(1+i)^n - 1}\right]$$

Series Present Worth:

To Find P
Given A

$(P/A, i, n)$

$$P = A\left[\frac{(1+i)^n - 1}{i(1+i)^n}\right]$$

Arithmetic Gradient

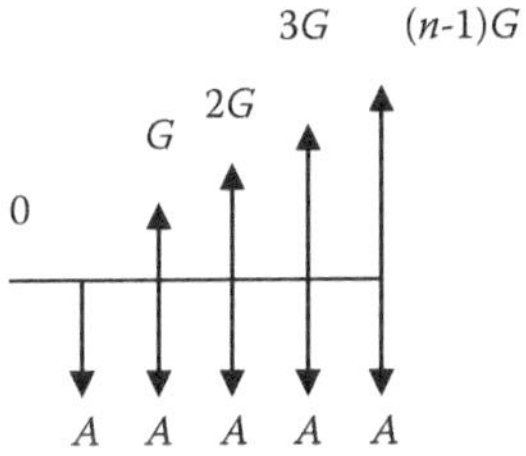

Arithmetic Gradient Uniform Series:

To Find A
Given G

$(A/G, i, n)$

$$A = G\left[\frac{(1+i)^n - in - 1}{i(1+i)^n - i}\right]$$

or

$$A = G\left[\frac{1}{i} - \frac{n}{(1+i)^n - 1}\right]$$

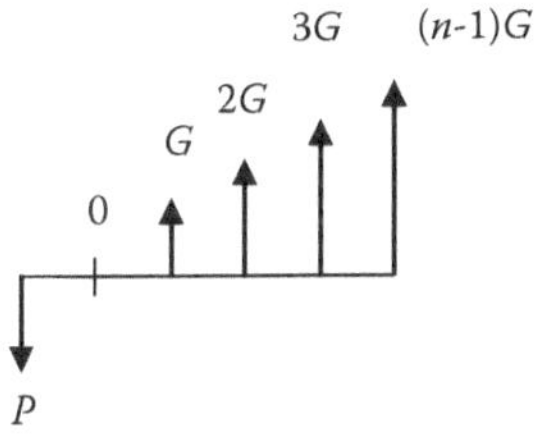

Arithmetic Gradient Present Worth:

To Find P
Given G

$(A/G, i, n)$

$$P = G\left[\frac{(1+i)^n - in - 1}{i^2(1+i)^n}\right]$$

Geometric Gradient

Geometric Series Present Worth

To Find *P*	(*P*/*A*, g, I, n)	$P = A_1\left[n(1+i)^{-1}\right]$
Given A_1, g	When $i = g$	
To Find P	(*P*/*A*, g, *i*, n)	$P = A_1\left[\dfrac{1-(1+g)^n(1+i)^n}{i-g}\right]$
Given A_1, g	When $i = g$	

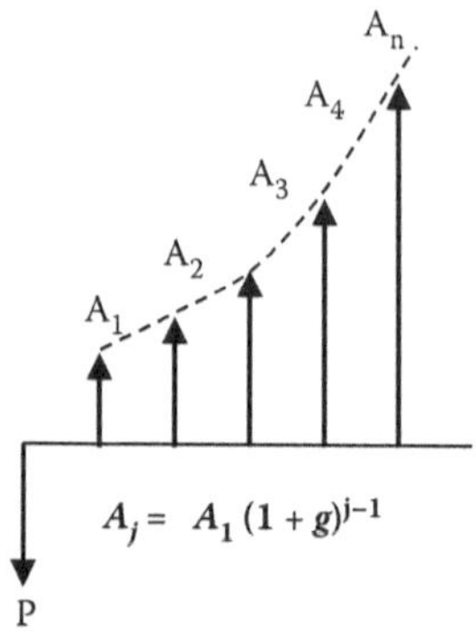

Compound Interest

i = Interest rate per interest period.

n = Number of interest periods.

P = The present sum of money.

F = The future sum of money. The future sum F is an amount, n interest periods from the present, that is equivalent to P with interest rate i.

A = The end-of-period cash receipt or disbursement in a uniform series continuing for n periods, the entire series equivalent to P or F with interest rate i.

G = Uniform period-by-period increase or decrease in cash receipts or disbursements; the arithmetic gradient.

g = Uniform *rate* of cash flow increase or decrease from period to period; the geometric gradient.

r = Nominal interest rate per interest period.

m = Number of compounding sub periods per period.

In most cases the interest period id one year. The period under consideration could be something else.

To Compute	Excel Formula
P	-PV(i, n, A, F, Type)
A	-PMT(i, n, P, F, Type)
F	-FV(i, n, A, P, Type)
n	NPER(i, A, P, F, Type)
i	RATE(n, A, P, F, Type, guess)
P	NPV(i, CF_1:CF_n)
i	IRR (CF_0:CF_n)

While we would love to dedicate this book to our loved ones, we ask for their forgiveness. Socrates stated that the only true wisdom is in knowing nothing. Thus, we hope that through this book, you end up learning something. This book is dedicated to you and your pursuits, without which the world is idle.

Contents

Part II Decision Making

Preface

Production engineering encompasses manufacturing and is a more global term. While it is generally defined as the process of converting raw materials into finished products by subjecting the former to certain value adding processes and operations, it is also used to represent output from processing industries such as the petrochemical industry, pharmaceutical industry, and mining industry. Since the output (production) from manufacturing industry represents a preponderance of industrial output, the word "production" in this book is analogous to the word "manufacturing." The emphasis of this book, therefore, is on manufacturing industry economics.

It is well known that manufacturing (and thus, production) creates wealth. It accomplishes this by creating value. Wealth creation is accomplished through creation of equity and profits (for the owner), creation of jobs (for the employee), and creation of value (for the end customer) not to mention the profit generated for the retailer through price markups. This principle can be substantiated by carefully examining the ascent to wealth of numerous countries globally that have successfully trodden down this path in the past. One only needs to look at the definitely successful and arguably breathtaking economic ascent to global wealth and power of countries such as Japan, South Korea, and more recently, China. India is currently planning a trek down a similar path by offering to serve as the global manufacturing and production hub of the future.

Beginning in the mid-1990s, US manufacturing jobs were lost to foreign competition en-masse primarily due to low wages offered by competitors. However, the manufacturing industry in the United States has recently begun a long awaited revival. But it is not 'your dad's manufacturing.' There is more to manufacturing than metals processing. Modern manufacturing is a function of traditional manufacturing methods and a host of new techniques such as flexible manufacturing systems, nanomanufacturing, and additive manufacturing to name a few. It also includes automation which has come to play an increasingly important role in the modern manufacturing economy. Automation is evident on the shop floor in the form of CNC, robotics, automated material handling systems, machine vision systems, and so on.

Through this book, we seek to introduce to students the economics of modern production methods. The focus is on examination of production and manufacturing techniques and methods from the cost perspective. The book seeks to quantify various costs encountered in a manufacturing setting. It deals with computation of direct and indirect costs for production and manufacturing operations as well as a variety of overhead operations included in such an environment. Costing of manufacturing methods such as casting, forging, turning, milling, and welding is addressed along with inventory analysis and costing. Related topics such as equipment replacement, comparison of alternatives, depreciation, buy versus make decisions, interest factors, and equivalence are addressed as well.

Each chapter presents basic information on the topic under consideration. The focus is on rigorous problem solving. It contains a variety of solved problems as well as about 15–20 unsolved problems for practice. This book is suitable for students taking courses such as Engineering Economy, Cost Analysis, Cost Engineering, and Cost Estimating. It will also be useful to students who plan to take the Fundamentals of Engineering (FE) Exam. Engineering Economy is about 10% of the FE exam and all topics related to engineering economy are covered in this book. This book packs significantly more value than any book on engineering economy on the market today.

The book is split into two parts. Part I deals with manufacturing costing only whereas Part II introduces techniques that can be used in managerial decision making to optimize yield and maximize value.

In Part I, the concept of the time value of money is discussed at length in Chapter 1. The importance of cost estimation and various cost estimation techniques are presented in Chapter 2. Selection of different materials and their costing is dealt with in Chapter 3. Chapter 4 addresses the all-important topic of process costing. In this chapter, we have addressed costing of traditional manufacturing processes such as turning, milling, casting, forging, shaping, and welding. We have also presented costing methodologies for additive manufacturing in this chapter. Overhead expenses are discussed in Chapter 5. The topic of allocating resources to capital equipment and operating costs are addressed in Chapter 6. Types of material handling systems and relevant costing are dealt with in Chapter 7. The concept of inventory analysis, Just in Time manufacturing, and pull versus push systems of production are presented in Chapter 8. Chapter 9 deals with the topic of administrative and facilities cost. Part I of the book is brought to conclusion by Chapter 10 which deals with computation of selling price.

Different techniques of decision making are presented in Part II. Chapter 11 presents the concept of minimum attractive rate of return and incremental analysis when comparing alternatives. The concept of break even and benefit cost analysis is discussed in Chapter 12. Different methods of depreciation are presented in Chapter 13. The important topic of equipment replacement is addressed in Chapter 14. Chapter 15 deals with the crucial concept of make versus buy analysis. The importance of this concept global economic terms cannot possibly be overemphasized. Chapter 16 discusses issues related to vendor selection and vendor cost verification. A case study examining the role played by various factors such as interest rates, labor rates, equipment reliability, and market factors on choice of assembly method is presented in Chapter 17.

We hope the material presented in this book will prove invaluable to you in your future endeavors.

Anoop Desai
Aashi Mital

Authors

Dr. Anoop Desai is currently an associate professor of Mechanical Engineering at Georgia Southern University. He received his PhD in Industrial Engineering from the University of Cincinnati in 2006. His areas of research interest include Product Development, Design for Manufacturing, Six Sigma and Quality Control, and Engineering Economic Analysis.

Dr. Desai is the author of *Product Development: A Structured Approach to Consumer Product Design, Development and Manufacture, Fundamentals of Work measurement: What Every Engineer Should Know* and *Global Sustainability and Innovation (Edited).*

Dr. Desai is also the author of twenty-five journal papers mainly focusing on the area of Design for X, environment conscious design and manufacturing, and the economics of production processes. He is one of Elsevier Science's most cited authors.

Aashi Mital holds a MA in American History and minor in Classical Archaeology through the University of Cincinnati. She is the CEO and founder of Pivotal Solutions Consulting. Her innovative methodologies involving resource utilization, strategic initiatives, and impact growth have caused several businesses to reinvent their approaches in today's multidisciplinary environment. Being among the most sought out consultants in the region, she is known for her ability to generate and execute strategies to improve operations and processes, critical thinking, problem solving, and a deeper understanding of multifaceted problems.

She has a strong background in resource management, community development, and marketing, all of which have contributed to publications covering a sundry of interdisciplinary topics, such as geopolitical patterns and historical forecasting, business and engineering economics, and human productivity and sustainability.

Some of her latest works include *Product Development: A Structured Approach to Consumer Product Development, Design and Manufacture,* as well as chapters "Changes in US Manufacturing Worker Productivity with Working Hours" and "A comparison of US Manufacturing Workers' Productivity with Similar Workers in Selected Developed and Fast Developing Economies" found within *Human Work Productivity: A Global Perspective.* She has also worked with various publications, such as International Journal of Industrial Engineering: Theory, Applications and Practice (IJIETAP), the Organization of American Historians (OAH) and Archaeological Magazine.

Ms. Mital is a recipient of the Craig Waff Researcher's Award, CPA Innovative STEM Educator's Award, and MHS Preservation Award.

Part I

Manufacturing Costing

1

Interest Factors, Equivalence, Present Worth, Annualized Equivalent Worth, and Future Worth Analysis

1.1 What Is Time Value of Money?

Money has value over time. If you were asked, "Would you rather have $10,000 today or $10,000 fifteen years from now?" which would you choose? When all things are considered, it is obvious. A sum of $10,000 today has more value than the same nominal sum of money fifteen years from now. This concept is intuitive to most people. It is easy to appreciate that $10,000 would have a greater amount of purchasing power today than it would fifteen years in the future.

The value of money over time is expressed by the term "rate of interest" or "interest rate" and is commonly used in cash flow analysis. Cash flow analysis deals with cash flows that occur over extended periods of time. Given the fact that money has value over time, the nominal values of those cash flows cannot simply be added together or subtracted. The equivalent value of that cash flow needs to be arrived at by discounting it over time. The net worth can then be computed by assimilating all the equivalent values together. In this chapter, we shall look at how this done.

The chapter will cover the following concepts in detail:

- Interest: Simple interest and compound interest
- Time value of money
- Present worth analysis
- Annualized equivalent worth
- Future worth analysis

1.2 What Is Interest?

Conceptually, interest can be defined as the arithmetic difference between the final amount of money and the beginning amount of money. Money is a valuable commodity. Every time money is lent to a borrower, the lender charges a percentage as interest to assure the safe return of the original sum (also referred to as principal), along with a charge to "rent" the money. Thus, interest can be looked upon as interest paid and interest earned.

Think about it. If you approach a bank to borrow money, you expect to pay the principle, plus an additional amount of money as *interest paid* (that is dependent on the term of loan and the interest rate). On the other hand, if you lend money to the bank by investing in certificates of deposit (CDs), you expect to get back your principal plus a certain amount in the form of *interest earned*. Thus, interest paid on borrowed funds can be described by a simple equation as follows:

$$\text{Interest} = \text{Amount owed at time "t"} - \text{Principal} \qquad \text{(Equation 1)}$$

Interest paid over a specific time unit is expressed as a percentage of the principal, also referred to as the interest rate.

$$\text{Interest rate (\%)} = \frac{\text{Interest accrued over time}}{\text{Principal}} \times 100 \qquad \text{(Equation 2)}$$

The unit of time over which interest has been calculated is referred to as "interest period." One year is the most commonly used interest period. Yet, for shorter maturity loans, the interest period could be expressed in terms of months, quarters or even semi-annual time periods. The rule of thumb is that if only the interest rate is given and nothing else, the interest period is automatically assumed to be one year. For instance, if there is an interest rate of 5%, it is assumed to be over one year. Consider the following example to fully understand the concept of interest.

EXAMPLE 1.1

John purchased a car for $15,000. In order to buy the car, John borrowed the money from his favorite bank. He must pay $17,000 one year later in order to repay the loan. Determine the amount of interest paid. Also, what is the rate of interest charged by the bank?

Solution

The amount of money borrowed is termed as the principal. In the present case, this amount is $15,000.

The final amount to be repaid to the bank at the end of one year is $17,000.

Using Equation 1, the amount of interest paid = $17,000 – $15,000 = $2000.

We use Equation 2 to compute the interest rate as follows:

$$\text{Interest rate (\%)} = \frac{2000}{15{,}000} \times 100 = 13.33\% \text{ per year} \qquad \textbf{Solution}$$

It should be borne in mind that inflation can have a harmful effect on the purchasing power of money and can raise the interest rate. Inflation represents decreased purchasing power of a currency, which is often combated by rising interest rates.

In the world of investments, government bonds are perhaps the safest debt instruments that investors can purchase. Buying a government bond implies that the investor is lending a certain sum of money to the government at a specific interest rate. Most government bonds have a 3%–4% real rate of return implicit in their interest rates. This means that a 7.5% interest rate on a bond points to inflation expectations in the 4.5%–3.5% range going forward.

1.3 Terminology

Throughout this book, we will utilize concepts that have a foundational basis in the time value of money. To that end, it is necessary to introduce the fundamental terms that are used in economic analysis. They are discussed in this section.

P: This is referred to as present worth (PW) or present value (PV). The present worth is the value of money at the present moment (designated as t = 0). The present worth or present value, being a measure of money, is designated in terms of dollars.

F: This is referred to as future worth (FW) or future value (FV). The future worth is the value of money at some time in the future moment (designated as t = n, the value of "n" will be provided in the problem statement). The future worth or future value, once again, being a measure of money, is designated in terms of dollars.

n: "n" is the number of interest periods. This, being a measure of time, is designated in terms of years or months or quarters, etc.

i: "i" is the rate of interest and is designated as a percentage value. It is generally measured on an annual basis unless specified otherwise.

t: "t" is time and is specified in terms of years or months, quarters, etc.

A: "A" is a series of consecutive amounts of cash flow that occur at the end of each time period. In order to qualify as a series with a value of "A," the cash flows need to have the same nominal value every time period and the same direction. This is important to remember. Thus, a cash flow series can be said to exist if the cash flows have the same nominal value every year and are all either positive or negative, but not both simultaneously.

1.3.1 Simple Interest

Simple interest is always computed ONLY on the original principal, never on the final lump sum. Keep in mind that this type of interest is used very rarely. Using the aforementioned terminology, simple interest will always be computed on the value of "P" only. The formula for computing simple interest on a principal "P" is presented in Equation 3:

$$\text{Amount of simple interest earned} = P \times i \times n. \quad \text{(Equation 3)}$$

The future value at the end of "n" years can be computed by first computing the total amount of simple interest earned on a principal of "P" at an interest rate of "i" over "n" years using Equation 3. This amount is then added to the original principal "P" in order to arrive at the future value. Thus, the future value using simple interest is computed using Equation 4.

$$\text{Future value (F)} = P + Pin. \quad \text{(Equation 4)}$$

This can be rewritten as follows:

$$F = P(1 + in). \quad \text{(Equation 5)}$$

In order to more clearly understand the concept of simple interest, consider the following example.

EXAMPLE 1.2

Steve loans Jack $10,000 for four years with an interest rate of 5%. Steve and Jack are very good friends, so Steve decides to charge only simple interest on his loan to Jack. How much money will Jack pay at the end of four years? How much interest will Steve earn on his money?

Solution

In the present case, the present value of money denoted by P is equal to $10,000.

The interest rate is 5%, thus the value of $i = 5\% = 0.05$.

Since the loan has been made for a period of four years, the value of $n = 4$ years.

In order to compute the future value, we use Equation 4.

$$\text{Future value (F)} = P(1 + in)$$

Substituting the values of P, i, and n in the equation, we obtain the value of F as follows:

$$F = \$10{,}000(1 + (0.05 \times 4))$$

Thus, $F = \$10{,}000(1 + 0.2) = \$10{,}000(1.2) = \$12{,}000$. This is the amount that Jack has to pay back to Steve.

Table 1.1
Computation of Future Worth Based on Compound Interest

Year	Present Value ($)	Interest ($)	Future Value ($)
1	10,000	500	10,500
2	10,000	500	11,000
3	10,000	500	11,500
4	10,000	500	12,000: Final answer

The total amount of interest earned by Steve is simply the difference between the future value of the money and the present value. Thus, the amount of interest earned is given by: $12,000 – 10,000 = $2000 over four years. This being simple interest, the amount equates to $500 per year.

Alternatively, the amount of interest earned can also be computed by substituting the respective values in Equation 3. Thus, the interest earned over four years is equal to $P \times i \times n = \$10{,}000 \times 0.05 \times 4 = \$2000$ over four years.

This amount is in agreement with the amount arrived at by subtracting the present value from the future value. Thus, both methods of computation are equally valid and the solution is robust. Most of the problems associated with the concepts presented in this chapter can be solved using several approaches all of which are equally sound. It does not matter what approach is used to solve any given problem, just as long as it is steeped in basic logic. Table 1.1 illustrates the amount of interest charged per year and the resulting worth at the end of that year. Please study this table carefully. It will be used to draw a comparison with compound interest when we approach that topic.

1.3.2 Compound Interest

Compound interest uses the basic concept of compounding. This is the type of interest is the most commonly used. In this method, any balance that is owed but not paid back is added to the principal. In the previous example, the interest of $500 that was owed but not paid back at the end of the first is added to the principal in order to compute the present value at the beginning of the second year. Thus, the principle of compound interest can be likened to "interest upon interest" compounding. The following formulas are used to compute future value and interest earned using compound interest.

$$\text{Future value (F)} = P \times (1+i)^n. \qquad \text{(Equation 6)}$$

Interest earned can always be computed by deducting the present value from the future value.

1.3.3 Compound Interest: The Most Powerful Force in the Universe

Einstein said that compound interest was the most powerful force in the universe. Investment stalwart Warren Buffett repeatedly touts the power of compound interest, especially over a long period of time. He often attributes his investment success to this power and his ability to compound healthy rates of return over extended periods of time. Let us consider an example to illustrate this latent power.

EXAMPLE 1.3

Consider a situation in which you, as a student, are negotiating your salary at the end of a series of grueling job interviews. Assuming that you have pretty good to strong powers of negotiation, you make the interviewer an offer as follows: You will accept one penny for the first day of actual work. The second day you work, you will get two pennies. The third day, you will get four pennies and so on. Assuming that you only get paid for the number of days you actually work, how much would you take home at the end of the year? What would be the sum total of your salary at year end?

Solution

We can use the formula given in Equation 6 to solve this problem. Insofar as the value of P is concerned, it is given to be one penny. This is the amount of money we start with. The rate of interest is obviously 100% since the salary doubles every day. In order to compute the number of working days, this is calculated as follows:

The total number of weeks in one year = 52 weeks.
Number of working days per week = 5 days.

Thus, the theoretical maximum number of working days = 52 weeks × 5 days/week = 260 days.

Assuming time off for vacations and so on, the actual number of working days could be about 220 days or so per year.

Thus, we have the prerequisites to solve our problem.

$$P = \$0.01;\ i = 100\%;\ n = 260;\ F = ?$$

Substituting these values into Equation 6, we have the future value as follows:

$$F = \$0.01(1 + 1)^{260}$$

This implies that $F = \$0.01(2)^{260}$.
Thus, the future value is equal to $\$1.85 \times 10^{76}$.

This is quite a considerable amount of money. Putting it in perspective, the size of the entire US economy in 2015 as expressed by GDP (Gross Domestic Product) was $\$17.95 \times 10^{12}$.

Let us consider Example 1.2 again, but this time from the point of view of compound interest to further understand the concept of compound interest and appreciate how different it is from simple interest.

EXAMPLE 1.4

Let us assume that Steve and Jack had some sort of falling out and Steve decided to loan the same amount of money (P = \$10,000) to Jack at an interest rate (i = 5%) for a period of four years (n = 4 years), only this time he decides to charge compound interest instead of charging simple interest. Compute the future value that is owed by Jack to Steve as well as the total amount of interest paid by Jack at the end of four years.

Solution

We are already quite familiar with the basic terminology that is used to express the present value (P), the future value (F), the interest rate (i), and the term of the loan (n).

In order to compute the future value using compound interest, we use Equation 6.

Therefore, the future value at the end of the first year (year 1) is given by:

$$F_1 = P \times (1+i)^1 = \$10{,}000 \times (1.05) = \$10{,}500.$$

Since this is compound interest, the future value at the end of the first year becomes the present value at the beginning of the second year. The future value at the end of the second year becomes the present value at the end of the third year and so on.

Future value at the end of the second year (year 2) is computed by:

$$F_2 = P \times (1+i)^2 = \$10{,}000 \times (1.05)^2 = \$11{,}025.$$

The future value at the end of the fourth year (year 4) can be calculated by changing the value of "n" in Equation 6. The remainder of the values will remain unchanged.

Thus,

$$F_4 = P \times (1 + i)^4 = \$10{,}000 \times (1.05)^4 = \$12{,}155.06$$

The interest earned by Steve can be calculated as the difference between the future worth and present worth of the loan. This is given as: \$12,155.06 − \$10,000 = \$2155.06.

The above computations are presented in a tabular format as follows:

Year	Present Value ($)	Interest ($)	Future Value ($)
1	10,000	500	10,500
2	10,500	525	11,025
3	11,025	551.25	11,576
4	11,576	578.8125	12,155: Final answer

It is obvious by comparing the two tables that Steve stands to earn more by charging compound interest instead of simple interest.

In the next section, we will consider a tool that is useful when computing the present value, future value of money, and equivalence. The understanding of the time value of money is facilitated by means of a visual tool, often referred to in the parlance of finance as a cash flow diagram.

1.4 Cash Flow Diagram

At its core, a cash flow diagram is a depiction of future cash flows over time. Cash flows can be positive and negative depending on the perspective. Assume that a customer bought a car and financed it by taking a loan from the bank. The monthly payment on the loan is $250. From the point of view of the customer, this cash flow is obviously negative because it takes money away from them. On the other hand, the bank sees the same cash flow as positive because it puts money into their account. Thus, any situation can be generalized by characterizing any outflow of funds as negative cash flow and any inflow of funds as positive cash flow. It is important to bear in mind that because cash flows occur over time, they cannot just simply be added or subtracted when they occur at different points in time. This becomes clear when you consider the cash flows in regard to the time value of money as explained in the previous section. In order to add or subtract cash flows that occur at different points in time, their values needs to be adjusted based on the interest rate environment (rate of return environment). A graphical depiction of cash flows over time is referred to as a cash flow diagram.

Every cash flow diagram starts at point 0 (the present moment) and proceeds further. Time is represented on the X axis and cash flows on the Y axis. Thus, positive cash flows are depicted above the X axis and negative cash flows are depicted below the X axis. A simple cash flow diagram is depicted in Figure 1.1.

The above cash flow diagram represents cash flows that are expected to occur over four years in a 5% interest rate environment. The following observations can be made based on our previous discussion:

- Time, being represented on the X axis, starts at point 0 (now).
- The first year starting now ends at point 1 wherein a positive cash flow of $100 is expected to occur.

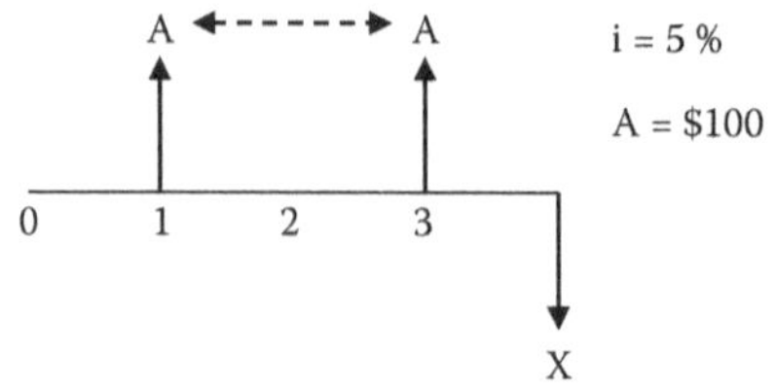

FIGURE 1.1

- Points 2, 3, and 4 on the X axis represent the ends of years 2, 3, and 4, respectively.
- A positive cash flow in the amount of $100 each is expected at the ends of years 1, 2, and 3.
- A negative cash flow in the amount of $100 is expected to occur at the end of year 4.

Now that the basic concept of a cash flow diagram has been described, we shall look the following concepts in the next section:

- Uniform series of cash flows and derivation of the formula to compute present value of a uniform series of cash flows
- Geometric series of cash flows and derivation of the formula to compute present value of a uniform series of cash flows
- Continuous compounding
- Nominal and effective interest rates

1.4.1 Uniform Series of Cash Flows

A uniform cash flow is defined as a fixed amount of nominal cash flow that occurs over "n" number of time periods. In most real life situations, cash flows are measured annually. In those cases, they are referred to as a uniform annual series. This, however, is not a hard and fast rule. Cash flows can be measured over months, quarters or any other consistent time frame. In either case, a uniform series is denoted by the symbol "A". In order to qualify as a uniform series, cash flows should exhibit the following characteristics:

1. Each cash flow should have the same *nominal* value.
2. Each cash flow should have the same direction. A uniform series can be comprised of either positive or negative cash flows, but not both simultaneously. Thus, an alternate positive and negative cash flow of $100 cannot be regarded as a uniform series.
3. Each cash flow in the series should happen at the end of *every* time period, be it years, quarters, months or days. A series of cash flows that occurs at the end of the first three years, skips the next two years, and resumes for the last two years cannot be characterized as a uniform series that lasts over seven years. It could, however, be considered to be two separate uniform series, the first over the first three years and the second over the last two years.
4. Regardless of the time periods and directions of the cash flows, if a uniform series is found to be present, the value of each cash flow is denoted by the symbol "A". An example of a uniform series is presented in Figure 1.2. It should be noted that "A" can assume any nominal value, for instance, A = $100 or A = $2000 depending on the cash flow.

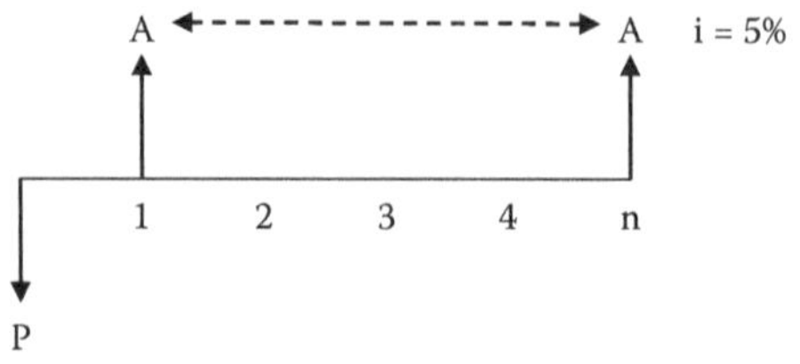

FIGURE 1.2

1.5 Computing the Present Value of a Uniform Series: Uniform Series Present Worth Factor (P/A) and Capital Recovery Factor (A/P)

Consider the uniform series presented in Figure 1.2. The present worth of this series is a function of the interest rate. Thus, cash flows that occur at the ends of years 1 and 2 cannot be added together because they have different present worths. Only the present worth of each of the cash flows can be added together. In order to derive an expression for the present worth of the series, it would be helpful if each value of "A" was considered to be a future worth "F." Using Equation 6, the present worth of an "F" can be computed as follows:

$$P = \frac{F}{(1+i)^n}. \qquad \text{(Equation 7)}$$

Substituting "A" for "F" we get the following equation:

$$P = \frac{A}{(1+i)^n}. \qquad \text{(Equation 8)}$$

Adding together a series over "n" years, we would obtain the following expression:

$$P = A\left[\frac{1}{(1+i)} + \frac{1}{(1+i)^2} + \frac{1}{(1+i)^3} + \frac{1}{(1+i)^4} + \cdots + \frac{1}{(1+i)^{(n-1)}} + \frac{1}{(1+i)^n}\right]. \qquad \text{(Equation 9)}$$

When $i \neq 0$, Equation 1.9 can be simplified to read as follows:

$$P = A\left[\frac{(1+i)^n - 1}{(i)(1+i)^n}\right]. \qquad \text{(Equation 10)}$$

In the parlance of finance, if A is known and P is to be computed, we use the following formula:

$$P = A(P/A, i, n). \qquad \text{(Equation 11)}$$

The factor P/A is referred to as the Uniform Series Present Worth factor.

Conversely, if P was known and A was unknown, the situation would be reversed and we have Equation 12.

$$A = P(A/P, i, n). \qquad \text{(Equation 12)}$$

The factor A/P is referred to as the Capital Recovery factor. It is algebraically denoted as follows:

$$A = P\left[i\frac{(1+i)^n}{(1+i)^n - 1}\right]. \qquad \text{(Equation 13)}$$

In order to more thoroughly understand the concept of a uniform series, let us consider a few examples.

EXAMPLE 1.5

Consider the cash flow diagram presented in Figure 1.3. Compute the present value of the cash flows.

Solution

The following observations can be made about the data presented in Figure 1.3:

- The cash flow is expected to occur over four years.
- The cash flow starts at year 1 and continues unabated until year 4.
- There is a positive cash flow per year, every year and its nominal value is \$250.
- In light of the observations stated above, the cash flow qualifies as a uniform series. The value of "A" = \$250 per year.
- The interest rate is 4% per year.

By definition, the present value of a cash flow diagram is the sum total of the discounted present value of individual cash flows. We can solve this problem by using Equations 10 and 11 alternatively. Either way, we should arrive at the same answer.

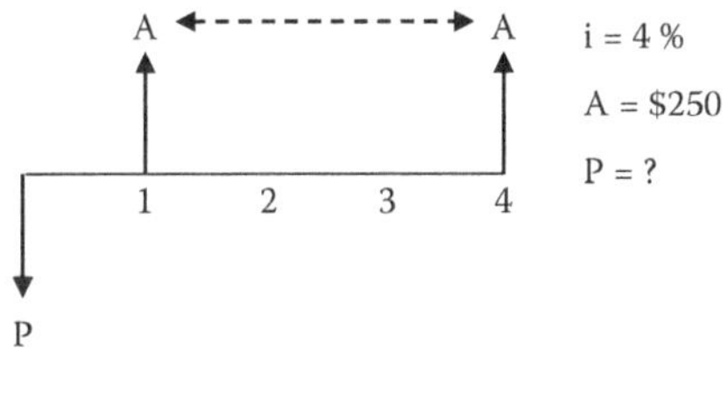

FIGURE 1.3

Method 1: Using Equation 10

The present value of the uniform series can be computed by using Equation 10 as follows:

$$P = A\left[\frac{(1+i)^n - 1}{(i)(1+i)^n}\right]. \qquad \text{(Equation 10)}$$

In the present case, the value of A = \$250 and the value of "i" = 4% = 0.04. Since the series lasts over four years, the value of n = 4 as well. Remember to NEVER substitute the value "i" = 4 in the equation as this would be patently wrong and yield a grossly wrong solution.

Substituting the above values in Equation 10, we obtain the following:

$$P = \$250\left[\frac{(1+0.04)^4 - 1}{(0.04)(1+0.04)^4}\right]$$

Solving the above we obtain the following result:

$$P = \$250\left[\frac{0.16985}{0.04679}\right] = \$907.428$$

Thus, the present value of the uniform series is \$907.428.

Answer method 1

Method 2: Using Equation 11

The problem can also be solved using Equation 11 as follows:

$$P = A(P/A, i, n). \qquad \text{(Equation 11)}$$

Once again, the value of A = \$250, i = 4% = 0.04, and n = 4 years.

The factor (P/A, i, n) is read to mean: "The value of P/A for a specific interest rate "i" and for a certain number of years "n".

This value can be read directly from standard interest rate tables that are available in many textbooks, as well as online.

It is important to look under the correct interest rate to start with. In this case, we look under 4% and the column corresponding to P/A and scroll down to n = 4. Reading that value directly from the table we get (P/A, 4%, 4) = 3.630.

Substituting that value into the original equation we get the following:

$$P = \$250(3.630) = \$907.5$$ **Answer method 2**

It will be observed that both answers match. The slight difference in the decimal values is due to rounding error.

Let us consider an example in which the uniform series starts at a point other that "1". In the above problem, since the series started at "1," it was just a matter of simple substitution to solve the problem. Nevertheless, there are many instances wherein a business may not be able to obtain positive cash flow right off the bat. How can the present value of the uniform series be determined in these types of situations? Consider the following example:

EXAMPLE 1.6

Consider the cash flow diagram presented in Figure 1.4. Compute the present worth of the cash flows.

Solution

The following observations can be made about the data presented in Figure 1.4:

- The cash flow is expected to occur over five years.
- The cash flow starts at year 3 and continues unabated until year 7.
- There is a positive cash flow per year once it starts in earnest, and its nominal value is $1000.
- In light of the above observations, the cash flow does qualify as a uniform series. The value of "A" = $1000 per year.
- The interest rate is 3% per year.

It will be noticed from Figure 1.4 that there is no cash flow, either positive or negative, for the first two years. The series only starts at year 3 and lasts until year 7. *Thus, the value of "n" = 5 NOT 7.* This is a crucial point to remember. *The value of "n" for a uniform series corresponds to the number of time periods over which the series lasts and NOT the total number of time periods in the cash flow diagram.*

Once again, we can solve this problem using both Equations 10 and 1.11. Note that those equations are only applicable for uniform series that start at point 1 ONLY. So, how can we apply those formulas to other situations such as the one presented in this example? Let us take a look:

Method 1: Using Equation 10

Study point to remember:

$$P = A\left[\frac{(1+i)^n - 1}{(i)(1+i)^n}\right] \qquad \text{(Equation 10)}$$

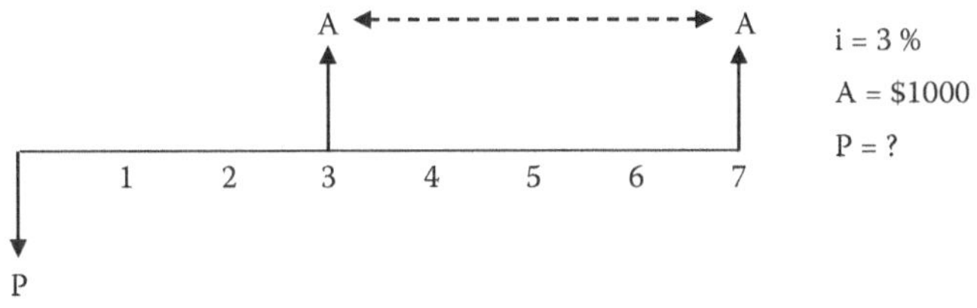

FIGURE 1.4

will always yield the present worth of the uniform series one year BEFORE the series commences.

Thus, in this case, it will yield the worth at the end of year 2 (since the series starts at year 3, one year prior to commencement is year 2).

Substituting the respective values of A, i, and n into the equation, we obtain the value of P as follows:

$$P_2 = \$1000\left[\frac{(1+0.03)^5 - 1}{(0.03)(1+0.03)^5}\right]$$

Thus,

$$P_2 = \$1000\left[\frac{(1.03)^5 - 1}{(0.03)(1.03)^5}\right]$$

This implies that

$$P_2 = \$1000\left[\frac{1.15927 - 1}{0.03477}\right]$$

Thus, $P_2 = \$1000[4.5806] = \$4580.6$.

However, this is not the present worth of the cash flow diagram. This is just the value at the end of year 2. In order to obtain the present worth (value at point 0), we need to discount the above value using the equation:

$$F = P(1 + i)^n$$

Considering the value of P_2 as F (since year 2 occurs in the future with respect to the present (point 0)), we have the following:

$$P_2 = P(1 + i)^2$$

Substituting the value in the equation, we have the present value of the cash flow diagram as follows:

$$P = \frac{P_2}{(1+i)^2}$$

Thus, $P = (\$4580.6/1.0609) = \$4317.65$

The present value of the cash flow diagram presented in Figure 1.4 is \$4317.65 **Solution 1**

Method 2: Using Equation 11

Study point to remember:

$$P = A(P/A, i, n) \qquad \text{(Equation 11)}$$

will always yield the present worth of the uniform series one year BEFORE the series commences.

Once again, as in method 1, it will yield the worth at the end of year 2 (since the series starts at year 3, one year prior to commencement is year 2).

The present value can be obtained by using the following modified equation in keeping with the method used to obtain solution 1.

$$P = A(P/A, i, n)\ (P/F, i, 2)$$

We multiply the (P/A, i, n) factor by the (P/F, i, 2) in order to account for the fact that the series starts at point 2.

We need to refer to the tables for an interest rate of 3%. Next, we read the value of (P/A) for $n = 5$ and we get the following:

$$(P/A, 3\%, 5) = 4.580$$

Also, the value of $(P/F, 3\%, 2) = 0.9426$.

Substituting the above values in the equation: $P = A\ (P/A, i, n)\ (P/F, i, 2)$ we get the following solution:

$$P = \$1000\ (4.580)\ (0.9426) = \$4317.10 \qquad \textbf{Solution 1}$$

Once again, it will be observed that the solutions are an almost perfect match within the margin of error (due to rounding error).

Present worth analysis is most commonly used to value an asset based on its future expected cash flows. In the world of investments, the purchase price of any asset should ideally be as low as possible, but how low is low? Is it zero? In practice, this is not the case. In such situations, the present worth is computed by discounting the value of future cash flows by the expected rate of return. The purchase price should be at most equal to the present worth if not less. The higher the expected rate of return, the lower should be the purchase price.

Let us consider another example with respect to the application of uniform series and present worth calculation.

EXAMPLE 1.7

An asset is expected to produce a regular income of $50,000 per year over a period of 10 years. What price should be paid for the asset if the investor wishes to obtain a rate of return equal to 20% on his investment per year?

Solution

In any word problem, it is always a good idea to first draw the cash flow diagram in order to enable quick visualization of the data. Thus, in the present case, we have the following diagram (Figure 1.5):

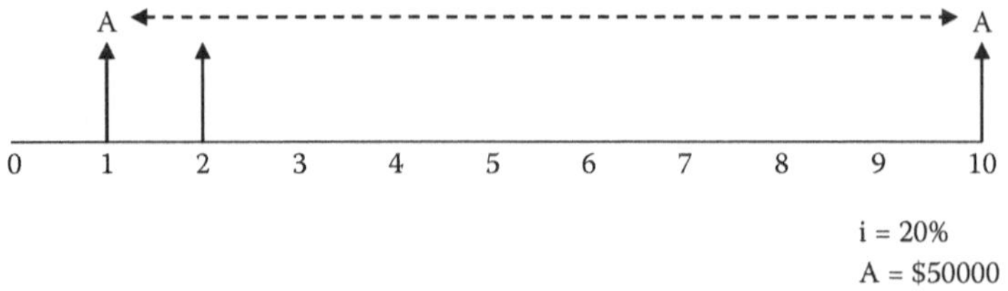

FIGURE 1.5

It will be observed that based on the description of the cash flow provided in the problem, it clearly constitutes a uniform series with a value of A = \$50,000 and n = 10 years. The value of i = 20% as this is the expected rate of return on the investment.

Students often complain about the value of i = 20% being too unrealistic for real world rates. We acknowledge this. As of the writing of this book in 2017, the mortgage rate on a 30 year fixed rate mortgage is about 4%. Yet, the "rate of return" in this problem is not the interest rate. Rather, it is the rate of return expected on an investment, which ideally speaking should be as high as possible.

The price to be paid for the asset should be at most equal to the present value of future cash flows as discounted by the rate of return.

Thus, in this case, purchase price ≤ (Present value of uniform series of \$50,000 per year over 10 years).

The present value of the cash flow is obtained by using Equation 10 as follows:

$$P = A\left[\frac{(1+i)^n - 1}{(i)(1+i)^n}\right]. \qquad \text{(Equation 10)}$$

Here, the pertinent values are as follows:

A = \$50,000
i = 20%
n = 10 years

Substituting the above values in Equation 10, we obtain the following:

$$P = \$50{,}000\left[\frac{(1+0.2)^{10} - 1}{(0.20)(1+0.20)^{10}}\right] \qquad \text{(Equation 14)}$$

Thus,

$$P = \$50{,}000\left[\frac{5.1917}{1.2383}\right]$$

This yields the present value of the cash flow to be equal to \$209,630.13.

This implies that the investor should pay at most an amount equal to $209,630.13 in order to obtain a 20% return on his investment per year.

Solution

It is evident that a lower purchase price would yield a higher rate of return and vice versa.

In the following section, we will examine how to compute the present value of a gradient and derive corresponding equations.

1.6 Computing the Present Value of an Arithmetic Gradient Series: Arithmetic Gradient Factor (P/G) and (A/G)

An arithmetic gradient series can be defined as a series of cash flows that can either increase or decrease by the same amount each period. The amount by which the cash flow changes (either increases or decreases) is referred to as a gradient.

For instance, maintenance on a computer numeric controlled (CNC) work cell could increase by a constant amount of $2500 every year beginning with $2500 for the first year. This implies that the first year maintenance could be $2500 for the first year, $5000 for the second year, $7500 for the third year, and so on. This series of cash flows would constitute an arithmetic gradient series.

In order to derive the equations for arithmetic gradient factors (P/G) and (A/G), consider Figure 1.6 below.

The present worth of the cash flow depicted in Figure 1.6 can be computed by using the following equation:

$$P = G(P/F, i, 2) + 2G(P/F, i, 3) + 3G(P/F, i, 4) + 4G(P/F, i, 5) + \cdots + (n - 2)(P/F, i, (n - 1)) + (n - 1)(P/F, i, n).$$

Substituting the formula for P/F into the above equation and factoring out the term "G," we have the following:

$$P = G\left[\frac{1}{(1+i)^2} + \frac{2}{(1+i)^3} + \frac{3}{(1+i)^4} + \cdots + \frac{(n-2)}{(1+i)^{n-1}} + \frac{(n-1)}{(1+i)^n}\right].$$

(Equation 15)

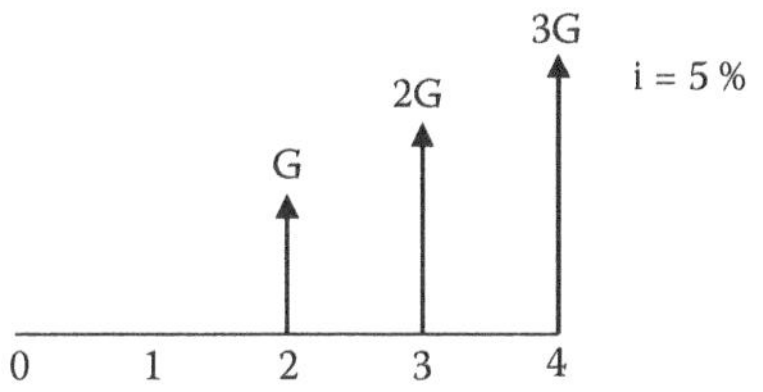

FIGURE 1.6

Multiplying both sides of the above equation by the term (1 + i) gives us the following equation:

$$P(1+i)=G\left[\frac{1}{(1+i)}+\frac{2}{(1+i)^2}+\frac{3}{(1+i)^3}+\cdots+\frac{(n-2)}{(1+i)^n}+\frac{(n-1)}{(1+i)^{n-1}}\right].$$

(Equation 16)

Subtract Equation 13 from Equation 1.14 and we obtain the following:

$$iP=G\left[\frac{1}{(1+i)}+\frac{1}{(1+i)^2}+\frac{1}{(1+i)^3}+\cdots+\frac{1}{(1+i)^n}\right]-G\left[\frac{n}{(1+i)^n}\right].$$

(Equation 17)

Simplifying Equation 15, we obtain the value of (P/G) as follows:

$$(P/G,i,n)=\frac{1}{i}\left[\frac{(1+i)^n-1}{i(1+i)^n}-\frac{n}{(1+i)^n}\right]. \quad \text{(Equation 18)}$$

The equivalent uniform annual series for an arithmetic gradient is computed by multiplying the present worth in Equation 16 by the formula for (A/P, i, n). Thus we obtain the following:

$$(A/G,i,n)=\left[\frac{1}{i}-\frac{n}{(1+i)^n-1}\right]. \quad \text{(Equation 19)}$$

Let us consider a few problems to better understand the application of the formulas we derived in the previous section.

EXAMPLE 1.8

Compute the present worth of the cash flow diagram presented in Figure 1.7.

Solution

A few observations can be made from Figure 1.7 as follows:

- The first year cash flow is equal to $100.

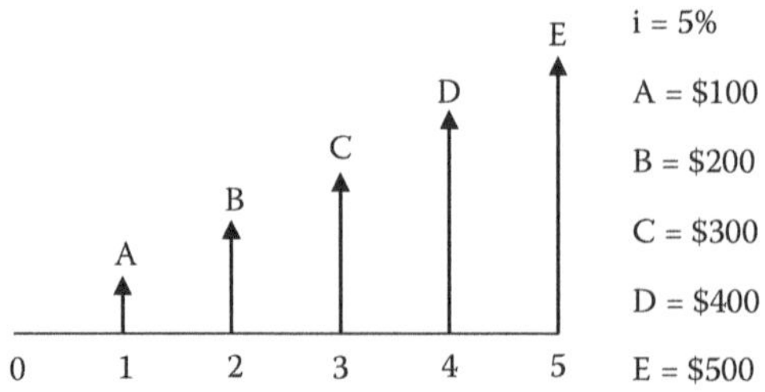

FIGURE 1.7

- The cash flow increases every year by an amount equal to \$100. Thus, the second year cash flow is equal to \$200, the third year yields \$300, and so on all the way to the fifth year.
- The cash flow lasts until year 5.
- The cash flow increases uniformly by an amount equal to \$100 every year and thus can be characterized as an arithmetic gradient series.
- The interest rate is i = 5%.

Based on the above observations, we can conclude that the value of A = \$100, G = \$100, n = 5 years, and i = 5%.

The value of G = \$100 since this is the amount of money by which the cash flow rises every year. The value of A = \$100 since this is the base minimum amount of cash flow over which the gradient (G) is applied. Once again, as with uniform series, since the cash flow lasts for five years, i.e., the aforementioned relationship holds over five years, the value of n = 5 years.

The present value of the cash flow presented in Figure 1.7 is obtained by using the following equation:

$$P = A(P/A, i, n) + G(P/G, i, n). \qquad \text{(Equation 20)}$$

We will use two methods to solve this equation and then compare the answers obtained by both methods to check the veracity of the solution.

Method 1: Using the Formula for (P/A) and (P/G)

From our prior discussion, we have seen that P = \$100; G = \$100, i = 5%, and n = 5 years.

Substituting these values into the actual equation for (P/A) and (P/G), we obtain the following:

$$P = A\left[\frac{(1+i)^n - 1}{(i)(1+i)^n}\right] + G\left[\frac{(1+i)^n - in - 1}{i^2(1+i)^n}\right] \qquad \text{(Equation 21)}$$

Substituting the respective values into Equation 18, we have the following:

$$P = \$100\left[\frac{0.2762}{0.0638}\right] + 100\left[\frac{0.02628}{0.00319}\right]$$

Solving the above equation, we have the present value of the cash flow to be equal to P = \$100 (4.329) + 100 (8.238) = \$1256.72. **Solution 1**

Method 2: Using Values for (P/A) and (P/G) from Interest Rate Tables

We can solve this problem by simply substituting the numeric value for (P/A) and (P/G) from the interest rate tables for i = 5%. We choose i = 5% because that is the interest rate given in the problem.

From the interest rate table for 5% and n = 5 years, we can obtain the value of (P/A) = 4.329 and (P/G) = 8.237.

Substituting these values into Equation 18, we obtain the following:

$$P = \$100\ (4.329) + 100\ (8.237) = \$1256.6 \qquad \textbf{Solution 1}$$

Once again, the solutions are a very good match and if not for the rounding error for the (P/G) factor, would have been a perfect match.

EXAMPLE 1.9

Compute the present worth for the following cash flow diagram (Figure 1.8):

Solution

A few observations can be made from Figure 1.7 as follows:

- The first year cash flow is equal to \$0. The second year cash flow is also equal to \$0.
- The third year cash flow is \$500.
- Thereafter, the cash flow increases every year by an amount equal to \$500. Thus, the fourth year cash flow is equal to \$1000, the fifth year yields \$1500, and so on, all the way to the sixth year.
- The cash flow lasts until year 6.
- The cash flow increases uniformly by an amount equal to \$500 every year after year 3 and thus can be characterized as an arithmetic gradient series. The arithmetic gradient lasts for four years ($A = \$500$, $G = \$500$, $n = 4$).
- Alternatively, the arithmetic gradient can be characterized to exist over five years, in which case, $A = \$0$, $G = \$500$, and $n = 5$ years.
- The interest rate is $i = 3\%$.

We will compute the present worth of both scenarios using interest rate factor values for respective variables. Next, we will check the veracity of the solutions by comparing them.

Method 1: A = \$500, G = \$500, n = 4 years

The present worth for an arithmetic gradient series is computed using Equation 18:

$$P = A(P/A, i, n) + G(P/G, i, n).$$

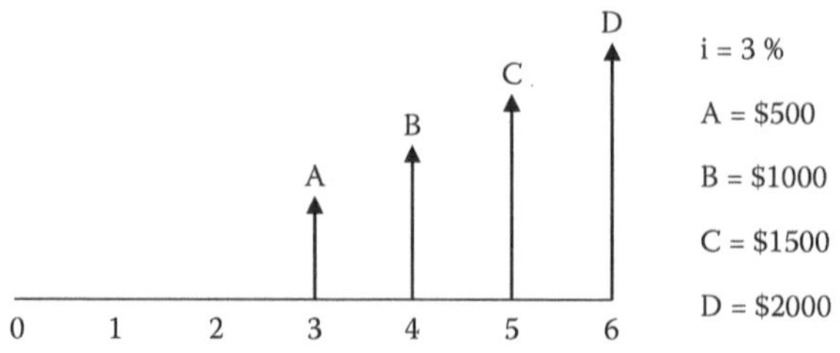

FIGURE 1.8

The values of the (P/A) and (P/G) for n = 4 years and i = 3% are as follows:

$$(P/A, 3\%, 4) = 3.717.$$

$$(P/G, 3\%, 4) = 5.438.$$

Substituting the values in Equation 18, we get the following:

$$P = \$500(3.717) + \$500(5.438) = \$4577.50.$$

This is not the final solution. As with a uniform series, application of the basic formula will only yield the value of the cash flow at the end of the year prior to the beginning of the series. In this case, the series starts at the end of year 3 (denoted by point 3). Thus the above value is the value of the cash flow at the end of year 2 (one year prior to the beginning of the series). It needs to be multiplied by the (P/F) factor for n = 2 years in order to compute the actual present worth.

Thus, we have the following equation:

$$P_2 = \$4577.50$$

$$P = P_2(P/F, 3\%, 2).$$

The value of (P/F, 3%, 2) = 0.9426.

Substituting this value, we obtain the present worth of the cash flow as:

$$P = (\$4577.50)\ (0.9426) = \$4314.75. \qquad \textbf{Solution 1}$$

Method 2: A = \$0, G = \$500, n = 5 years

As before, we use Equation 18 to solve this problem.

The values of the factors are obtained from the interest rate tables as follows:

$$(P/A, 3\%, 5) = 4.580.$$

$$(P/G, 3\%, 5) = 8.889.$$

This scenario has the series starting at point 2. Thus the present worth using Equation 18 will only yield the value at the end of year 1 (one year prior to the beginning of the series). We need to multiply that value by (P/F, i, 1) to compute the true present worth.

The value of (P/F, i, 1) = 0.9709.

Substituting these values, we can compute the present worth as follows:

$$P = [\$0\ (4.580) + \$500\ (8.889)]\ (0.9709)$$

Thus,

$$P = (4444.5)\ (0.9709) = \$4315.16 \qquad \textbf{Solution 2}$$

This problem can also be solved by substituting the values of the variables directly into the respective equations if one so desires. There is no express need to read the values from interest rate tables to solve the problem.

So far we have considered only problems wherein the gradient rises with the passage of time. How do we deal with cases wherein the gradient actually decreases with time? Consider the following example of such a case.

EXAMPLE 1.10

Compute the present worth for the cash flow diagram presented in the following Figure 1.9:

Solution

A few observations can be made from Figure 1.9 as follows:

- The first year cash flow is equal to \$1000.
- Thereafter cash flow decreases every year by an amount equal to \$100. Thus, the second year cash flow is equal to \$900, the third year yields \$800, and so on, all the way to the fourth year.
- The cash flow lasts until year 4.
- The interest rate is i = 5%.

Since the cash flow starts at year 1, this is a relatively straightforward problem with the following variables:

$$A = \$1000, G = -\$100, n = 4 \text{ years}, i = 5\%.$$

Note that when the cash flow decreases with the passage of time, the value of the gradient is negative. The value of "A" will be the highest amount of cash flow from which "chunks" of cash flow are taken out in accordance with the value of the arithmetic gradient.

In order to solve the problem, we substitute the aforementioned values into Equation 18.

Thus,

$$P = \$1000(P/A, i, n) - \$500(P/G, i, n).$$

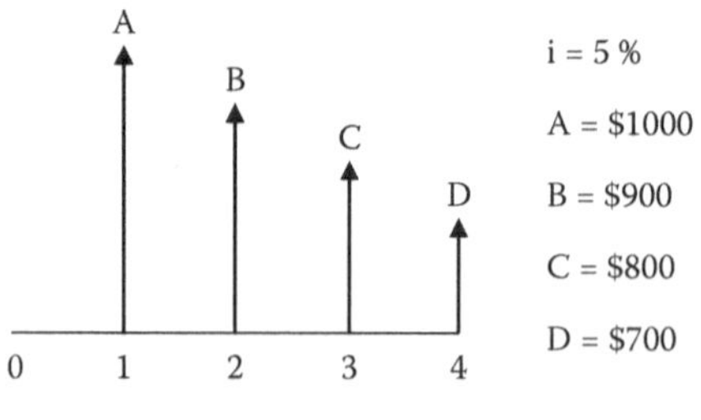

FIGURE 1.9

This gives us the following equation:

$$P = \$1000(P/A, 5\%, 4) - \$500(P/G, 5\%, 4).$$

Substituting values into Equations 10 and 16, we get the following:

$$P = \$1000(3.546) - \$500(5.103).$$

Thus, the present value of the cash flow is $P = \$3546 - \$2551.5 = \$994.5$.

1.7 Computing the Present Value of a Geometric Gradient Series: Arithmetic Gradient Factor (P/G) and (A/G)

A geometric gradient series can be defined as a series of cash flows that can either increase or decrease by the same percentage each period. The percentage by which the cash flow changes (either increases or decreases) is referred to as a gradient.

The following variables need to be defined in order to derive the equation for the present value of a geometric series:

g = Constant rate of change. This is different from the arithmetic gradient in which G was the actual amount of change.

A_1 = Initial cash flow at the end of the first year (defined by point 1 on the cash flow diagram).

P = Present worth of the geometric gradient series.

The present worth of a geometric gradient series can be computed by using the following equation:

$$P = \frac{A_1}{(1+i)} + \frac{A_1(1+g)}{(1+i)^2} + \frac{A_1(1+g)^2}{(1+i)^3} + \cdots + \frac{A_1(1+g)^{n-1}}{(1+i)^n}.$$

Thus,

$$P = A_1\left[\frac{1}{(1+i)} + \frac{1+g}{(1+i)^2} + \frac{(1+g)^2}{(1+i)^3} + \cdots + \frac{(1+g)^{n-1}}{(1+i)^n}\right].$$

The above equation can be further simplified and written in its final form as follows:

$$P = A_1\left[\frac{1-(i+g)^n(1+i)^{-n}}{(i-g)}\right]. \qquad \text{(Equation 22)}$$

The above equation holds if i ≠ g.

Let us consider an example to fully understand the concept of geometric gradient.

EXAMPLE 1.11

It costs $250 during the first year to maintain a slightly used car. The cost of maintenance rises uniformly at the rate of 15% every year. Thus, it would cost $287.50 in maintenance related expenses during the second year, $330.63 during the third year, and so on. What is the present worth of the cost of the first ten years of maintenance if the interest rate was 5%?

Solution

It will be observed from this problem that maintenance costs follow a geometric gradient. The rate of growth is 15% per year as specified in the problem. In the case of descriptive problems such as this one, it is always a good idea to draw the cash flow diagram in order to better visualize the nature of the cash flow.

The cash flow diagram representative of the data given in the problem is presented in the following Figure 1.10.

The present worth of the maintenance expenses can be obtained by using Equation 19. The values of the different variables pertinent to the equation are given below:

$A_1 = \$250$. This is the first year expense and the beginning of the geometric gradient.
g = 15%. This is the rate of growth of maintenance expenses per year.
n = 10 years as given in the problem statement.
i = 5%, again as specified in the problem.

It will be observed that in this case, i ≠ g. (5% ≠ 15%). Thus, we use Equation 19 to obtain the present worth of the geometric gradient series.

Substituting the values of the variables into Equation 19, we obtain the following:

$$P = A_1 \left[\frac{1-(1+g)^n(1+i)^{-n}}{(i-g)} \right]. \qquad \text{(Equation 22)}$$

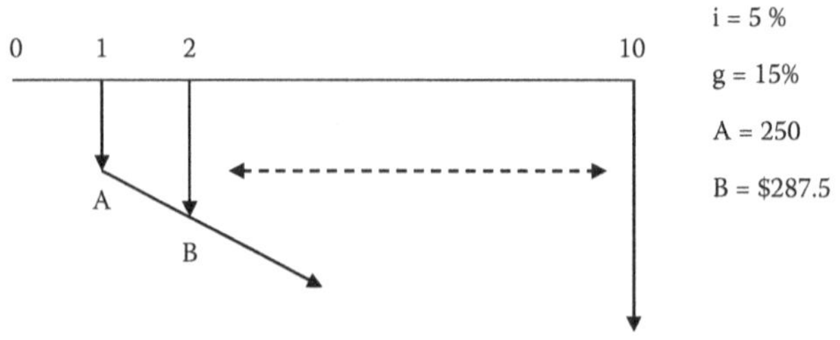

FIGURE 1.10

Thus,

$$P = \$250\left[\frac{1-(1+0.15)^{10}(1+0.05)^{-10}}{(0.05-0.15)}\right]$$

Solving the equation, we obtain the following:

$$P = \$250\left[\frac{1-(4.0455)(0.6139)}{-0.10}\right]$$

This implies that the present value, P = \$250[0.35816/−0.10].

Solving the above equation, we obtain the present value of the maintenance expenses as follows:

$$P = \$895.40.$$ **Solution**

Let us consider another problem to fully grasp the concept of geometric gradient series.

EXAMPLE 1.12

An ethanol extraction plant has adopted some new technology at the initial cost of \$10,000 to help improve yields. The technology is in the form of several devices which are expected to have a useful life of 10 years at the end which they will be sold at a salvage value of \$1000. In order to maintain these devices, it costs \$2000 the first year. Maintenance expenses rise at the rate of 7% every year and continue until the devices are scrapped at salvage value (at the end of ten years). Compute the equivalent present worth of the new technology if the interest rate is 5% per year.

Solution

Before we start solving this problem, we first need to define the concept of salvage value. Salvage value is the amount of money obtained by selling an asset at the end of its useful life. On the cash flow diagram, it is often represented as a positive cash flow if it is an income. Sometimes, as in the case of environmentally friendly disposal of assets, salvage value can actually be negative (i.e., it costs money to dispose of the asset). Such cases occur infrequently.

Once again, because this is a descriptive problem, we draw the cash flow diagram in order to better visualize the nature of the cash flow. It is presented in the following Figure 1.11.

It will be observed from the figure that the cash flow diagram has three distinct components as follows:

- Initial payment of \$10,000 occurring at point 0 (present moment in time). This is an expense since it is necessary to spend money in order to obtain the technology.

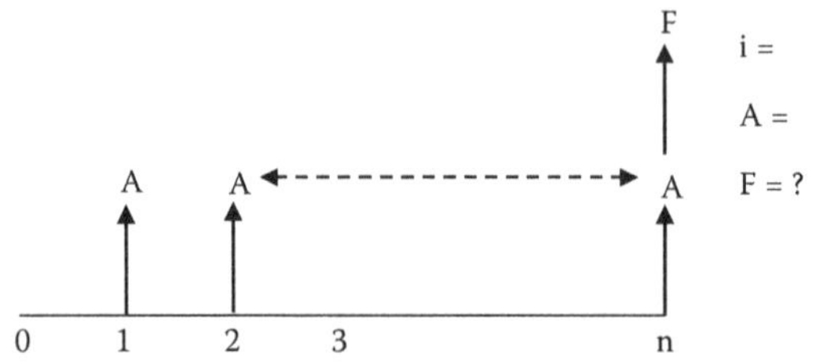

FIGURE 1.11

- Maintenance expenses in the form of a geometric gradient series.
- Salvage value ($1000), in the form of a one-time positive cash flow that occurs at the end of 10 years.

As far as the geometric series is concerned, the following variables need to be defined:

A_1 = $2000. This is the first year expense and the beginning of the geometric gradient.
g = 7%. This is the rate of growth of maintenance expenses per year.
n = 10 years as given in the problem statement.
i = 5%, again as specified in the problem.

The net present worth of the asset is given by the following equation:

$$P = -\$10{,}000 - \$2000\left[\frac{1-(1+0.07)^{10}(1+0.05)^{-10}}{(0.05-0.07)}\right] + \$1000(P/F,5\%,10).$$

In order to compute the value of (P/F, 5%, 10), we can use $P = F\,(1+i)^{-n}$
Solving this equation, we obtain the following values:

$$P = -\$10{,}000 - \$2000\left[\frac{1-1.2076}{-0.02}\right] + \$1000\,(0.6139).$$

Thus,

$$P = -\$10{,}000 - \$2000(10.38) + \$613.9.$$

Thus, the net present value = −$30,146.10. **Solution**

The negative sign indicates that the net present value is in the form of an expense.

Next, we will deal with the concept of future worth followed by nominal and effective interest rates and conclude the chapter with the concept of equivalence.

1.8 Computing the Future Worth of a Uniform Series and the Expression for Compound Amount Factor (F/A)

Future worth analysis deals with situations in which we are interested in computing the value of a cash flow at some time in the future. This is contrasted with present worth analysis wherein the value of a series of cash flows is computed at the present moment in time ($t = 0$). Thus, by definition, future worth analysis computes the value of a series of cash flows at $t > 0$. The basic form of future worth analysis is presented by the compound interest equation:

$$F = P(1 + i)^n \qquad \text{(Equation 6)}$$

Similarly, once the present worth has been computed, calculation of future worth is actually quite a simple exercise. It is obvious from the above equation that in order to compute the value of "F," the following values must be known: Present worth, interest rate, and number of time periods.

Consider a uniform series of cash flows where the nominal value of each cash flow is equal to "A". Assume that the interest rate is "n" and the present value is "P" as depicted in the following Figure 1.11.

From Equation 10, we know that the present value of this series of cash flows would be equal to

$$P = A\left[\frac{(1+i)^n - 1}{(i)(1+i)^n}\right]. \qquad \text{(Equation 10)}$$

In order to compute the future value of this cash flow, we would substitute the value of "P" from Equation 10 into Equation 6.

Thus, we have the following expression:

$$F = A\left[\frac{(1+i)^n - 1}{(i)(1+i)^n}\right](1+i)^n$$

Canceling the term $(1 + i)^n$ from both the numerator and denominator, we obtain the following expression for the future value of a uniform series:

$$F = A\left[\frac{(1+i)^n - 1}{i}\right]. \qquad \text{(Equation 23)}$$

Conversely, if the future worth was known and it was required to convert it into a series of uniform cash flows, all we need to do is to obtain the inverse of Equation 20. This can be written as follows:

$$A = F\left[\frac{i}{(1+i)^n - 1}\right]. \qquad \text{(Equation 24)}$$

Study point to remember: The future worth of a series of uniform cash flows as represented by the formula:

$$F = A\left[\frac{(1+i)^n - 1}{i}\right], \qquad \text{(Equation 23)}$$

is always obtained at the point in time that the series ends, i.e., at point "n" in time.

Let us consider a few examples to better understand the aforementioned principles.

EXAMPLE 1.13

Compute the future worth of the following series of cash flows:

Solution

From the above Figure 1.12, the following facts are obvious:

- The cash flow represents a uniform series due to the fact that it occurs at every year, each annual cash flow has the same nominal value, and all the cash flows are positive.
- The value of "A" for the uniform series is \$10,000.
- The series lasts for 10 years. Thus, the value of n = 10.
- The value of the interest rate, from the figure, is i = 5%.

The future worth of the uniform series can be computed by using Equation 20 as follows:

$$F = A\left[\frac{(1+i)^n - 1}{i}\right]. \qquad \text{(Equation 23)}$$

Substituting the respective values in Equation 20, we obtain the following:

$$F = \$10{,}000\left[\frac{(1+0.05)^{10} - 1}{0.05}\right]. \qquad \text{(Equation 25)}$$

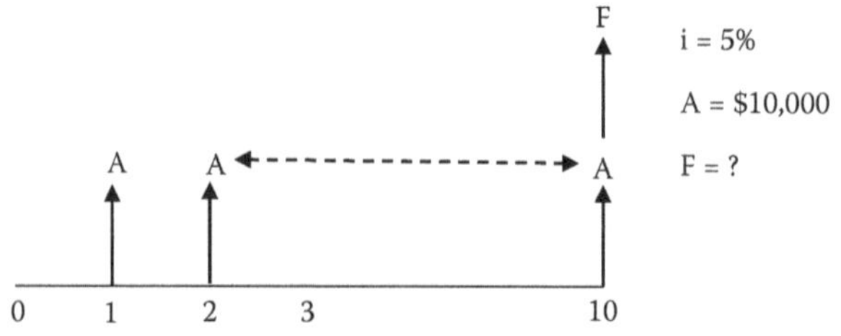

FIGURE 1.12

Solving this equation, we get the following values:

$$F = \$10,000\left(\frac{0.6288}{0.05}\right) = \$125,778.92. \qquad \textbf{Solution 1}$$

Alternative Method

As in the case of present worth analysis, problems related to future worth analysis can be solved using values obtained directly from interest rate tables. This is demonstrated as follows:

We use the equation: $F = A\ (F/A,\ i,\ n)$ in order to compute the future value of the uniform series.

In order to compute the value of (F/A) factor for an interest rate of 5% and $n = 10$ years, we look up that value in the interest rate table for $i = 5\%$. Scrolling down to $n = 10$, we read the value as $(F/A) = 12.578$.

Substituting this value, we obtain the future worth of the cash flow as follows:

$$F = \$10,000(12.578) = \$125,780. \qquad \textbf{Solution 2}$$

Comparing solution 2 with solution 1, it will be noticed that both solutions converge. Once again, either method is valid and can be used to solve problems of this nature.

EXAMPLE 1.14

An investor has decided to buy a house which is expected to provide $1000 in annual cash flow for a period of ten years. At the end of ten years, the house will be conservatively sold for the same price that the investor paid for it ($100,000). Compute the net future value of this investment if the interest rate is 5%.

Solution

As is the case with all descriptive problems, the first step in solving this problem is to draw the cash flow diagram. It is presented in the following Figure 1.13:

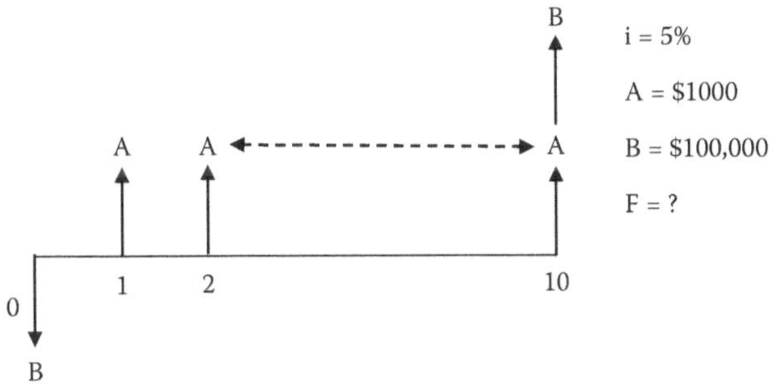

FIGURE 1.13

The following observations can be made from the cash flow diagram above:

- \$100,000 is spent in acquiring the house at the outset. This represents an outflow of money and is represented by a downward pointing arrow at point "0."
- Since the annual benefit is expected to be \$1000 for a period of ten years, this constitutes a uniform series and is depicted as such on the cash flow diagram.
- The house will be sold at the end of ten years for \$100,000. This represents an inflow of money and is therefore represented as an upward pointing arrow on the cash flow diagram at point "10."
- The interest rate is 5% and is depicted as such on the cash flow diagram.

The net future worth can be calculated by using the following formula:

$$F = \$100{,}000 + A(F/A, i, n) - \$100{,}000(F/P, i, n).$$

When solving problems related to future worth, it is necessary to note that the frame of reference shifts towards the future and the time at which the future worth is to be calculated. Contrast this with present worth always being calculated at point "0". Thus, in this case, the future worth is being computed at the end of year 10. This is the reason that the cash flow (+\$100,000) occurring at that point in time is not multiplied by any factor.

From the interest rate table, we observe the following values for substituting in the equation:

$$(F/A, 5\%, 10) = 12.578,$$

$$(F/P, 5\%, 10) = 1.629.$$

Alternatively, the same values can be arrived at by substituting the values of "i" and "n" in the respective formulas for (F/A) and (F/P) as follows:

$$(F/A, 5\%, 10) = \left[\frac{(1+i)^n - 1}{i}\right] = \left[\frac{(1+0.05)^{10} - 1}{0.05}\right] = \left[\frac{0.628894}{0.05}\right] = 12.578.$$

$$(F/P, 5\%, 10) = (1.05)^{10} = 1.629.$$

It will be observed that these values are in agreement with the values read from the interest rate table.

Substituting all relevant values in the equation, we get the following:

$$F = \$100{,}000 + \$1000\ (12.578) - \$100{,}000\ (1.629)$$

Thus,

$$F = \$100{,}000 + \$12{,}578 - 162{,}900 = -\$50{,}322.$$

This implies that the investor would lose $50,322 if they decided to go ahead with this particular deal.

In the next part of this chapter, we will consider how to convert present value to an annual cost/benefit as the case may be.

1.9 Annual Cash Flow Analysis: Converting Present Value to a Uniform Series (A/P)

The main objective of annual cash flow analysis is to convert a present amount of money (present value) into a uniform series of cash flows. The following cash flow diagram illustrates such a situation (Figure 1.14).

In order to accomplish this objective, we use the factor (A/P, i, n) also termed as the capital recovery factor. The formula to compute the value of (A/P) has already been discussed in a previous section of this chapter. We will consider it here again as follows:

$$A = P\left[i\frac{(1+i)^n}{(1+i)^n - 1}\right]. \qquad \text{(Equation 13)}$$

In order to fully understand this concept, let us consider an actual example.

EXAMPLE 1.15

Consider the situation presented in problem 1.14. Instead of computing the future value to the investor, compute the equivalent uniform annual cost/benefit of buying and holding onto the property for ten years.

Solution

Since this problem has already been solved albeit from a different angle, most of the parameters will remain the same. However, the question

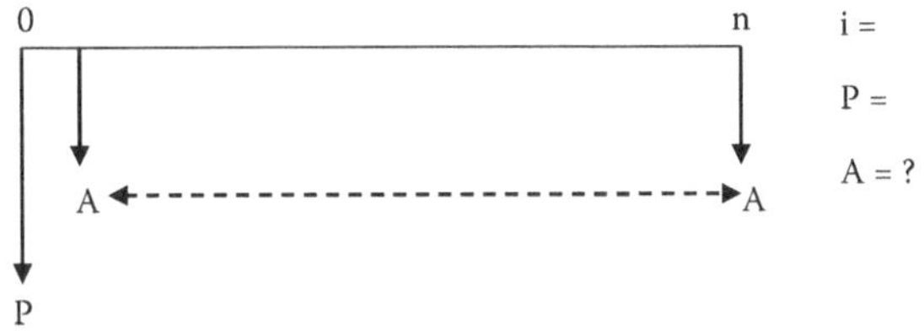

FIGURE 1.14

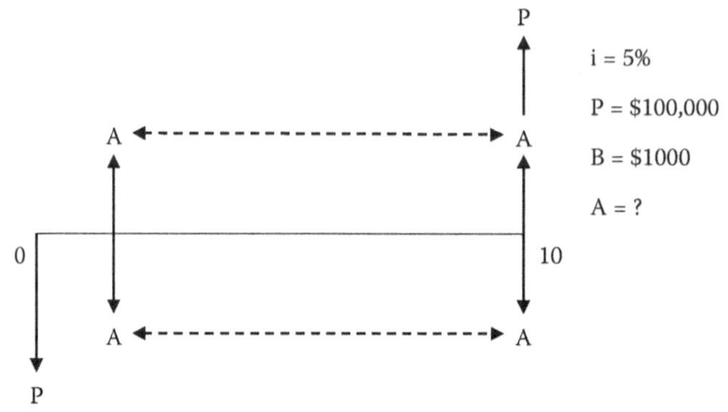

FIGURE 1.15

at hand, being different, will entail some changes to be made to the cash flow diagram. The modified cash flow diagram is presented below (Figure 1.15).

The easiest way to solve an annual cash flow analysis problem is to first compute the present worth (or future worth). The next step is to multiply that value by the factor (A/P) or (A/F), respectively. We will consider both methods here.

Method 1: Compute Present Value and use Capital Recovery Factor (A/P)

Step 1: Compute the present worth of the cash flow diagram.

$$P = -\$100{,}000 + \$10{,}000\ (P/A, i, n) + \$100{,}000\ (P/F, i, n).$$

The value of (P/A, 5%, 10) = 7.722.
The value of (P/F, 5%, 10) = 0.6139.
Substituting the values of different factors, we obtain the following expression.

$$P = -\$100{,}000 + \$1000(7.722) + \$100{,}000(0.6139)$$

Thus,

$$P = -\$100{,}000 + \$7{,}722 + \$61{,}390 = -\$30{,}888.$$

Step 2: Compute the value of annual cash flow by multiplying the present worth by the capital recovery factor (P/A).

$$A = P(A/P, i, n).$$

The value of (A/P, 5%, 10) = 0.1295

Substituting the respective values into the above equation, we have the following:

A = −\$30,000(0.1295) = −\$3999.96, which is approximately \$4000. This implies that the investor would lose the equivalent of \$4000 every year by buying and holding onto this investment for ten years. This loss is the equivalent of cost since it entails a negative cash flow.

Method 2: Compute the Future Value and use Compound Amount Factor (A/F)

Step 1: Compute the future worth of the cash flow diagram.

The future worth of this cash flow diagram was already computed in problem 1.94 and we know that it is −\$50,322.

Step 2: Compute the value of annual cash flow by multiplying the future worth by the compound amount factor (A/F).

$$A = F(A/F, i, n).$$

The value of (A/F, 5%, 10) = 0.0795.

Substituting the respective values into the above equation, we have the following:

$$A = -\$50{,}322(0.0795) = -\$4000.5$$

Once again, this solution is in agreement with the solution obtained using method 1 and would be identical if not for rounding errors in calculation. Nevertheless, the essence of the solution remains the same.

In the next section of this chapter, we will consider the concepts of nominal interest rate, effective interest rate, and continuous compounding.

1.10 Nominal Interest Rate and Effective Interest Rate

Nominal interest rate per year (i) is defined as the annual interest rate without taking into account the effects of any compounding. This interest rate is what is given in the problem statement and assumes that it is compounded annually. In real life, interest can be compounded any number of ways, such as being compounded monthly, semi-annually, quarterly, etc. In such cases, we need to examine the concept of effective interest rate.

Effective interest rate (i_a) is defined as the annual interest rate arrived at by considering the effect of compounding that has occurred over the course of the year. Thus, an interest rate of 5% per year compounded quarterly will have a different (higher) effective value than a nominal interest rate of 5% compounded annually.

The effective annual interest rate can be derived from the nominal annual interest rate using the following formula:

$$i_a = \left(1 + \frac{i}{m}\right)^m - 1. \qquad \text{(Equation 26)}$$

In the equation above, "m" is the number of compounding periods per year. For example, if the interest rate is compounded monthly, the value of "m" would be 12 (there are 12 months in a year). If interest is compounded quarterly, the value of "m" would be 4, due to the fact that there are four quarters in a year, and so on.

Let us consider an example in order to make the complete distinction between nominal interest rate and effective interest rate.

EXAMPLE 1.16

The interest rate charged by a bank to its customers is 5% per year compounded monthly. What is the nominal interest rate and what is the effective interest rate charged by the bank?

Solution

The value of the nominal interest rate "i" in the problem is 5%. Due to compounding that occurs on a less than annual basis (monthly, in this case), the effective interest rate will be substantially different from the nominal interest rate. We compute the value of effective interest rate using Equation 22 as follows:

$$\text{Effective interest rate } i_a = \left(1 + \frac{i}{m}\right)^m - 1. \qquad \text{(Equation 27)}$$

Since there are 12 months in a year, the value of "m" in the present case would be equal to 12.

Substituting different values into Equation 22, we obtain the following:

$$i_a = \left(1 + \frac{0.05}{12}\right)^{12} - 1. \qquad \text{(Equation 28)}$$

This amounts to: $i_a = (1 + 0.004167)^{12} - 1$.

Thus, the effective interest rate would be $i_a = 1.05116 - 1 = 0.05116 =$ 5.116%.

It will be observed that 5.116% > 5%. Thus, the effective interest rate is actually higher than nominal interest rate for cases where compounding occurs at less than one year intervals.

EXAMPLE 1.17

Compute the future value after three years of a sum of \$10,000 deposited into a savings account that earns interest at the rate of 3% per year compounded quarterly. How much would the future value be if interest was compounded annually?

Solution

From the problem description, we have the values of respective parameters as follows:

$$P = \$10{,}000;\ i = 3\%,\ F = ?$$

Case 1

In this case, we will consider the effects of monthly compounding. Since there are 12 months in a year, there would be $(12 \times 3) = 36$ months in a three year period. In this case, $n = 36$. As far as the value of "i" is concerned, it would have to be divided by 12 in accordance with Equation 1.22.

Thus, in this case, $i = 3/12 = 0.25\%$.

Substituting these values into the equation: $F = P(1 + i)^n$, we obtain the following:

$$F = \$10{,}000(1 + 0.25\%)^{36}$$

Thus,

$$F = \$10{,}000(1 + 0.0025)^{36}$$

Thus, the future value is equal to \$10,940.5. **Solution 1**

Case 2

In this case, the interest is only compounded annually. Thus, the calculation is a matter of simple substitution as follows:

$$F = \$10{,}000(1.03)^3$$

Thus,

$$F = \$10{,}927.27.$$ **Solution 2**

Note that \$10,940.5 > \$10,927.27.

Once again, it will be observed that the same nominal interest rate compounded more frequently will yield a higher effective interest than its counterpart that is compounded only once every year.

1.11 Continuous Compounding

Unlike compounding that occurs on a conventional basis, such as annual compounding, quarterly compounding, monthly compounding, etc., continuous compounding, as the name suggests, occurs continuously. We will derive the notations for continuous compounding in this section.

We are already familiar with the single payment compound amount formula to compute the future worth. It is written as follows:

$$F = P(1 + i)^n \qquad \text{(Equation 6)}$$

In order to increase the frequency of compounding, the above formula can be rewritten as follows:

$$F = P\left(1+\frac{r}{m}\right)^{mn}$$

It will be observed from the nature of the above equation that if the numbers of compounding sub periods per year (m) are increased without limit, the value of "m" will become very large. Consequently, since the term "r/m" contains "m" in the denominator, its value will become very small and will eventually approach zero.

The above condition can be written mathematically as follows:

$$F = P\lim_{m\to\infty}\left(1+\frac{r}{m}\right)^{mn}$$

Setting r/m = x, mn = (1/x) (rn). As we have seen before, as m approaches infinity, x approaches zero. Thus, the above equation is now written as follows:

$$F = P\left[\lim_{x\to 0}(1+x)^{1/x}\right]^{rn}$$

We know from calculus that $\lim_{x\to 0}(1+x)^{1/x} = 2.71828 = e$.

Substituting this value into the equation for F, we obtain the following expression:

$$F = Pe^{rn} \qquad \text{(Equation 23)}$$

Thus, for continuous compounding, $(1 + i) = e^r$.

The value of effective interest rate per year for continuous compounding is as follows:

$$i_a = e^r - 1. \qquad \text{(Equation 24)}$$

Let us consider a few examples to fully understand the concept of continuous compounding.

EXAMPLE 1.18

A bank pays 4% nominal interest compounded continuously. How much interest would a deposit of \$10,000 earn after five years?

Solution

Interest earned on the deposit = Future value − Present value.

We know from the data given in the problem that present value, P = \$10,000.

We also know that n = 5 years, i = 4% compounded continuously.

Thus, the future value for continuous compounding can be computed by using Equation 1.23 as follows:

$$F = Pe^{rn}. \qquad \text{(Equation 23)}$$

Substituting respective values into the equation, we obtain the following:

$$F = (\$10{,}000)e^{(0.04)(5)}$$

Thus,

$$F = (\$10{,}000)e^{0.2} = \$10{,}000(1.22) = \$12{,}214.$$

The interest earned on the deposit = F − P = \$12,214 − \$10,000 = \$2214.

Solution

EXAMPLE 1.19

A hedge fund offers investors \$20,000 per deposit at the end of 15 years. They offer to pay interest at the rate of 10% compounded continuously. The final payment will be made in the form of one lump sum at the end of 15 years. Determine the pricing of each deposit being offered to investors now.

Solution

When interest is compounded continuously, we know the expression for future value from Equation 1.23 as:

$$F = Pe^{rn}. \qquad \text{(Equation 23)}$$

In this case, the value if F = \$20,000; r = 10%; n = 15 years; P = ?

Rewriting Equation 1.23 to solve for the present value, P, we get the following:

$$P = Fe^{-rn}. \qquad \text{(Equation 29)}$$

Substituting respective values into Equation 1.24, we obtain the following results:

$$P = (\$20{,}000)e^{-(0.10)(15)}$$

Thus,

$$P = (\$20{,}000)\ (0.223) = \$4462.60.$$

This implies that the present value of each deposit is $4462.60 based on the conditions given in the problem. **Solution**

1.12 Capitalized Cost

The concept of Capitalized Cost is used to address cost situations in which a service needs to be offered for extended periods of time. This is typical of services offered by the government. Services that are offered and consequently need to be maintained for long periods of time include sanitation, public transport, etc. In the private sector, when such services are offered, they are referred to as "services that need to be offered for indefinitely long time horizons."

Succinctly put, capitalized cost is defined as the present sum of money that needs to be set aside at a particular interest rate in order to yield the funds that would be necessary to provide the service in question for an indefinite period of time.

If the principal amount of money is considered to be equal to an amount "P", the interest for that period would be equal to "$i \times P$".

Thus, the amount at the end of the period = "$P + (i \times P)$".

Thus, for a present sum of money equal to the amount "P," the end of period withdrawal equal to the amount "A" would have to be equal to the value "iP". Withdrawals can continue forever without ever decreasing the value of "P." Thus, for infinite time periods, we have the following relationship:

$$\text{If } n = \infty,\ A = i \times P.$$

Capitalized cost, the present sum of money that needs to be set aside, is equal to "P." Thus the equation for capitalized cost can be derived using the relationship above. This equation is presented below:

$$P = A/i. \qquad \text{(Equation 30)}$$

In order to better understand the concept of capitalized cost, consider Examples 1.20 and 1.21.

EXAMPLE 1.20

Calculate the amount of funds that should be set aside to pay \$7500 per year for maintenance on a street light. Interest is assumed to be 5% per annum. The maintenance needs to be performed over an infinite period of time. As such, the principal should remain undiminished.

Solution

The only condition given in this problem is that the maintenance task needs to be performed over an infinite time horizon. Thus, we use the concept of capitalized cost to solve the problem.

From the data given in the problem, we know that the value of "A" is equal to \$7500 per year since this constitutes a uniform series of payments. We also know that the interest rate is 5%.

Using Equation 25, the present value of money can be computed as follows:

$$P = \$7500/0.05 = \$150{,}000.$$

Thus, the amount of money to be set aside to pay for annual maintenance should be \$150,000. **Solution**

EXAMPLE 1.21

An oil and gas pipeline is expected to cost \$25 million. It is expected to last for 75 years. It will be maintained regularly so as to be kept in service indefinitely. Assuming an interest rate of 5%, compute the capitalized cost for this project.

Solution

In order to compute the capitalized cost for this project, we first need to calculate the value of "A".

The present cost of the pipeline is given as \$25 million.

If the pipeline lasts for 75 years, we can compute the end of year payments by converting the present value into a uniform series using the (A/P, i, n) expression. This will yield the value of "A" that is necessary to solve the problem. We can compute the value of "A" as follows:

Thus,

$$A = P(A/P, i, n).$$

Substituting the values of $i = 5\%$ and $n = 75$ years, we obtain the following:

$$A = \$25{,}000{,}000(0.0513).$$

Thus,

$$A = \$1{,}282{,}500.$$

Now we compute the capitalized cost using Equation 25 as follows:

$$P = A/i.$$

Thus,

$$P = \$1,282,500/0.05 = \$25,650,000.$$

The capitalized cost for the project would be \$25,650,000. **Solution**

1.13 Conclusion

This chapter presented crucial concepts related to the computation of present worth, future worth, and equivalent uniform annual cash flows. We also drew a clear distinction between lump sum payments and a series of payments in terms of calculating their present worth and future worth. The topic of nominal interest rate was addressed. The difference between nominal interest and effective interest was clearly delineated with the help of examples. The concept of continuous compounding was explained with the help of several problems. Remember that the concepts presented in this chapter will form the backbone of most engineering economic analyses, as well as financial analyses. Make absolutely sure to learn these concepts well. Most of the formulas derived in this chapter and the underlying concepts will be revisited numerous times in the following chapters.

PROBLEMS

1. If \$15,000 is deposited into an account earning 7% interest per year, how much would be in the account after:
   a. 5 years
   b. 10 years
   c. 15 years
   d. 25 years
2. If a computer will cost \$5000 seven years from now, how much should be deposited today in order to pay for it if you can earn 3% interest per year? *(Answer: \$4,065.45).*
3. Max borrowed \$10,000 from his bank and repaid them \$25,000. What is the interest rate the bank charged him if the repayment occurred after:
   a. 5 years
   b. 7 years

c. 15 years
d. 20 years

4. An investment can earn a return of 10% per year conservatively. $20,000 in invested into such an account. How many years would it take to earn the following interest:
   a. $2000
   b. $7000
   c. $10,750
   d. $15,756
5. Alan invested $7000 in the stock index futures. For the first five years, his average return was 15% per year and was 25% per year for the next seven years. How much money would be in his account after twelve years? *(Answer: $67,136.29)*
6. $10,000 is invested at an interest rate of 5% per year compounded semi-annually. How much more interest was earned by investing in this arrangement instead of using an annual compounding system? Would it be wiser to invest the same amount of money at an interest rate of 4.85% compounded monthly?
7. Compute the nominal and effective interest rate on 6 month CDs (certificates of deposit) if they are being offered for $10,000 and the bank will repay the investor $10,200 at the end of six months.
8. A sum of money invested at 5% per year compounded quarterly will take how many years to double? *(Answer: 14 years).*
9. If the nominal annual interest rate is 7% compounded monthly, compute the effective annual interest rate. *(Answer: 7.229% per year).*
10. Jim deposits $75,000 into an account for six months and obtains $80,000 at the end of that period. Interest was compounded continuously. What were the nominal rate of interest and the effective rate of interest?
11. Warren wishes to grow his money quickly. What is your advice to him in terms of selecting from the following options:
    a. 7.8% per year compounded continuously
    b. 8% per year compounded quarterly
    c. 8.1% per year compounded monthly
    d. 8.2% per year compounded semi-annually
12. A manufacturing company spends $100,000 annually for a required inspection program. A new online monitoring device has been invented recently and will eliminate the need for such a program. How much should the company pay for the device if the interest rate is 7% per year? The company wishes to consider a 7 year

time horizon in order to make its decision. *(Answer: No more than $538,900).*

13. Brad wishes to buy a car that costs $10,000. He will finance this purchase by obtaining a loan from his bank at an interest rate of 5% per year over a period of 3 years. How much money should he save for his down payment?
14. Sally deposits $7000 in her savings account every year for 7 years. She makes no deposits for the next seven years. How much money will be in her account at the end of fourteen years if the interest rate is 8% per year compounded quarterly? *(Answer: $109,106.28).*
15. A manufacturing company buys a machine for $100,000. The deal is structured so that the machine will be fully paid off at the end of seven years. The company agrees to make seven equal annual payments in order to pay off the machine. An interest rate of 5% per year is charged by the bank. After the first four payments were made, the CEO of the company was able to obtain an arrangement wherein the balance on the loan would be paid off at the end of the following year. Compute the value if this final payment. *(Answer: $41,425.81).*
16. Jennifer, a young engineer (age 22), wishes to have $5 million by the time she is 60 years old. She is confident that she can obtain an interest rate of 10% per year. How much should she invest every year in order to achieve her goal?
17. A new car can be obtained for $30,000. Brad wishes to finance no more than $20,000 over four years to purchase the car. If he can obtain an interest rate of 5% on his investment, how much should he save each year?
18. A debt of $15,000 can be repaid at an interest rate of 6% per year. The following is a schedule of payments. Compute the value of the payment at the end of the fifth year (denoted by X). *(Answer: $3191.49).*

Year	Payment ($)
1	$1500
2	$3000
3	$4500
4	$6000
5	X

19. The maintenance cost on an SUV is estimated to be $750 the first year. Beginning with the second year, the maintenance cost is expected to increase by $250 every year. Assuming the interest rate to be 5%

per year, how much money should be set aside to meet maintenance expenses if you plan to sell the vehicle after ten years?

20. Compute the present value of the following cash flow (Figure 1.16).

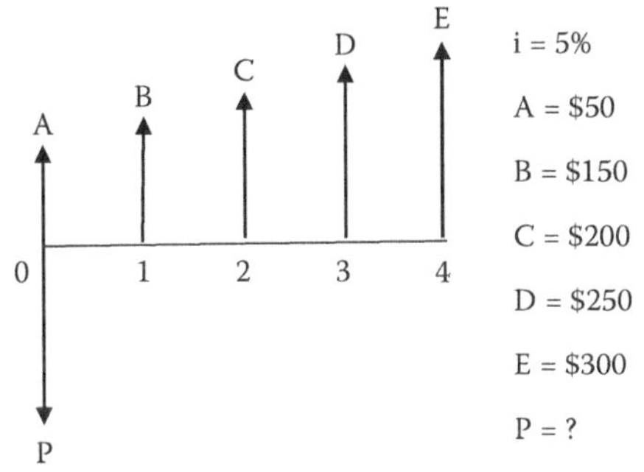

FIGURE 1.16

21. The following cash flows have a present value of 0. Compute the value of D (Figure 1.17).

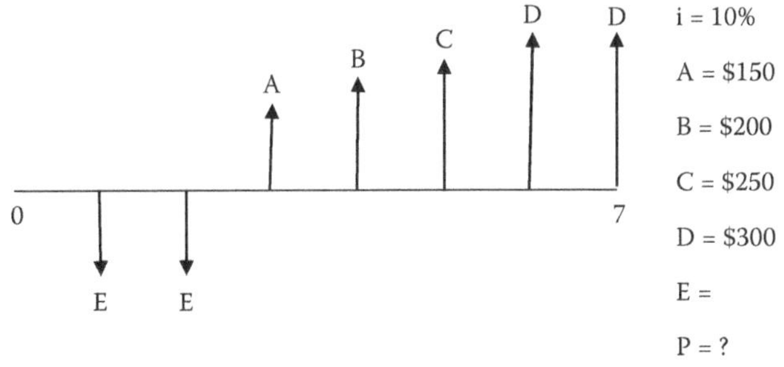

FIGURE 1.17

22. Compute the present worth of the following cash flows (Figure 1.18).

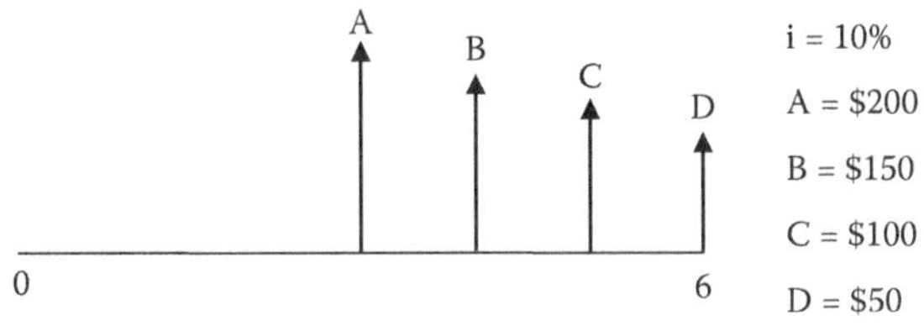

FIGURE 1.18

23. A series of cash flows begin at $15,000 the first year and steadily increases over the next few years until year 10. The interest rate is 5% per year. What is the present value of these cash flows if they increase at the rate of 7% per year?
24. An industrial engineer deposits 10% of her salary every year into an account that pays 5% per year compounded semi-annually. She expects her salary to increase by 7% annually. Her current salary is $100,000. How much money will she have accumulated if she works for 25 years?
25. Kim wishes to be a millionaire by the time she retires in 35 years. How much money should she save every year if she can obtain an interest rate of 5% per year compounded continuously?
26. Two machines are equally capable of performing a task. If the interest rate is 7%, which machine would you recommend based on annual cost of operation?

	Machine A	Machine B
Initial cost ($)	10,000	12,000
Salvage value ($)	2000	2400
Annual maintenance cost ($)	200	250
Useful life (years)	5	5

27. Convert the following into a uniform series of benefits (Figure 1.19).

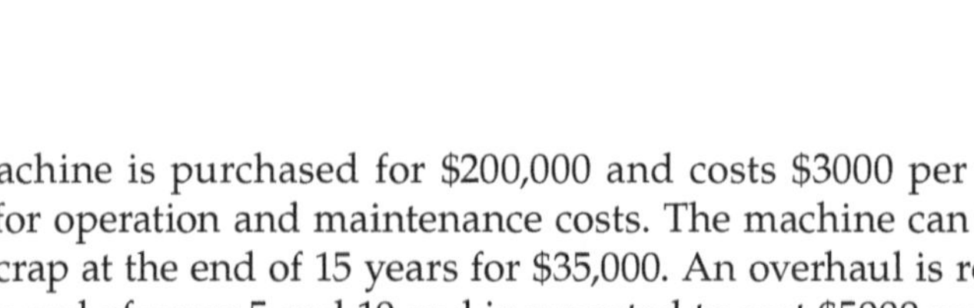

FIGURE 1.19

28. A machine is purchased for $200,000 and costs $3000 per year to pay for operation and maintenance costs. The machine can be sold for scrap at the end of 15 years for $35,000. An overhaul is required at the end of years 5 and 10 and is expected to cost $5000 each time. Compute the equivalent uniform annual cost to buy and operate the machine if the interest rate is 5% per year. *(Answer: $21,310.42 per year).*
29. Compute the present value of the following cash flow (Figure 1.20).

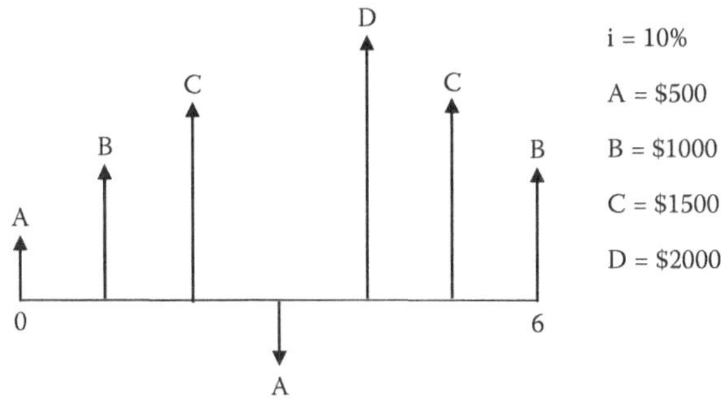

FIGURE 1.20

30. ABC, Inc. is the owner of a residential apartment complex. The company plans to install a new sprinkler system. The system is expected to save $1500 per year in maintenance costs and will need to be replaced every ten years. At the end of its useful life, it can be sold for 10% of its original cost. As an owner of the building, how much would you pay for the sprinkler system?
31. A recent graduate wants to grow his money as swiftly as possible. Which of the following options would you recommend to ensure he meets his goals?
    a. 9.2% per year compounded monthly
    b. 8.4% per year compounded semi-annually
    c. 8.3% per year compounded quarterly
    d. 8% per year compounded continuously
32. It costs $100 per year to maintain your future cemetery plot. If the interest rate is 6%, how much would need to be set aside to pay for maintenance on the plot without touching the principal amount?
33. What is the uninflated present worth of $2000 in 3 years if the average rate of inflation is 4% and interest is 12%?
34. The owner of a local pizzeria borrowed $85,000, agreeing to repay the loan in quarterly payments over 6 years. If the interest rate is 10% compounded quarterly, how much is each payment?
35. The Diamond Credit Card Company charges 19% interest per year, compounded monthly. What is the effective annual interest rate charged by the company?
36. You are thinking about retirement and considering investing money as soon as possible so you will have $700,000 in 20 years. If the interest rate is 9% and the interest is compounded annually, calculate the amount that you must invest now. *(Answer: $124,901.63).*

37. An employee at the Blue Ribbon Cookie Company receives a bonus package of $1500 at the end of the first year, $2500 at the end of the second year, and so on for the first eight years of employment. What is the present worth of the bonus package at 5% interest? *(Answer: $30,664.5).*
38. A young pharmacist plans to buy a home, but has been advised that her monthly house and property tax payment should not exceed 30% of her disposable monthly income. After researching the market, she determines she can obtain a 27-year home loan for 8.95% annual interest per year, compounded monthly. Her monthly property tax payment will be approximately $150. What is the maximum amount she can pay for a house if her disposable monthly income is $1800?
39. An engineering firm needs a rapid prototyping machine. The company gives you two options. The first is where you purchase the machine for $60,000, pay a maintenance contract of $1500 annually, and expect to resell the machine after 8 years with a salvage value of $12,000. The second option is to lease the machine at $8000 annually, pay no maintenance, but receive no salvage. Assume you will take in $9500 per year in income from this machine. Also assume that there is an additional option where you do not buy the machine, but invest the money in a bank at 7.3% interest. Which option would be the best for the firm?
40. If you were to invest your paycheck at 8% per year compounded semi-annually, how many months would it take to triple?
41. Expenses for water treatment at a state park are expected to be $90,000 now, $35,000 in the first year, and $12,000 per year for every year thereafter. At an interest rate of 11% per year, how much would the capitalized cost of the treatment be?
42. Interest payments on a bond are $400 every four months. If the value of the bond is $20,000, what is the interest rate of the bond?
43. Congratulations! You just won $5,000,000 in the lottery. Unfortunately, the commission will pay you in installments of $250,000 each year for the next 20 years, starting at the end of the first year. What is the present worth of your winnings? Assume a nominal annual interest rate of 12%. *(Answer: $1,867,250).*
44. You have accumulated $7000 in credit card debt. The credit card company charges 19% nominal annual interest compounded monthly. You can only afford to pay $300 per month. How many months will it take you to pay off the debt and how much money will you have paid in interest?

45. It is likely that train tickets will increase 9% annually over the next 4 years. The cost of a ticket by the end of the first year will be $210. How much money would a student need to place in a savings account to pay for their travel home at the end of the year for the next 4 years? Assume the savings account pays 4% annual interest.
46. Compute the Present Worth of the following cash flow diagram (Figure 1.21).

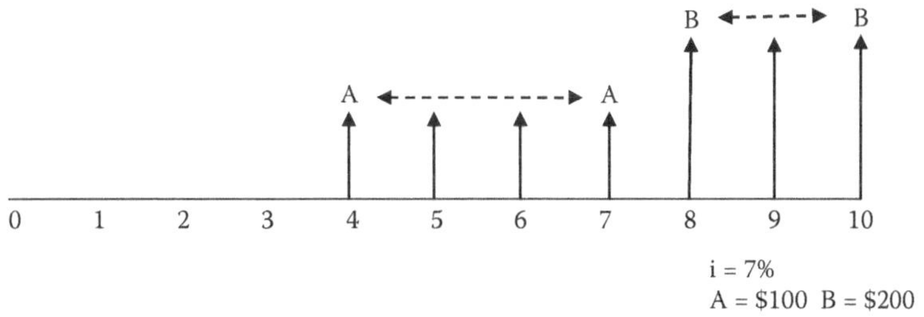

FIGURE 1.21

47. Compute the Present Worth of the following cash flow diagram (Figure 1.22).

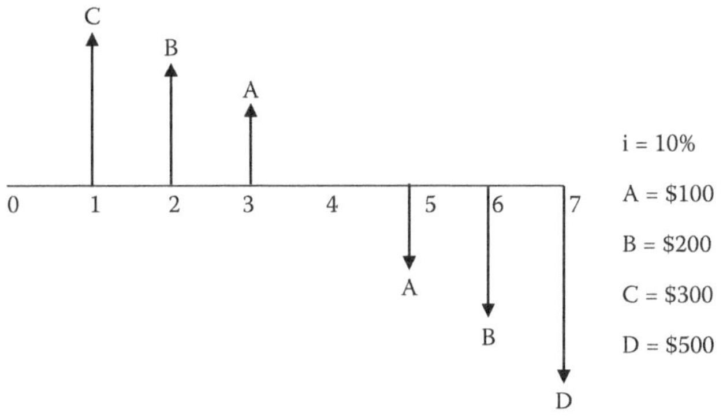

FIGURE 1.22

48. Compute the Present Worth of the following cash flow diagram (Figure 1.23).

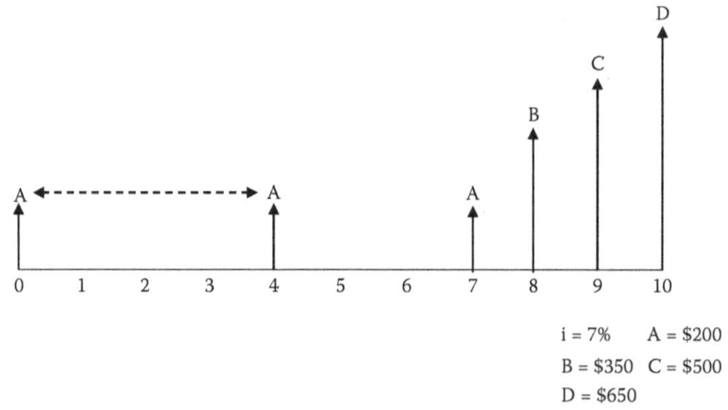

FIGURE 1.23

49. Compute the Future Worth of the following cash flow diagram (Figure 1.24).

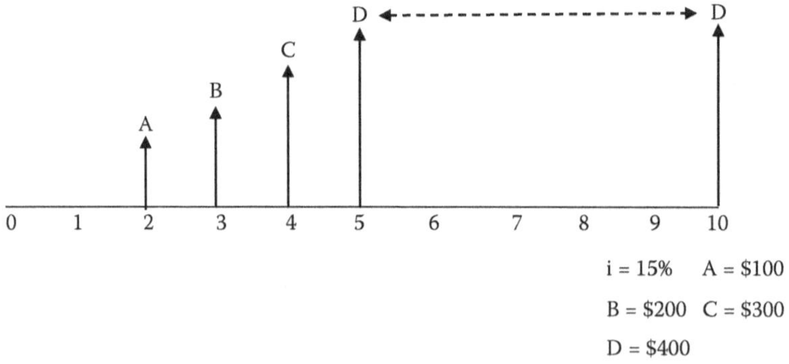

FIGURE 1.24

50. Compute the Future Worth of the following cash flow diagram (Figure 1.25).

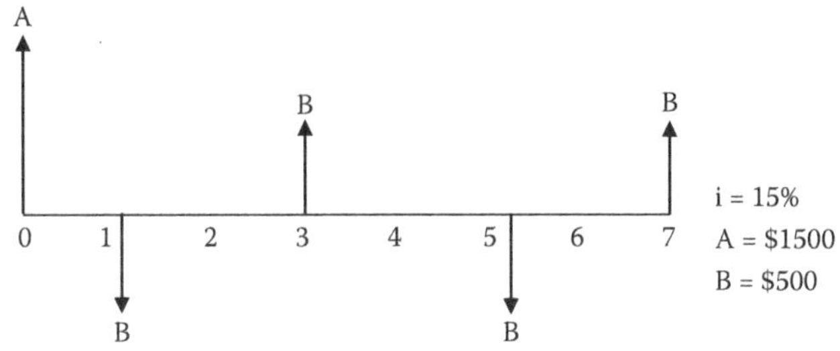

FIGURE 1.25

51. Convert the following Cash flow diagram into a Uniform Series over seven years (Figure 1.26).

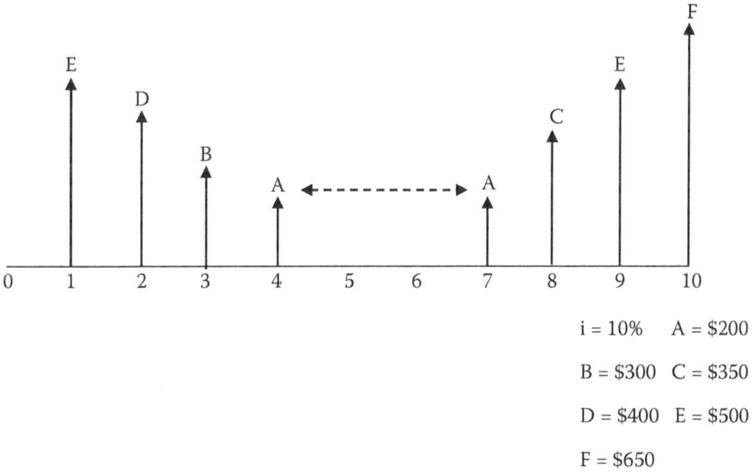

FIGURE 1.26

2

Cost Estimation

2.1 What Is Cost Estimation?

It is necessary to accurately estimate production costs prior to production in order to establish a viable and successful production operation. Cost estimation predicts how much a product would cost. Since profitability is a function of product cost, cost estimation plays a vital role in economic evaluation.

Cost estimation is often confused with costing. While cost estimation is an estimate of how much a product is likely to cost, costing, on the other hand, is an account of the actual costs that contributed to the production. Thus, cost estimation is a "before production" process whereas costing is an "after production" process. Estimation of any kind is an inherently complex process. This is especially true of cost estimation. It requires knowledge of a variety of parameters. Some of these parameters are listed as follows:

- An intimate understanding of the product design.
- Knowledge of the different materials that will fulfill the functional requirements intended of the product. Similarly, knowledge of related manufacturing and production processes is crucial as well.
- It is important to have knowledge of the costing structure of each production process.
- Quality assurance and quality control costs money. In order to arrive at an accurate cost estimate, it is necessary to be familiar with the various costs inherent in controlling product quality at different stages of its production.
- Once a product has been introduced into service, it costs money to regularly maintain it. Knowledge of maintenance costs is important.
- A product that does not satisfy its intended function or at worst, fails and causes bodily injury or harm to the customer is likely to be the center of product liability suits. While rare, such cases do occur. Knowledge of costs involved in the litigation process is an important aspect of cost estimation.

- We live in an era of increased environmental awareness. Governments around the world have imposed strict restrictions and passed regulations pertaining to environmentally friendly product disposal of end of life products. It is essential to have first-hand knowledge of costs associated with this process.

The primary function of any business is to make a profit. The role played by accurate cost estimates in this capacity cannot possibly be overemphasized. Given the fact that profit is the difference between revenue and cost, it is obvious that in order to maximize profit, a business needs to simultaneously maximize revenue and minimize cost. This is where cost estimation plays an important role. An accurate estimation of costs likely to be incurred during a production operation would enable a business owner to look for ways that would reduce and hopefully minimize costs. This would lead to a jump in productivity and consequently improve profitability of the business. The following methods are commonly used to improve business profitability:

- Raise selling price per unit of product
- Increase the volume of sales
- Optimize the product mix
- Minimize cost per unit of product

Raising the selling price per unit of product is easier said than done. Selling price cannot be altered at will without considering the competition. The same is true to a large extent when it comes to increasing sales. Cost minimization, however, can be achieved by making the operation leaner and more efficient. Wasteful operations cost money and should be eliminated. Impractical design specifications are expensive to manufacture and should be modified. Too much packaging is a waste of money and can be done away with. Consider the following example, which illustrates the effects of unit sales price, cost per unit, sales volume, and product mix.

Consider a situation in which a company produces two products: X and Y. Results of different strategies such as reigning selling price, lowering cost, increasing sales volume, etc., have been tabulated below (Table 2.1). It will be observed from the above table that the greatest increase in profit is achieved by a 10% increase in selling price per unit. But as we have already pointed out, this task can only be achieved through great difficulty due to the very nature of a competitive marketplace.

The second best improvement in profits is achieved by reducing costs by 10%. This task is easily achievable since it is an internal matter and as such cannot be impeded by external competition. The crucial role of cost reduction in the long-term profitability of a company is aptly illustrated in the table above. It should be borne in mind though that cost reduction can be achieved only by improving productivity. Productivity is often improved by investing

TABLE 2.1

Effect of Changing Costs on Profitability

Item	Present Situation	10% Price Increase	10% Sales Increase	2 (Sales of Y) Decrease Sales of X by Same Number	Reduce Cost by 10%
Sales (Units)					
Product X	160	176	176	120	160
Product Y	40	44	44	80	40
Cost ($)					
Variable Cost ($/unit)					
Product X	114	114	125.4	85.5	102.6
Product Y	24	24	26.4	48	21.6
Fixed Cost ($)					
Product X	38	38	38	38	34.2
Product Y	8	8	8	8	7.2
Total Cost ($)					
Product X	152	152	163.4	123.5	136.8
Product Y	32	32	34.4	56	28.8
Gross Profit ($)					
Product X	8	24	12.6	−3.50	23.2
Product Y	8	12	9.6	24	11.2
Total Profit	16	36	22.2	20.50	34.4
Increase in Profit					
Product X	—	16	4.6	−11.5	15.2
Product Y	—	4	1.6	16	3.2
Total Increase		20	6.2	4.50	9.4
Product X%		200%	57.5%	−143.75%	190%
Product Y%		50%	20%	200%	80%
Total Percentage		125%	38.8%	28.1%	115%

in more efficient and leaner processes and equipment. In the next section of this chapter, we will consider the different types of cost estimates.

2.2 Types of Cost Estimates

Five types of cost estimates are commonly used. They are listed as follows:

1. Overall factor method
2. Material cost method

3. Function method
4. Cost indexes
5. Cost capacity method

2.2.1 Overall Factor Method

According to the overall factor method, the estimated cost of an item can be expressed by the following equation:

$$C = F \times AM \qquad \text{(Equation 1)}$$

wherein:

C: Estimated cost of item.

F: Factor of total cost estimate, such as cost per mile, cost per square foot, cost of component per pound of casting, etc. This factor may be an individual factor or an overall factor that represents the sum of individual cost components such as direct labor, direct materials, indirect costs, etc.

AM: Amount of major cost item. This can also include cost of major equipment.

This method can also be used to estimate the cost of a plant. In that case, "C" would represent total plant cost with the remainder of the variables remaining the same.

Consider the following example:

EXAMPLE 2.1

A casting weighs 45 kg. What is the estimated cost of the casting if the cost factor is $2.00 per kg of casting?

Solution

Using the formula from Equation 1, we obtain the cost of the casting as follows:

$$\text{Estimated cost } (C) = F \times AM \qquad \text{(Equation 1)}$$

In this case, F = $2/kg and AM = 45 kg.

Substituting these values into Equation 2.1 we obtain the following:

$$\text{Estimated cost } (C) = \$2/\text{kg} \times 45\ \text{kg} = \$90 \text{ per casting}$$

Solution

EXAMPLE 2.2

A solar panel manufacturing company is considering expanding its facilities using state of the art technology. If the delivered equipment costs $3 million and the overall cost factor of this type of plant is 4.21, compute the total cost of the plant.

Solution

The total cost of the plant can be estimated by using the equation for the factor method.

In this case, the value of the factor is 4.21 and the major cost item is worth \$3 million. Thus the cost of the plant is estimated using Equation 1 as follows:

$$C = F \times AM \qquad \text{(Equation 1)}$$

Substituting respective values into the equation, we obtain the following:

$$C = 4.21 \times \$3\text{ million} = \$12.63\text{ million} \qquad \textbf{Solution}$$

EXAMPLE 2.3

A new CNC machining center is expected to have a delivered cost of about \$900,000. The cost factor for installation of foundation, concrete, steel, insulation, etc., is 0.51. The construction factor is 0.55 and the indirect cost factor is 0.22. Compute the total cost if:

1. All cost factors are applied to the cost of the delivered equipment.
2. The indirect cost factor is applied to the total direct cost.

Solution

1. Since the cost factors are applied to the cost of delivered equipment, we need to add all factors together and compute the total cost using Equation 1 as follows:

$$\text{Overall cost factor} = 1 + 0.51 + 0.55 + 0.22 = 2.28$$

Thus,

$$\text{Total cost} = 2.28 \times \$900{,}000 = \$2{,}052{,}000 \qquad \textbf{Solution 1}$$

2. In this case, since indirect cost factor is applied to total direct cost, we first need to compute the total direct cost.

From the data presented in the problem, we know the following:

$$\text{Indirect cost factor} = 0.22$$
$$\text{Construction factor} = 0.55$$

Installation of foundation, concrete, etc., factor = 0.51.
Thus,

$$\text{Total direct cost factor} = 1 + 0.55 + 0.51 = 2.06.$$

This can be used to compute the total direct cost as follows:

$$\text{Total direct cost} = 2.06 \times \$900{,}000 = \$1{,}854{,}000.$$

The indirect cost factor will now be applied to the total direct cost in order to compute the total cost as follows:

$$\text{Total cost} = \text{Total direct cost} \times \text{Indirect cost factor}.$$

Thus,

$$\text{Total cost} = \$1{,}854{,}000 \times 1.22 = \$2{,}261{,}880 \qquad \textbf{Solution 2}$$

2.2.2 Material Cost Method

The material cost method is used to estimate the total cost of the product using the following ratio:

$$\text{Cost} = \frac{\text{MC}}{\text{MCS}}. \qquad \text{(Equation 2)}$$

wherein:

MC: Material cost of item under consideration
MCS: Material cost share of item under consideration

The material cost shares of various items are presented in Table 2.2. Consider the following illustrative example.

TABLE 2.2

Material Cost Shares of Various Items

Item	Material Cost Share (%)	Share Classification
Automobile	70	Large
Railway freight car	68	Large
Chemical plant	64	Large
Railway passenger car	58	Large
Construction equipment	54	Large
Diesel engines	50	Large
Machine tools	48	Large
Amplifiers	45	Large
Clocks, watches, etc.	25	Medium
Electrical instrumentation	25	Medium
Measuring tools	20	Medium
Glass products	14	Small
China products	7	Small

Source: Adapted from Wierda, LF. 1988. *'Detailed Cost Estimation Concept and Database Structure for DIDACOE'*, Report: IM K 177, Delft University of Technology, The Netherlands, p. 122 [18] and Hubka, V. 1982. *'Bewertein und Entscheiden beim Konstruieren'*, Konstruktionsmethoden-Reading, Heurista WDK-8, pp. 164 [8].

EXAMPLE 2.4

The material cost of a truck is $3000. Compute the total cost of this truck.

Solution

In order to solve this problem, we need to use the value of material cost share percentages from the above table.

In the present case, the truck can be considered to be a class of automobile. As such its material cost share from the above table is 70%.

Using Equation 2, we can estimate the cost of the truck by substituting the respective values as follows:

Cost = $3,000/0.70 = $4,285.71 **Solution**

2.2.2.1 *Material Cost Method Applied to Different Industry Groups*

The material cost method can also be applied to different manufacturing industry groups. The production costs of about 20 different manufacturing industry groups were analyzed. It was realized that direct materials cost comprise the largest share of total costs. The material cost shares for different industry groups are presented in Table 2.3. It is expressed as direct material cost share of total annual sales.

TABLE 2.3

Direct Material Cost as a Percentage of Sales for Various Items

SIC Code	Industry Group	Direct Materials	Direct Labor	Indirect Payroll	Other
20	Beverages	0.65	0.08	0.05	0.22
21	Tobacco	0.55	0.07	0.02	0.36
22	Textile mills	0.56	0.20	0.05	0.19
23	Apparel	0.48	0.23	0.07	0.22
24	Wood products	0.52	0.21	0.05	0.22
25	Furniture and fixtures	0.45	0.24	0.09	0.22
26	Paper products	0.51	0.17	0.07	0.25
27	Printing and publishing	0.33	0.20	0.16	0.31
28	Chemicals	0.43	0.09	0.07	0.41
29	Petroleum products	0.75	0.04	0.02	0.19
30	Rubber and plastics	0.45	0.20	0.08	0.27
31	Leather	0.46	0.25	0.07	0.22
32	Stone, clay, and glass	0.41	0.21	0.08	0.30
33	Primary metals	0.56	0.17	0.06	0.21
34	Fabricated metals	0.46	0.21	0.09	0.24
35	Non-electrical machinery	0.42	0.21	0.12	0.25
36	Electrical equipment	0.42	0.19	0.14	0.25
37	Transportation equipment	0.57	0.15	0.08	0.20
38	Instruments	0.34	0.17	0.14	0.35
39	Miscellaneous manufacturing	0.44	0.21	0.10	0.25
Average for all manufacturing industries		0.51	0.16	0.10	0.23

EXAMPLE 2.5

The cost of raw materials to make 1000 kg of cast iron is \$250. Compute the expected selling price per ton of cast iron.

Solution

The factors presented in the above table are expressed as direct material cost share of annual sales. Thus, using the above values will help us determine the selling price (since revenue = selling price × sales volume).

From the table we can see that the direct materials cost share in the primary metals industry is 0.56. Cast iron can be classified as a primary metal. Also, remember that 1000 kg = 1 ton. Substituting the values into Equation 2, we obtain the following:

$$C = 250/0.56 = \$357.14 \text{ per ton}$$ **Solution**

2.2.2.2 1-3-9 Rule for Product Cost Estimation

The following is the significance of the numbers under the 1-3-9 rule for product cost estimation:

1: Material cost

3: Manufacturing cost

9: Selling price

It is to be noted that the material cost includes a 10% scrap factor and a 10% tooling cost.

Thus, the material cost = actual material value of the component + 20% addition for scrap costs and tooling costs.

Consider the following example.

EXAMPLE 2.6

Aluminum costs \$1.25 per kg. An aluminum part weighs 5 kg. Compute the estimated material cost, part cost, and selling price.

Solution

We can compute the material cost using the 1-3-9 rule as follows:

$$\text{Material cost} = \text{Material cost } (\$/\text{kg}) \times \text{Weight of the part (kg)} \times 1.2.$$

Notice the presence of the 1.2 factor in accordance with its inclusion in the material cost portion of the 1-3-9 rule.

Thus,

$$\text{Material cost} = \$1.25/\text{kg} \times 5 \text{ kg} \times 1.2 = \$7.5.$$

Thus, the part cost based on the 1-3-9 rule is 3 × \$7.50 = \$22.50.

The selling price = 9 × Material cost = 9 × \$7.5 = \$67.50. **Solution**

2.2.3 Function Method

The function method, though similar to the factor method, involves more variables. Also the expressions are not necessarily linear. The function provides cost estimates for the constituent variables. For instance, the expression for the cost of a turbofan engine can be written as follows:

$$C = 0.13937 \times a^{0.7345} \times b^{0.00775}. \qquad \text{(Equation 3)}$$

wherein:

C: Cost of the engine development, as expressed in $\$1 \times 10^6$
a: Maximum engine thrust as expressed in pounds
b: Number of engines produced

Consider the following example.

EXAMPLE 2.7

Compute the development cost of an engine if the engine thrust is 35,000 lbs and the company seeks to produce 250 engines.

Solution

The total development cost of the engine can be compute by using Equation 3 as follows:

$$C = 0.13937 \times a^{0.7345} \times b^{0.00775}$$

a: Maximum engine thrust as expressed in pounds = 35,000 lbs
b: Number of engines produced = 250 engines

Substituting these values into the equation, we obtain the following:

$$C = 0.13937 \times (35{,}000)^{0.7345} \times (250)^{0.00775}$$

Thus, the engine development cost = 0.13937 × 2175 × 1.043 = \$316.4 million. **Solution**

The main limitation of this formula is that it is highly specific to the application at hand, namely engine development, and as such cannot be used in a generic sense. Furthermore, the operating range for constituent variables is not provided either. It should be noted, however, that for any application, a function to estimate costs can be derived based on empirical observations. A mathematical model can be developed using such data. The principal drawback of this approach, as stated before, is their limited utility value.

Consider another model for estimating cost of a manufacturing operation. The following model was developed to estimate costs of a turning operation.

Its biggest drawback was that it was limited to parts that had a mean diameter ranging from 2 inches to 50 inches.

$$C = (R \times N_d/60) \times [(K_c \times L \times D_m/4.77) + S_t/(X \times Q)] + M. \quad \text{(Equation 4)}$$

wherein:

C: Cost of turned component expressed in \$/unit.
R: Hourly labor rate plus overhead expressed in \$/hour.
N_d: Complexity factor of the component.
K_c: Machine-material complexity factor. This is the product of K_1, K_2, and K_{mat}.
K_1: Machining factor.
K_2: Machine type factor.
K_{mat}: Material type factor.
L: Total machined length expressed in inches.
D_m: Mean diameter expressed in inches.
S_t: Machine set up time expressed in minutes.
X: Machine tooling capacity.
Q: Batch size expressed as number of units.
M: Material cost per unit including estimated scrap losses, expressed as \$/unit.

The following is an explanation of the various factors involved in the above formula:

The hourly labor rate is pretty straightforward and does not require much explanation.

N_d: The component complexity factor is comprised of the number of set ups in addition to the type and number of discontinuities on the machined surface. The discontinuity ratings for turned surfaces are presented in the following table. The discontinuity factors I expressed as the sum of the products of the number of discontinuities and the rating of each discontinuity. A discontinuity is defined as any surface feature that is not part of a smooth surface. For instance, a recess, threads, chamfers, etc., are examples of discontinuities.

K_1: K_1 is defined as a machining factor. It is a function of the tolerance and surface finish of the component being machined.

K_2: K_2 is defined as the machine type factor. It distinguishes between different machines and their capability to handle various machining situations. The factors for different machines are also detailed in a subsequent table.

K_{mat}: K_{mat} is a function of the rate of metal removal per unit horsepower. Different materials will be removed at different rates. These values are presented in Table 2.4 as well.

TABLE 2.4

Ratings Factor for Various Machined Features

Component	Rating Factor
Number of set ups	≥1
Unmachined surfaces	0
Machined surfaces	1
Threaded surfaces-not previously machined	1
Threaded surfaces-previously machined once	2
Threaded surfaces-previously machined twice	3
Shoulders or faces < 1/16 inch	0
Shoulders or faces ≥ 1/16 inch	1
Chamfers < 1/8 inch	0
Chamfers ≥ 1/8 inch	1
Radii < 1/8 inch	0
Radii ≥ 1/8 inch	1
Drilled holes	1
Reamed holes	1
Bored holes	1
Recesses ≤ ¼ inch	1
Recesses > ¼ inch	3

Source: Adapted from Mahmoud, MAM and Pugh, S. 1979. 'The costing of turned components at the design stage', *Information for Designers. Fourth International Symposium*, University of Southampton, pp. 37–42 [10].

X: X, in this case, represents machine tooling capacity. It represents the number of tools that the machine can handle at any time.

D_m: D_m represents the mean diameter and it stands for the average of all turned diameters.

L: L represents the total machined length. It includes features such as chamfers, radii, shoulders, faces, etc. (Tables 2.5–2.7).

TABLE 2.5

Factor K_1 for Different Ranges of Tolerance

Machining Factor K_1	Range of Tolerance (inches)	Type of Surface Finish
1.0	>0.010	Rough
1.5	0.005 < x < 0.010	Medium
2.0	<0.005	Fine, threading

Source: Adapted from Mahmoud, MAM and Pugh, S. 1979. 'The costing of turned components at the design stage', *Information for Designers. Fourth International Symposium*, University of Southampton, pp. 37–42 [10].

TABLE 2.6

Machine Tooling Capacity and Machining Factor for Different Machines

Machining Factor K_1	Machine Description	Machine Tooling Capacity: X
1.00	Center lathe	7
0.75	Capstan 2S lathe, 4S	11,12
0.80	Combination turret	14
0.33	NC lathe	12

Source: Adapted from Mahmoud, MAM and Pugh, S. 1979. 'The costing of turned components at the design stage', *Information for Designers. Fourth International Symposium*, University of Southampton, pp. 37–42 [10].

TABLE 2.7

Material Type Factor for Different Materials

Material Type Factor: K_{mat}	Material
1.00	Mild steel
0.60	Duralumin
0.35	Cast brass
0.70	Cast iron
0.50	Die cast aluminum
0.60	Bronze

Source: Adapted from Mahmoud, MAM and Pugh, S. 1979. 'The costing of turned components at the design stage', *Information for Designers. Fourth International Symposium*, University of Southampton, pp. 37–42 [10].

Consider an example to help illustrate the significance of the above equation and related factors.

EXAMPLE 2.8

Determine the cost per unit of the component presented in Figure 2.1. The component is to be rough turned and manufactured on a numerical controlled (NC) lathe. The values of various factors, set up time, batch size, material cost, and labor and overhead rate are presented in the following table.

Item	Value
Labor rate + overhead rate	$45/hour
Machining factor, K_1	1.0
Tooling capacity factor, X	12

(Continued)

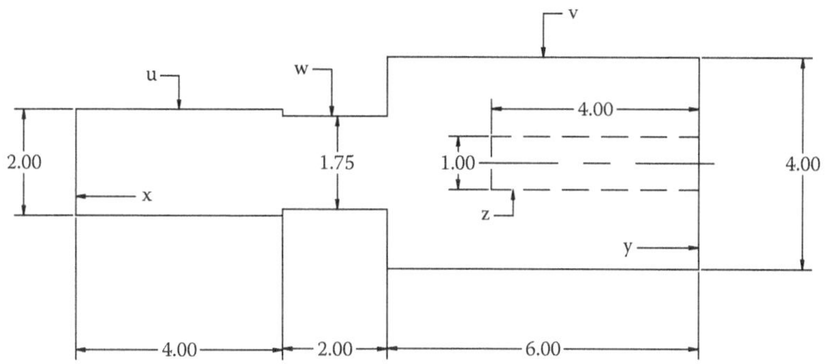

FIGURE 2.1
Turning layout for Example 2.8.

Item	Value
Material factor for mild steel, K_{mat}	1.00
Machine set up time, S_t	40 minutes
Batch size	300 units
Material cost including scrap	\$5/unit
Machine type factor, K_2	0.33

Solution

The following can be calculated the above figure:

$$\text{Mean diameter } (D_m) = \frac{2 + 1.5 + 4}{3} = 2.5 \text{ inches.}$$

The total length is calculated as follows:

$$\text{Total length} = \text{length of the part} + \text{face length} + \text{internal length.}$$

From the part drawing, we have the following values:

$$\text{Length to be turned} = 4'' + 1'' + 6'' = 11''$$

$$\text{Face length} = 2'' + 4'' = 6''.$$

$$\text{Internal length} = 4''.$$

Thus

$$\text{Total length} = 11 \text{ inches} + 6 \text{ inches} + 4 \text{ inches} = 21 \text{ inches.}$$

$N_d = 1 \times 2$ set ups (one set up each for surfaces x and y) + (1 × 1 hole (z)) + (1 × 2 machined surfaces (u and v) + 3 × 1 recess (w).

Thus,

$$N_d = 8.$$

$$K_c = K_1 \times K_2 \times K_{mat}.$$

From the respective tables, we can read the corresponding values and substitute them into the equation in order to obtain the value of the machine material complexity factor as follows:

$K_1 = 1$ for rough turning
$K_2 = 0.33$ for NC lathe
$K_{mat} = 1$ for mild steel

Thus,

$$K_c = 1 \times 0.33 \times 1 = 0.33.$$

The cost of machining the part can be estimated by using Equation 4 as follows.

$$C = (R \times N_d/60) \times [(K_c \times L \times D_m/4.77) + S_t/(X \times Q)] + M. \quad \text{(Equation 4)}$$

Substituting values into the above equation, we obtain the following:

$$\begin{aligned} C &= (\$45/\text{hour} \times 8/60) \times [(0.33 \times 21 \text{ inches} \times 2.5 \text{ inches}/4.77) \\ &\quad + 40/(12 \times 300)] + \$5/\text{unit}. \\ &= \$6 \times (3.63 + 0.0111) + \$5. \\ &= \$21.85 + \$5. \\ &= \$26.85/\text{unit}. \end{aligned}$$

2.2.4 Cost Indexes

A cost index is a ratio of two quantities: present cost of the item to the cost of the item at some point in the past. Because it is an index of similar variables, it has no dimension. Its main function is to measure the change of cost over a period of time. Given the obvious sensitivity of the index to factors such as time, technological up gradation, etc., it is essential to update the index over time.

Table 2.8 illustrates examples of several indexes used to estimate values of different variables in various sectors of the economy.

The cost of an item at time "t" can be obtained by using the general concept of cost indexes. This is expressed through Equation 5 as follows:

$$\frac{C_t}{C_0} = \frac{I_t}{I_0}. \quad \text{(Equation 5)}$$

wherein:

C_t: Estimated cost at time "t"
C_0: Cost at time t_0
I_t: Index value at time "t"
I_0: Index value at time t_0

TABLE 2.8

Examples of Various Types of Cost Indexes

Index	Source of Information
Construction	
Overall chemical plant	Chemical engineering
Equipment, machinery, and supports	
Buildings	
Engineering and supervision	
Construction labor costs	
Engineering News Record	
Construction costs	Engineering news record can be accessed at: www.construction.com
Building costs	
Common labor costs	
Skilled labor costs	
Material costs	
FRED Indexes	
Construction index	Federal Reserve Economic Data (FRED)
House price index	
Construction index	
Import price index	
Equipment Indexes	
Marshall and Swift overall index	Marshall and Swift
Marshall and Swift specific industries index	

As an example, values of different indexes from the Federal Reserve Economic Data (FRED) database are presented in Table 2.9. The data presented in the table begins with year 2010 and continues until year 2015.

In order to fully understand the application of cost indexes to real world problems, consider Example 2.9.

TABLE 2.9

Examples of two FRED Cost Indexes' Values Over Six Years

Year	Crude Oil Index (2005: Baseline 100)	Consumer Price Index (1982–1984: Baseline 100)
2010	160.6	220.5
2011	194.4	227.2
2012	192.3	231.3
2013	196.8	234.8
2014	140.2	236.5
2015	79.3	238.0

EXAMPLE 2.9

An engineering manager wishes to estimate the cost of skilled labor for a construction job. Upon doing some background research, he finds that a project similar in scope and magnitude was successfully completed four years ago. The skilled labor cost on that project amounted to $450,000. The skilled project index at the time was 3550. The same index stands at 5175 today. Can you help the manager estimate the cost of skilled labor for the new project?

Solution

From the problem statement, the following inferences can be drawn:

C_0 = Cost of skilled labor four years ago = $450,000
C_t = Estimate of cost of skilled labor at the present moment in time
I_0 = Value of index four years ago = 3550
I_t = Value of index today = 5175

The cost estimate for skilled labor at the present moment in time can be obtained by using Equation 5 as follows:

$$\frac{C_t}{C_0} = \frac{I_t}{I_0}. \qquad \text{(Equation 5)}$$

Substituting the respective values into the equation, we obtain the following:

$$C_t = \$450{,}000 \times (5175/3550).$$

Thus,

$$C_t = \$450{,}000 \times (1.4577) = \$655{,}985.91.$$

Thus, it is estimated that the cost for skilled labor will amount to about $655,985.91 at the present time. **Solution**

2.2.5 Relationship between Cost and Size: Cost Capacity Equations

Unit size can be used to estimate the cost of a product. Costs of different designs can be compared either using the actual cost as a basis for evaluation or on a relative basis. Similarly, design variables for equipment, plants, etc., are determined during the early stages of construction. Examples of such variables include physical size, speed, weight, and thrust. An example of this is the cost capacity equation.

According to the cost capacity equation, the cost of a component, system or plant is related to its capacity. It is often referred to as the sizing model or the power law. The common form of the cost capacity equation is represented as follows:

$$C_2 = C_1\{Q_2/Q_1\}^x. \qquad \text{(Equation 6)}$$

wherein:

C_1: Cost at capacity Q_1
C_2: Cost at capacity Q_2
x: Correlating exponent

The value of the exponent "x" for various components, systems, and plants can be derived from various sources such as journals and handbooks in chemical engineering. Examples of such publications include chemical engineer handbooks and various technical journals. A partial listing of exponents for various units is presented in Table 2.10.

It should be noted that when the exponent value for a particular item is not known, it should be assumed as being $x = 0.6$.

TABLE 2.10

Cost Capacity Equations: Exponent Values for Different Components

Component	Size Range	Exponent Value
Activated sludge plant	1–100 MGD	0.84
Aerobic digester	0.2–40 MGD	0.14
Blower	1000–7000 ft/min	0.46
Centrifuge	40–60 inches	0.71
Chlorine plant	3000–350,000 tonnes/year	0.44
Clarifier	0.1–100 MGD	0.98
Reciprocating compressor	5–300 hp	0.90
Compressor	200–2100 hp	0.32
Cyclone separator	20–8000 cu.ft/min	0.64
Dryer	15–400 sq.ft	0.71
Sand filter	0.5–200 MGD	0.82
Heat exchanger	500–3000 sq.ft	0.55
Hydrogen plant	500–20,000 scfd	0.56
Laboratory	0.05–50 MGD	1.02
Aerated lagoon	0.05–20 MGD	1.13
Centrifugal pump	10–200 hp	0.69
Reactor	50–4000 gallons	0.74
Sludge drying beds	0.04–5 MGD	1.35
Stabilization pond	0.01–0.2 MGD	0.14
Stainless tank	100–2000 gallons	0.67

MGD: Millions of gallons per day; scfd: standard cubic feet per day; hp: horsepower.

It will be observed from Table 2.8 that the value of the exponent "x" should ideally lie between 0 and 1. The following rule is applicable to all exponents.

- If $x < 1$, cost advantages for larger sizes are provided by economies of scale.
- If $x = 1$, a linear relationship exists.
- If $x > 1$, this indicates the presence of diseconomies of scale. This indicates that a larger size is more expensive than that of a linear relationship.

The cost at time "t" can be determined by writing the following equation:

$$C_t = C_0 \left\{ \frac{I_t}{I_0} \right\} \qquad \text{(Equation 7)}$$

wherein:

C_t: Estimated cost at present time "t"
C_0: Cost at previous time t_0
I_t: Index value at time t
I_0: Index value at time t_0

The cost at time "t" and capacity level 2 may be derived by combining Equations 5 and 6. This is written as follows:

$$C_2 = C_1 \left\{ \frac{Q_2}{Q_1} \right\}^x \left\{ \frac{I_t}{I_0} \right\}. \qquad \text{(Equation 8)}$$

Consider the following example in order to better understand the underlying concepts of Equation 7.

EXAMPLE 2.10

An activated sludge plant is being designed to handle a flow of 2 million gallons per day (MGD). The design and construction cost for this project was \$3 million in 2012. Estimate its cost in 2017 if the flow rate was 3 MGD. The cost index in 2012 was 150. It has been updated to 258 for 2017.

Solution

It will be observed from the problem statement that this problem involved both capacity as well as time-adjusted costs. Thus, Equations 2.6 and 2.7 by themselves would be insufficient to solve this problem. We would need to use the hybrid form of the equation, namely Equation 2.8, to solve this problem.

From Table 2.8, we notice that the value of the exponent, x, for an activated sludge plant is 0.84.

Thus, the estimated cost is given by the equation:

$$C_2 = C_1 \left\{\frac{Q_2}{Q_1}\right\}^x \left\{\frac{I_t}{I_0}\right\}. \qquad \text{(Equation 8)}$$

The following are the values that need to be substituted:

C_1: Cost in 2012. This was \$3 million
Q_2: 3 MGD
Q_1: 2 MGD
x: 0.84
I_t: 258
I_0: 150

Substituting the above values into Equation 8, we obtain the following:

$$\begin{aligned}\text{Cost in } 2017 &= \$3{,}000{,}000 \times \{3\ \text{MGD}/2\ \text{MGD}\}^{0.84} \times \{258/150\}\\ &= \$3{,}000{,}000 \times (1.406) \times (1.72)\end{aligned}$$

Thus, the cost in 2017 would be \$ 7,254,960. **Solution**

The cost size relationship allows one to compare the costs of different designs relative to each other. This comparison can also be done on an absolute basis. The cost capacity relationship indicates that there exists a direct correlation between the size of equipment or plant capacity and its corresponding cost. However, this correlation is not necessarily linear in nature. Cost capacity relationships are often used in chemical industries. In this section, we have examined the relationship between cost and size. It can be written as follows:

$$C = A \times V^B. \qquad \text{(Equation 9)}$$

wherein:
C: Cost per unit of part of size "V", expressed in \$/unit
V: Size of the part
B: Cost-size exponent
A: Cost-size constant

An expression for estimating cost for the investment casting process was developed by Michaels and Wood (1989). It can be expressed as follows:

$$\text{Relative cost of investment casting} = 5V^{0.6}. \qquad \text{(Equation 10)}$$

wherein:
V: Volume of investment casting, expressed in cubic inches.
0.6: Cost size exponent for investment casting process. This is actually the value of "B" for that process.
5.0: Relative cost of one cubic inch of investment casting.

Let us consider an example to illustrate the practical significance of this cost estimation technique.

EXAMPLE 2.11

A turbine blade is being manufactured using the investment casting method. The objective is to triple the volume of this blade. Currently, the volume of the blade is 5 cubic inches. It is sought to triple this size to 15 cubic inches. What would be cost of a 15 cubic inch blade compared to the cost of a one cubic inch blade?

Solution

We can use Equation 9 in order to solve this problem.

Thus, the cost of a 15 cubic inch blade can be computed by using the cost size exponent for investment casting and the comparative size relationships as follows:

$$C = \text{Cost of a 5 cubic inch blade} \times (15/5)^{0.6}$$

In this case, the volume of the blade has been tripled. This is expressed by the ratio: 15/5 in the above equation.

Thus, the cost of the 15 cubic inch blade would be 1.933 times that of the cost of a 5 cubic inch blade.

Similarly, comparing the cost of a 15 cubic inch casting to a one cubic inch casting, we have the following relationship:

$$\text{Cost of a 15 cubic inch casting} = \text{Cost of a one cubic inch casting} \times (15/1)^{0.6}$$

Solving the above equation, we get the following:

$$\text{Cost of a 15 cubic inch casting} = \text{Cost of a one cubic inch casting} \times 5.077$$

This implies that the cost of a 15 cubic inch casting is about five times the cost of a one cubic inch casting. **Solution**

2.3 Conclusion

Cost estimation is one of the important facets of manufacturing costing. It plays a vital role in production planning and planning of operations. While estimates are arrived at prior to actual production, cost estimation offers a window into the different costs that are likely to be encountered during manufacturing.

Several techniques related to cost estimation were discussed at length in this chapter. Most of these techniques deal with "scaling up" prevalent cost data in order to fit the end goal. Examples were drawn not just from manufacturing but from chemical engineering as well to facilitate the estimation process. To this end, several numerical examples were solved to help clarify fundamental concepts related to each method of cost estimation.

PROBLEMS

1. The cost of a recently purchased 500 HP compressor was $20,000. How much would an 800 HP compressor cost? *(Answer: $23,246).*
2. A manufacturing process has a capacity of 15,000 units per day. The cost for implementing this process is about $1,000,000. The cost for a plant with a capacity of 250,000 units per day was $7 million. Compute the value of the exponent in the cost capacity equation. *(Answer: 0.69).*
3. An engineering manager wishes to estimate the cost of some chemical equipment in 2017. The corresponding cost of a unit one-third its size in 2010 was $100,000. The value of the exponent in the cost capacity equation is 0.33. Use the equipment cost index to arrive at the cost estimate.
4. The cost of some mineral extraction equipment is $3 million. The overall cost factor of this type of facility is 2.75. Compute the estimated plant cost.
5. The cost of construction of a desulfurization system at a 500 MW power plant is estimated to be $300 million. A similar system at a smaller plant costs $45 million. The value of the exponent in the cost capacity equation is 0.67. Compute the size of the smaller plant.
6. A cyclone system has a capacity of 75,000 cubic feet per minute. It is estimated to cost $600,000. An actual value of $300,000 for a 40,000 cubic feet per minute system was used in the cost capacity equation. Compute the value of the exponent.
7. A mass spectrometer can be purchased today for $110,000. This cost is expected to increase based on the specialty laboratory equipment cost index over the next ten years. If the specialty equipment inflation rate is estimated to be 3% per year for the next four years and 7% per year for every subsequent year, compute the cost of the spectrometer in ten years.
8. Equipment necessary for a manufacturing operation cost $250,000 when the FRED equipment cost index was 210. Compute the estimated cost of a similar system in 2015 when the index was at 245.9. *(Answer: $303,425.38).*

9. The cost of a piece of equipment was $100,000 when the relevant cost index was 1501.7. What was the index value when the same equipment was estimated to cost $125,000? *(Answer: 1877.13)*.
10. A compressor and pump system cost $120,000 in 2011. Applying the FRED equipment cost index, compute the estimated cost for a similar system when the index value is 275. *(Answer: $145,246.48)*.
11. Estimate the cumulative average number of direct labor hours for a product requiring 7 hours for the first unit, if 28 units are to be produced with an improvement of 25% between doubled production quantities.
12. A piece of equipment cost $15,000 in 2009 and must soon be replaced. The present equipment is to be replaced in 2019 when the price index is estimated to be 173 with 2009 as the base year. What is the cost of the unit in 2019 if the coefficient for this equipment is 0.7?
13. The net annual operating cost for a plant was estimated to be $3 million annually. The estimate was based on the $300,000 per year cost of a 1-MGD plant. If the exponent in the cost capacity equation is 0.60, what was the size of the larger plant?
14. The cost of a high quality 400-horsepower compressor was $33,000 when recently purchased. What would a 550-horsepower compressor be expected to cost? *(Answer: $36,540.19)*.
15. Estimate the cost in 2016 of a 1050-horsepower air compressor if a 300-horsepower unit cost $265,000 in 2007. The exponent in the cost capacity equation is 0.45. The equipment cost index increased by 42% between the years.
16. The price of a jet engine is nearly $13 million in 2017. What was the price of the same engine in 2007 if the engine and engine cost indexes are 177.5 and 205.0 for 2007 and 2017, respectively? *(Answer: $11,256,097.56)*.
17. The cost for construction at a 730-megawatt plant was estimated to be $434 million. If a smaller plant has a cost of $80 million and the exponent in the cost capacity equation is 0.78, what was the size of the smaller plant that served as the basis for the cost projection?

3

Materials Costing

3.1 The Role of Materials' Cost in Product Costing

Every product is comprised of some combination of materials. In keeping with the needs of the consumer, materials have evolved over time. During the dawning years of the twentieth century, relatively simple materials such as steel were used for different purposes. The passage of time has given way to not only technological sophistication, but a greater degree of material complexity as well. As a matter of fact, it would not be a stretch to state that technological complexity and material complexity are symbiotic in nature. In view of the central role played by materials in product functionality, structural rigidity, usability, manufacturability, assembly, disassembly, and end of life disposal, it is obvious that material costing should be studied as an important element of product cost.

In terms of the costing structure itself, the amount of funds invested in obtaining the appropriate material for the product are defined as direct material cost. The role of direct material cost as a part of total product cost has evolved over time as well. For instance, direct labor cost comprised the major chunk of total product cost during the early part of the twentieth century. This is no longer the case. An analysis of contemporary product costing structure reveals that material cost now constitutes the largest element of product cost. The typical costing structure for a typical large equipment manufacturing company is depicted in Figure 3.1.

Adopting many approaches can minimize direct labor cost. Direct material, on the other hand, is a different story. Researchers have tried to devise different approaches to reduce direct material cost. These approaches are explained below:

- The first approach consists of teaming up with the designer in order to select an appropriate material that is not only cost effective, but functional as well. This implies that product functionality would not be compromised at the expense of lower costs.
- The second approach is to optimize the operation sequence by minimizing scrap. Attaching a monetary value minimizes scrap

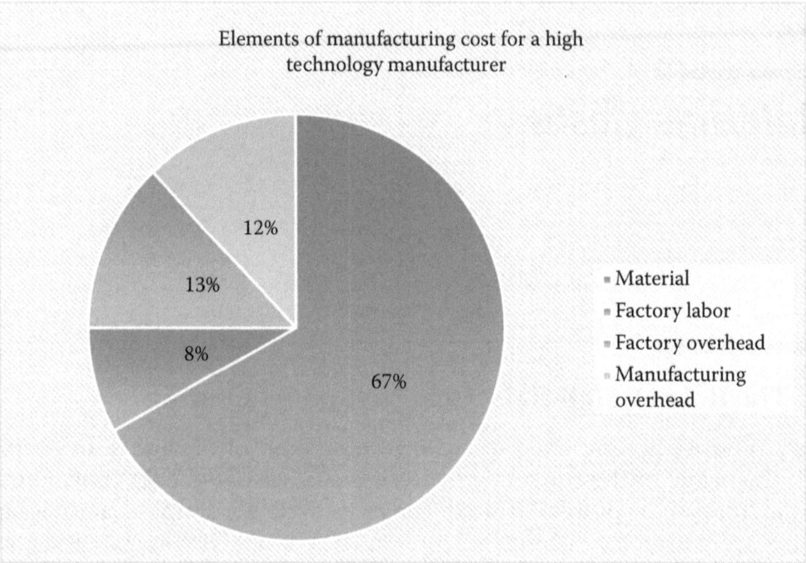

FIGURE 3.1
Components of manufacturing cost.

and defines it as waste. An optimal operating sequence is bound to lead to lower scrap rates, higher efficiency, and consequently lower product costs.

This chapter presents the aforementioned two methods used to compute materials costs.

3.2 Material Cost Selection System Based on Performance

The basic premise of material selection based on performance is that the intended material should satisfy the design requirements set forth by the designer. As such, it should meet requirements related to structural rigidity, product functionality, ease of manufacturing, assembly and disassembly, and ease of maintenance. As a secondary requirement, it should also satisfy requirements related to environment conscious disposal. Environment conscious waste disposal is an increasingly important design objective because of strict government regulations concerning environmental degradation.

It should be obvious at the outset that different materials have different design properties. Thus, engineering design parameters such as thickness, cross sectional area, etc., will be different for different materials. Thus, a range of materials can satisfy the same functional requirements. Each material possesses a unique engineering design signature. For instance, specific construction functionality can be satisfied by cast iron as well as steel. However, in light of the different properties of each material, it is easy to appreciate that the dimensions of engineering design, such as thickness and cross sectional area, will be different for each material. This implies that the amount of cast iron used in a structure would be quite different from the amount of steel used in the same structure to satisfy the same functional requirement. Depending on the material cost, it could be more economical to use one material instead of the other.

Material cost is generally expressed in terms of \$/lb. The total cost is expressed using Equation 1 as follows:

$$\text{Cost} = \text{C (m)} \times \text{M}. \qquad \text{(Equation 1)}$$

wherein:

C (m): Material cost expressed in \$/lb
M: Material weight expressed in lbs. (pounds) per unit

Furthermore, material weight can be expressed as a function of material density. Material density is unique to each material.

$$\text{Density} = \frac{\text{Mass}}{\text{Volume}}$$

Cross-multiplying to solve for mass, we obtain the following expression:

$$\text{Mass} = \text{Density} \times \text{Volume}.$$

Substituting this relationship into Equation 1, we obtain the modified form of that equation as follows:

$$\text{Cost} = \text{C (m)} \times \text{Density} \times \text{Volume}.$$

A modified form of this equation is represented by Equation 2.

$$\text{Cost} = \text{C (m)} \times \rho \times \text{Volume}. \qquad \text{(Equation 2)}$$

wherein:

ρ: Density of the material expressed in lb./in^3
V: Volume of material expressed in cubic inches

Let us take Equation 2 one step further. Basic design relationships are comprised of the interplay between cross sectional area, length, volume, and density.

TABLE 3.1

Basic Design Relationships Used for Structural Cross Sections

Shape of Cross Section	Type of Loading	Design Relationship Stress	Design Relationship Deflection
Solid cylinder	Bending under center load	$S = mc/I$	$\delta = PL^3/48 \times EI$
Solid bar, rod or cylinder	Tension and compression	$S = P/A$	$\delta = PL/AE$
Solid rectangular plate	Bending under center load	$S = mc/I$	$\delta = PL^3/48 \times EI$

Volume can be expressed as the product of cross sectional area and length. Using this relationship, Equation 2 is further modified as follows:

$$\text{Cost} = C\ (m) \times \rho \times A \times L. \qquad \text{(Equation 3)}$$

wherein:

A: Cross sectional area expressed in square inches

L: Length of the design element, expressed in inches

Table 3.1 depicts the basic design relationships used for structural cross sections.

The nomenclature for the terms in Table 3.1 are presented below:

I: Moment of inertia expressed as inch^4.

m: Moment, expressed as inch-pounds or N-m.

c: Distance from neutral axis to the outermost layers. This is expressed in inches.

S: Material strength as expressed in pounds per square inches (psi).

A: Cross sectional area expressed in square inches.

P: Amount of load applied to the part, expressed in pounds.

The design relationships depicted in Table 3.1 can be further modified using material properties such as modulus of elasticity and material strength. Expressions can be developed for specific material properties such as yield strength, density, modulus of elasticity, and material cost. Knowledge of the aforementioned parameters helps the designer in choosing the appropriate material as well as costing it.

Detailed design and cost relationships for strength and deflection are presented in Table 3.2.

EXAMPLE 3.1

An engineer manager is looking for the best material to construct a solid bar. The bar should be able to support a load of 55,000 lbs. over a length of 15 inches. The manager may choose from a variety of materials. The pertinent design data is presented in the following table. Based on this information, which material should be chosen?

Material	Yield Strength, S (kpsi)	1/S	Density, ρ (lbs./in³)	ρ/S	Elastic Modulus, E (Mpsi)	Material Cost C (m), ($/lbs.)
Magnesium						
AS91C-T6	21	0.0476	0.066	0.00314	6.5	2.00
AK60A-T5	41	0.244	0.066	0.0161	6.5	2.20
Aluminum						
2024-T861	66	0.01515	0.100	0.001515	10.5	1.00
3003-H18	29	0.0344	0.099	0.003406	10.2	0.95
7175-T66	76	0.01315	0.101	0.00133	10.4	1.25
Titanium						
Ti-6Al-6V-2Sn	140	0.00714	0.165	0.001178	16	6.00
Ti-13V-11Cr-3Al	160	0.00625	0.175	0.00109	16	6.25
Steel						
1015	50	0.02	0.285	0.0057	30	0.30
4140 Q&T	252	0.003968	0.285	0.001133	30	1.25
304 Stainless steel	85	0.0176	0.287	0.00505	28	2.00
Cast Iron						
Gray cast iron	20	0.05	0.250	0.0125	15	0.20
Ductile iron	50	0.02	0.250	0.005	30	0.25

kpsi: kilo-pound per square inch, Mpsi: mega-pound per square inch.

Solution

In the context of the present problem, "best material" can be defined in a variety of ways as follows:

1. Lowest cost material that will meet the prescribed strength requirement
2. Material with the lowest volume
3. Material with the lowest weight

Each of the aforementioned criteria are solved in the following sections:

Lowest Cost Material That Will Meet the Prescribed Strength Requirement

In order to search for the lowest cost material that will meet the prescribed strength requirement, we use the following equation:

$$\rho \times C\,(m)/S.$$

The values for each material corresponding to the above expression are calculated and the material corresponding to the lowest value is chosen.

TABLE 3.2

Material Properties for Design and Material Costing

			Design Equation for Area		General Material Relationship		General Cost Performance Relationship	
Shape of Cross Section	**Type of Load**	**Design Area**	**Strength**	**Deflection**	**Strength**	**Deflection**	**Strength**	**Deflection**
Solid bar, rod or cylinder	Compression or tension	A	P/S	PL/δE	1/S	1/E	ρ × C (m)/S	ρ × C (m)/E
Solid bar	Bending	A						
• Circular cross section • Rectangular cross section	Single load	Π.D^2/4	[P.L(Π)$^{0.5}$/S]$^{2/3}$	[P.L^3/3δE]$^{1/2}$	(1/S)$^{2/3}$	(1/E)$^{1/2}$	ρ × C (m)/S$^{2/3}$	ρ × C (m)/E$^{1/2}$
W: Width of cross section H: Height of cross section	Single load	W × H	[3PLW/2S]$^{1/2}$	[P.L^{3}W^2/4 δE]$^{1/3}$	(1/S)$^{1/2}$	(1/E)$^{1/3}$	ρ × C (m)/S$^{1/2}$	ρ × C (m)/E$^{1/3}$

Thus, for magnesium AS91C-T6, the value of the above expression is calculated by substituting the respective values into the expression.

For this alloy,

$$C\ (m) = \$2/\text{lb.}$$

$$S = 21\ \text{kpsi},\ \rho = 0.066\ \text{lbs./in}^3$$

Thus, for magnesium AS91C-T6, the value of the expression is given as follows:

$$0.066 \times \$2/\text{lb.}/21\ \text{kpsi} = 0.0062857.$$

A comprehensive list of values for all materials is presented in the following table:

Material	**$\rho \times$ C (m)/S**
Magnesium	
AS91C-T6	0.00628
AK60A-T5	0.00354
Aluminum	
2024-T861	0.00152
3003-H18	0.00307
7175-T66	0.00166
Titanium	
Ti-6Al-6V-2Sn	0.00707
Ti-13V-11Cr-3Al	0.00683
Steel	
1015	0.00171
4140 Q&T	0.00141
304 Stainless steel	0.00675
Cast Iron	
Gray cast iron	0.00250
Ductile iron	0.00125

Comparing the values from the above table, it is clear that ductile iron, with a value of 0.00125, is the best choice.

The next task is to determine the actual cost of the material. The cost of the material is computed using Equation 3 as follows.

$$\text{Cost} = C\ (m) \times \rho \times A \times L. \qquad \text{(Equation 3)}$$

In the equation stated above, all values are known except the cross sectional area.

The design objective in this problem is to construct a solid bar. The cross sectional area for a solid bar can be computed by using the expression P/S from Table 3.2.

In this case, the magnitude of the load to be applied is 55,000 lbs. Thus, the value of P = 55,000 lbs.

The value of "S" is read from the data given in the problem itself. Thus, the value of "S" for ductile iron = 50 kpsi.

Substituting these values into the equation for cross sectional area, we obtain the following:

$$A = P/S = 55{,}000 \text{ lbs.}/50 \text{ kpsi} = 1.1 \text{ in}^2$$

Substituting this value into Equation 3, we obtain the material cost for ductile iron as follows:

$$\text{Cost} = \$0.25/\text{lb.} \times 0.250 \text{ lb./in}^3 \times 1.1 \text{ in}^2 \times 15 \text{ inches.}$$

Thus,

$$\text{Material cost} = \$1.03125/\text{unit.}$$

Next, we will consider the second constraint, namely, volume.

Material with the Lowest Volume

If the design objective is to minimize material volume, the material with the highest yield needs to be selected. This will invariably be the material with the lowest (1/S) value.

From the data presented in the problem, the lowest 1/S value corresponds to 4140 Q&T steel.

In order to obtain the volume of the material, we use the following equation:

$$\text{Volume} = \text{Area} \times \text{Length.}$$

Cross sectional area can be calculated, as before, using the following equation:

$$A = P/S.$$

Substituting this value into the equation for volume, we obtain the following expression:

$$\text{Volume} = (P/S) \times \text{Length.}$$

Thus,

$$\begin{aligned} \text{Volume} &= (55{,}000 \text{ lbs.}/252 \text{kpsi}) \times 15 \text{ inches.} \\ &= 0.218 \text{ in}^2 \times 15 \text{ inches.} \\ &= 3.27 \text{ in}^3 \end{aligned}$$

Furthermore, mass of the product can be computed by using the following equation:

$$\text{Density} = \frac{\text{Mass}}{\text{Volume}}$$

Thus,

$$\text{Mass} = \text{Density} \times \text{Volume}.$$

Therefore, mass of the object can be computed by substituting the respective values into the above equations as follows:

$$\text{Mass} = 0.285 \text{ lbs./in}^3 \times 3.27 \text{ in}^3.$$

This gives us the value of mass = 0.932 lbs.

Next, we will consider the second constraint, namely, weight.

Material with the Lowest Weight

If the design objective is to minimize material weight, the material with the lowest ρ/S value should be selected.

Referring to the table from Example 3.1, the lowest ρ/S value is 0.00109 and corresponds to titanium Ti-13V-11Cr-3Al alloy.

The mass of the object can be calculated by using the following equation:

$$\text{Mass} = \text{Density} \times \text{Volume}.$$

Expressing volume as the product of cross sectional area and length, we have the following modified equation:

$$\text{Mass} = \text{Density } (\rho) \times \text{Cross Sectional Area} \times \text{Length}.$$

As before, cross sectional area is computed using the expression: P/S.

For titanium Ti-13V-11Cr-3Al alloy, the cross sectional area for the given load constraint = 55,000 lbs./160 kpsi = 0.34375 in^2.

Substituting this value into the equation for mass, we obtain the mass of titanium Ti-13V-11Cr-3Al alloy for the given application as follows:

$$\text{Mass} = 0.175 \text{ lb./in}^3 \times 0.34375 \text{ in}^2 \times 15 \text{ inches}.$$

Thus,

$$\text{Mass} = 0.902 \text{ lbs}.$$

The problem above proves that different materials will fit different descriptions. The main design objective needs to be properly defined before the appropriate material can be chosen. Let us consider one more example to help demonstrate this concept.

EXAMPLE 3.2

Which of the following materials should be chosen as the lowest cost material to construct a solid bar designed to support a load of 60,000 lbs.? The bar can only have a maximum extension of 0.10 inches over a length of 12.5 inches.

- Aluminum 7175-T66
- Steel 4140 Q&T
- Titanium Ti-13V-11Cr-3AL alloy

Solution

This problem can be solved using the following steps:

1. Computation of cross sectional area in keeping with the following conditions:
   - Load of 60,000 lbs
   - Maximum extension limit = 0.10 inches over a length of 12.5 inches

In order to compute the cross sectional area for the load constraint, we use the A = P/S relationship. In order to compute the cross sectional area for the extension constraint, we use the A = PL/δE relationship. An illustrative example for aluminum 7175-T66 is demonstrated below.

Computation of Cross Sectional Area for a Load of 60,000 lbs. for Aluminum 7175-T66

Using equation A = P/S and substituting the respective values from the data given in the problem, we have the following value for cross sectional area:

$$A = 60{,}000 \text{ lbs.}/76 \text{ kpsi} = 0.789 \text{ in}^2.$$

Computation of Cross Sectional Area for an Extension of 0.10 inches over a Length of 12.5 inches for Aluminum 7175-T66

Using equation A = PL/δE from Table 3.2 and substituting the respective values from the data given in the problem, we have the following value for cross sectional area:

$$A = 60{,}000 \text{ lbs.} \times 12.5 \text{ inches}/0.10 \text{ inches} \times 10.4 \text{ Mpsi}.$$

In this case, L = original length = 12.5 inches.

$$\delta = \text{value of extension} = 0.10 \text{ inches}.$$

$$E = \text{Elastic modulus} = 10.4 \text{ Mpsi for Aluminum 7175-T66}$$

Thus,

$$\text{The value of cross sectional area} = 0.7211 \text{ in}^2.$$

Following the method used above, the values for the remaining materials can be calculated. The values for the three materials are presented in the following table.

	Cross Sectional Area (in^2)	
Material	**Load, A = P/S**	**Extension A = PL/δE**
Aluminum 7175-T66	0.789	0.7211
4140 Q&T steel	0.238	0.2500
Ti-13V-11Cr-3Al	0.375	0.468

We choose the higher cross sectional area for each material in order to ensure functionality. Choosing a higher cross sectional area includes the added benefit of a built-in factor of safety, therefore enhancing functionality.

Based on this logic, we choose the following cross sectional area for each of the materials under consideration in this problem:

For Aluminum 7175-T66; A = 0.789 in^2
For 4140 Q&T Steel; A = 0.2500 in^2
For Ti-13V-11Cr-3Al; A = 0.468 in^2

As before, the unit cost for each material can now be calculated by using the equation:

$$\text{Cost} = C\,(m) \times \rho \times A \times L. \qquad \text{(Equation 3)}$$

We can substitute the respective values from the data given in the problem.

Thus,

$$\text{Cost for Aluminum 7175-T66} = \$1.25/\text{lb.} \times 0.101\ \text{lb./in}^3 \times 0.789\ \text{in}^2 \times 12.5\ \text{inches.}$$

This equates to a unit cost of $1.245/unit.

Similarly,

$$\text{Cost for 4140 Q\&T Steel} = \$1.25/\text{lb.} \times 0.285\ \text{lb./in}^3 \times 0.25\ \text{in}^2 \times 12.5\ \text{inches.}$$

This equates to a unit cost of $1.113/unit.

$$\text{The cost for Ti-13V-11Cr-3Al} = \$6.25/\text{lb.} \times 0.175\ \text{lb./in}^3 \times 0.468\ \text{in}^2 \times 12.5\ \text{inches.}$$

This equates to a unit cost of $6.398/unit.

Considering the cost structure as computed above, the obvious choice is the lowest cost material. In this case, it happens to be 4140 Q&T Steel corresponding to a unit cost of $1.113/unit.

So far, we have solved examples related to strength and deflection. Next, we will consider an example that involves bending.

EXAMPLE 3.3

An engineer has been assigned the task of designing a solid plate to support a 7500 lbs. load in bending over a span of 45 inches. The width of the plate is 15 inches. The two materials available to the engineer are aluminum 3003-H18 and 1015 steel. Which material should be selected?

Solution

The design task in this problem is to design a solid plate. The cost performance relationship for a solid plate can be calculated from Table 3.2. It is written as follows:

$$\rho \times C\ (m)/S^{1/2}$$

wherein each notation corresponds to predefined variables.

Substituting the respective values from the data presented in Example 3.1, we get the following values for each material:

For aluminum 3003-H18, the value of $\rho \times C\ (m)/S^{1/2}$ is computed as follows:

$$0.099\ \text{lb./in}^3 \times \$0.90/\text{lb.}/(29)^{1/2}$$

This equates to a value of 0.0165.

Similarly, for 1015 Steel, the value of $\rho \times C\ (m)/S^{1/2}$ is computed as follows:

$$0.285\ \text{lb./in}^3 \times \$0.30/\text{lb.}/(50)^{1/2}$$

This equates to a value of 0.01209.

Since $0.01209 < 0.01605$, it implies that 1015 steel is the more economic choice over aluminum 3003-H18.

In order to compute the actual cost of each material, once again, we start with computation of the cross sectional area necessary to ensure functionality under the given design conditions. Cross sectional area computation is presented as follows.

In this example, the design condition is one pertaining to strength only. There is no design condition of deflection mentioned in the problem statement.

For a plate with a width of cross section "w" and height of cross section "h", under conditions of strength the cross sectional area can be obtained by using the equation presented in Table 3.2.

The cross sectional area "A" $= [3PLW/2S]^{1/2}$

Computation of Cross Sectional Area for Aluminum 3000-H18

Substituting the respective values in the above equation for cross sectional area, we obtain the following value for aluminum 3000-H18.

Cross sectional area for a solid plate under conditions of strength:

$$A = [3PLW/2S]^{1/2}$$

Thus,

$$A = [(3 \times 7500 \text{ lbs.} \times 45 \text{ in} \times 15 \text{ in})/(2 \times 29 \text{ kpsi})]^{1/2}$$

wherein:

W: Width of the plate = 15 inches as given in the problem statement

L: Length (span) of the plate = 45 inches, also given in the problem statement

Thus, cross sectional area, "A" for aluminum 3000-H18 = 5.117 in^2.

Computation of Cross Sectional Area for 1015 Steel

Substituting the respective values into the above equation for cross sectional area, we obtain the following value for 1015 steel. The values of the variables P, L, and W remain unchanged since they correspond to the design constraints. The only variable that would change in this situation is the yield strength, S. This is because of the fact that it is a material property. As such, it is unique to the specific material, in this case, 1015 steel.

$$A = [(3 \times 7500 \text{ lbs} \times 45 \text{ in} \times 15 \text{ in})/(2 \times 50 \text{ kpsi})]^{1/2}$$

Thus,

Cross sectional area, "A" for 1015 steel = 3.897 in^2.

Using Equation 3, we can obtain the unit cost for each material as follows:

$$\text{Cost} = C(m) \times \rho \times A \times L. \qquad \text{(Equation 3)}$$

The following are the values for cross sectional area:

For

Cross sectional area "A", for aluminum 3000-H18 = 5.117 in^2

Cross sectional area "A", for 1015 steel = 3.897 in^2

Thus,

Cost per unit for aluminum 3000-H18 = \$0.90/lb. × 0.099 lb./in^3 × 5.117 in^2 × 45 inches.

This equates to a unit cost of \$20.52/unit.

The cost for 1015 steel = \$0.30/lb. × 0.285 lb./in^3 × 3.897 in^2 × 45 inches.

This equates to a unit cost of \$14.99/unit.

Since \$14.99/unit < \$20.52/unit, we choose the material corresponding to a unit price of \$14.99.

Thus, we choose 1015 steel for this design problem.

In the next section, we will consider the problem of material selection from the point of view of scrap rates.

3.3 Material Selection Based on Optimization of Operation Sequence through Minimization of Scrap

Every manufacturing and production operation is characterized by a certain amount of wasted material. Material can be wasted due to the following reasons:

- *Limitations of the manufacturing and production processes*: Every production process is defined by the amount of output obtained per unit of input. In production parlance, this is referred to as process yield. Process yield is inversely proportional to scrap rates. This implies that a process with a higher yield will produce a lower amount of scrap and vice versa.
- *Quality control*: Engineering managers have adopted various approaches to maximizing product quality. These include both reactive as well as proactive approaches. Examples of reactive approaches to quality control include control charts. On the other hand, design philosophies, such as six sigma, are examples of proactive approaches to quality control. In general, a higher level of quality implies a lower scrap rate and vice versa.

From the point of view of scrap rates and process yield, the total operation cost can be expressed as the sum of operation cost and the operation scrap cost. This relationship is depicted in Equation 4.

Total cost of an operation = Operation cost + Operation scrap cost (Equation 4)

The computation of operation scrap cost is expressed through Equation 5.

Operation scrap cost = {(100 − Yield)/Yield} × {Current operation cost + Cumulative unit cost}. (Equation 5)

wherein:

Yield is defined as the yield of the process under consideration. This is strictly a process characteristic and is expressed as a percentage.

Cumulative Unit Cost is defined as the cumulative operation cost of all operations in sequence prior to the current operation. Thus, a part could be subjected to nine different sequential operations. The cumulative unit cost for the seventh operation would be the sum of operation costs for the previous six operations.

Let us consider the effect of scrap rates on material costing through an illustrative example. Consider Example 3.4.

EXAMPLE 3.4

Consider a manufacturing operation in which a component undergoes processing through three different operations, namely: casting, turning, and plating. Unit costs are represented in the following table. They include unit material cost and unit operation cost as well as scrap rates for each process. Compute the unit scrap costand cumulative total unit cost as well as total unit scrap cost.

Item	Operation/Material Cost ($/Unit)	Scrap (%)	Yield (%)
Raw material	3.00	1.50	98.50
Casting	4.50	11.00	89.00
Turning	2.25	4.00	96.00
Plating	1.50	12.00	88.00

Solution

The first step in solving this problem is to compute the scrap cost per unit for each operation. This computation is presented below.

Computation of Unit Scrap for Each Operation

Unit Scrap Cost for Raw Material

Using Equation 5, we can obtain the scrap cost for the first process in the sequence as follows:

$$\text{Operation scrap cost} = \{(100 - \text{Yield})/\text{Yield}\} \times \{\text{Current operation cost} + \text{Cumulative unit cost}\}. \quad \text{(Equation 5)}$$

In this case, the yield for the process is 98.50%.

This implies that the value of the expression: 100 − Yield = 100% − 98.5% = 1.50%.

Thus,

$$\frac{100 - \text{Yield}}{\text{Yield}} = \frac{1.50}{98.50} = 0.0152.$$

This ratio is multiplied by the current operation cost, which in this case is equal to \$3/unit.

Since this is the first operation in sequence, cumulative total unit cost = \$0 at this stage.

Thus,

The scrap cost for the first operation = 0.0152 × \$3/unit = \$0.0456.

Thus,

Total operation cost = \$3.00 + \$0.0456 = \$3.0456.

Unit Scrap Cost for Casting

Using a similar approach using Equation 5, we compute the unit scrap cost for casting as follows:

Operation scrap cost = {(100 − Yield)/Yield} × {Current operation cost + Cumulative unit cost}. (Equation 5)

In this case, yield = 89%.

Thus,

(100 − Yield) = 11%.

Thus,

$$\frac{100-\text{Yield}}{\text{Yield}}=\frac{11}{89}=0.1236.$$

This ratio is multiplied by the sum of current operation cost and cumulative total unit cost.

The cumulative total unit cost at this stage = \$3.0456.

From the data presented in the table, current operation cost = \$4.50

Thus,

Operation scrap cost = 0.1236 × {\$4.50 + \$3.0456} = \$0.9326.

Thus,

Operation cost = \$4.50 + \$0.9326 = \$5.4326.

Unit Scrap Cost for Turning

In this case, yield = 96%.

Thus,

(100 − Yield) = 4%.

Thus,

$$\frac{100 - \text{Yield}}{\text{Yield}} = \frac{4}{96} = 0.04166.$$

This ratio is multiplied by the sum of current operation cost and cumulative total unit cost.

The cumulative total unit cost at this stage = \$3.0456 + \$5.4326 = \$8.4782.

From the data presented in the table, current operation cost = \$2.25

Thus,

$$\text{operation scrap cost} = 0.04166 \times \{\$2.25 + \$8.4782\} = \$0.4469.$$

Thus,

$$\text{Operation cost} = \$2.25 + \$0.4469 = \$2.6969.$$

Unit Scrap Cost for Plating

In this case, yield = 88%.

Thus,

$$(100 - \text{Yield}) = 12\%.$$

Thus,

$$\frac{100 - \text{Yield}}{\text{Yield}} = \frac{12}{88} = 0.13636.$$

This ratio is multiplied by the sum of current operation cost and cumulative total unit cost.

The cumulative total unit cost at this stage = \$3.0456 + \$5.4326 + \$2.6969 = \$11.1751.

From the data presented in the table, current operation cost = \$1.50

Thus,

$$\text{Operation scrap cost} = 0.13636 \times \{\$1.50 + \$11.1751\} = \$1.728.$$

Thus,

$$\text{Operation cost} = \$1.50 + \$1.728 = \$3.228.$$

The unit scrap costs and total operation costs as well as the cumulative cost at each stage are presented in the following table.

Scrap Cost ($/unit)	Operation Cost ($/unit)	Cumulative Cost ($/unit)
0.0456	3.0456	3.0456
0.9326	5.4326	8.4782
0.4469	2.6969	11.1751
1.728	3.228	14.4031

It will be appreciated from the above example that lower scrap rates imply higher yields. The higher the yield, the lower is the scrap cost per unit.

For instance, in the case of the plating operation, operation scrap costs amounted to as much as \$1.728 per unit. Notice the comparatively lower yield of this process at only 88%.

The scrap cost was consequently much higher for this process than for the turning operation for instance. In the case of the turning operation, the yield was quite healthy at as much as 96%. The scrap cost per unit was significantly lower in that case at just \$0.4469.

The above discussion implies that a strongly positive process yield can reduce overall product costs and vice versa.

3.4 Conclusion

In this chapter, we dealt with the issue of material costing in depth. Two main approaches to materials costing were discussed, namely: costing based on functionality and costing based on process yield. In the case of the former, different conditions of loading were examined in terms of material density, cross sectional area, type of structure, etc. Corresponding material costs were computed. In the case of the latter, the inverse correlation between process yield and unit scrap costs was clearly established. It should be noted in this context that the importance of high process yield and the consequent role played by quality control could not possibly be overemphasized.

PROBLEMS

1. An engineering manager is trying to design a solid bar to support a load of 70,000 lbs. If the length of the bar is 17 inches, which material should be selected from the following list of materials?

Material	Yield Strength, S (kpsi)	1/S	Density, ρ (lbs./in^3)	ρ/S	Elastic Modulus, E (Mpsi)	Material Cost C (m), (\$/lbs.)
Magnesium						
AS91C-T6	21	0.0476	0.066	0.00314	6.5	2.00
AK60A-T5	41	0.244	0.066	0.0161	6.5	2.20
Aluminum						
2024-T861	66	0.01515	0.100	0.001515	10.5	1.00
3003-H18	29	0.0344	0.099	0.003406	10.2	0.95
7175-T66	76	0.01315	0.101	0.00133	10.4	1.25

(Continued)

Material	Yield Strength, S (kpsi)	1/S	Density, ρ (lbs./in³)	ρ/S	Elastic Modulus, E (Mpsi)	Material Cost C (m), ($/lbs.)
Titanium						
Ti-6Al-6V-2Sn	140	0.00714	0.165	0.001178	16	6.00
Ti-13V-11Cr-3Al	160	0.00625	0.175	0.00109	16	6.25
Steel						
1015	50	0.02	0.285	0.0057	30	0.30
4140 Q&T	252	0.003968	0.285	0.001133	30	1.25
304 Stainless steel	85	0.0176	0.287	0.00505	28	2.00
Cast Iron						
Gray cast iron	20	0.05	0.250	0.0125	15	0.20
Ductile iron	50	0.02	0.250	0.005	30	0.25

2. Which of the following materials
   - Aluminum 7175-T66
   - Steel 4140 Q&T
   - 304 Stainless Steel
   - Aluminum 3003-H18

   should be chosen as the lowest cost material to construct a solid bar designed to support a load of 60,000 lbs.? The bar can only have a maximum extension of 0.10 inches over a length of 12.5 inches. Use the data presented in the table above.
3. A manufacturing engineer has been assigned the task of designing a solid plate to support a 10,500 lbs. load in bending over a span of 60 inches. The width of the plate is 25 inches. The two materials available to the engineer are aluminum 3003-H18 and 1015 steel. Which material should be selected? Use the data presented in the above table.
4. An engineering manager has been assigned the task of designing a solid plate to support a 9000 lbs. load in bending over a span of 40 inches. The width of the plate is 15 inches. The two materials available to the engineer are 304 stainless steel and aluminum 2024-T861. Which material should be selected? Use the data presented in the above table.
5. Consider a manufacturing operation in which a component undergoes processing through four different operations, namely: casting, turning, and plating. Unit costs are represented in the following table. They include unit material cost, unit operation cost as well as scrap rates for each process. Compute the unit scrap cost, cumulative total unit cost as well as total unit scrap cost.

Item	Operation/Material Cost ($/Unit)	Scrap (%)	Yield (%)
Raw material	3.00	2.00	98.00
Process 1	4.50	10.00	90.00
Process 2	2.25	8.00	92.00
Process 3	1.50	12.00	88.00
Process 4	5.00	5.00	95.00

6. Consider a manufacturing operation in which a component undergoes processing through five different operations, namely: casting, turning, and plating. Unit costs are represented in the following table. They include unit material cost and unit operation cost as well as scrap rates for each process. Compute the unit scrap cost and cumulative total unit cost as well as total unit scrap cost.

Item	Operation/Material Cost ($/Unit)	Scrap (%)	Yield (%)
Raw material	5.00	2.00	98.00
Process 1	6.50	17.00	83.00
Process 2	3.25	12.00	88.00
Process 3	2.50	10.00	90.00
Process 4	6.00	5.00	95.00
Process 5	4.00	15.00	85.00

7. Which of the following materials should be chosen as the lowest costing material WHEN constructing a solid bar designed to support a load of 75,000 lbs.? The bar can only have a maximum extension of 0.50 inches over a length of 15.60 inches.
   - Aluminum 7175-T66
   - Steel 4140 Q&T
   - Titanium Ti-13V-11Cr-3AL alloy
8. As an engineer working at Coarse Metals, Inc., you have been assigned the task of designing a solid plate that supports a 6500-pound load in bending over a span of 82 inches. The width of the plate is 17 inches. The materials available are aluminum 3003-H18 and 1015 steel. Which material would you select and why?

   Use the following information to answer the questions below.

A foundry decides to price its castings based on delivered weight per 250 pounds. They price every casting based on past records of a similar design.

Poured metal cost (per casting) = (Furnace labor and overhead + Cost of metal charged) × Casting poured weight.

Cost of metal (in finished casting) = Poured metal cost per casting less amount of re-melted metal × value of re-melted metal.

Cost per pound of delivered metal is item (k) divided by finished weight.

9. For an estimate, the casting poured weight is (8 pounds finished weight) 12 pounds and the cost of charged metal is \$0.13 per pound. Furnace labor and overhead are \$0.9 per pound. The amount of remelted metal is 5 pounds with a value of \$0.18 per pound. What is the material cost per pound?
10. What is the material cost per pound if:
    - The casting poured weight is (15 pounds finished weight) 7.61 pounds
    - The cost of charged metal is \$0.63 per pound
    - Furnace labor and overhead is \$0.27 per pound
    - The amount of remelted metal is 11 pounds with a value of \$0.28 per pound
11. A manufacturing engineer has been assigned the task of designing a solid plate to support a 3800 lbs. load in bending over a span of 78 inches. The width of the plate is 32 inches. The two materials available are aluminum 3003-H18 and 1015 steel. Which material should be selected? Use the data presented in the table for Problem 1.

Use the following information to answer the questions below.

A design having 43.64 cubic inches of metal volume, along with casting allowances included, is going to be purchased from Company B's foundry. You, a casting estimator, calculated the volume from various parts and received quotes from a job-shop foundry:

Pounds of Total Order	Price per Pound
50,000–90,000	\$0.750
90,000–150,000	\$0.390
150,000–500,000	\$0.180
500,000-	Negotiable

12. If iron's density is 0.3790 pounds per cubic inch, what are the unit prices for the midpoint of each range?
13. If iron's density is 0.7800 pounds per cubic inch, what are the unit prices for the midpoint of each range?

4

Process Costing

4.1 What Is Process Costing?

Traditionally, most manufacturing operations are comprised of transforming a piece of raw material into a finished product by subjecting the raw material to one or more material removal processes. The raw material itself is manufactured using processes such as casting and forging. These processes are also referred to as net shape manufacturing processes. Additionally, several other processes like welding and soldering are used to join different components of a product together. These processes are referred to as assembly processes or fastening processes.

Recent advances in nontraditional manufacturing processes, such as 3D printing and additive manufacturing, have upended the world of traditional subtractive manufacturing. Additive manufacturing by definition is the opposite of traditional manufacturing processes, like those of milling, turning, shaping, and drilling. These processes are characterized by material removal from a stock of raw material. They are commonly referred to as subtractive manufacturing processes. Additive manufacturing, on the other hand, is a process which involves addition of material to a base and the final part is built in layers, much like a building. Given the absence of material removal from a stock of raw material, additive manufacturing is the opposite of traditional manufacturing processes. Regardless of whether material is being added or removed, investment in terms of money, time, and resources is essential to enable any of the aforementioned processes.

It is important to learn about process costing because of the crucial role it plays in manufacturing operations. Beginning with shaping of the raw material, proceeding through machining, and finally culminating in product assembly, manufacturing processes are everywhere. A book on production economics would be incomplete without a chapter on process costing. In this chapter, we will discuss the costing of parts manufactured using the following operations:

Machining Operations

1. Turning Operation
2. Milling Operation

3. Shaping Operation
4. Drilling Operation

Metal Shaping Operations

1. Casting Operation
2. Forging Operation

Joining Operations

1. Welding Operation

Nontraditional Manufacturing Operations

1. Additive Manufacturing

4.2 Machining Operations

Throughout this chapter, we will define machining operations as operations which involve removal of material. Examples of such operations include turning, milling, shaping, and drilling.

In order to compute the cost of a part being manufactured using any machining operation, it is first necessary to compute machining time. Machining time is determined by breaking it down into two elements as follows:

1. *Fixed element time*: Amount of time that is independent of the number of units manufactured.
2. *Piece related time*: Amount of time that is directly related to the number of units manufactured. This can be further subdivided into two components as follows:
   i. *Productive elements*: Elements of the machining process which involve actual removal of material.
   ii. *Nonproductive elements*: Elements of the machining process which do not involve any removal of material.

4.2.1 Fixed Elements and Related Time

1. *Setup Time*: Setup time is defined as the amount of time necessary to prepare a machine for operation. It includes the time necessary to study a drawing, time necessary to obtain and install the correct tooling, and the time necessary to align the proper jigs and fixtures

for mass production as well as the amount of time necessary to actually produce and inspect the first piece for quality. It should be borne in mind that setup time occurs once per batch.

2. *Teardown Time*: In many ways, teardown time is the opposite of setup time. Thus, it is the amount of time necessary to remove tooling, jigs, and fixtures from the machine as well as the amount of time necessary to clean the machine and tools post production. Just like setup time, teardown time occurs only once per batch.
3. *Downtime:* To put it succinctly, downtime is defined as nonproductive time that occurs due to a variety of factors such as machine disrepair, lack of proper raw materials, delay in obtaining tooling, lack of an operator, etc. The greater the downtime, the lower the efficiency of the operation.

4.2.2 Piece Related Elements and Time

As we have already discussed, piece related elements and associated time can be classified as follows:

4.2.2.1 Productive Elements

1. *Machining Time*: Machining time is defined as the amount of time that the actual material removal process is taking place. This is the amount of time that the machine works on the component. It is counted from the moment in time when the tool commences the material removal operation to the time when the operation is complete. Most of the material in this chapter presented under the subheading: machining will deal with computation of machining time.

4.2.2.2 Nonproductive Elements

1. *Handling Time*: Handling time is defined as the amount of time taken by an operator to load and unload the part.
2. *Tool Return Time*: Tool return time is defined as the amount of time taken to return the tool to the beginning of the cut and is included as part of handling time.
3. *Unit Tool Changing Time:* This is defined as the amount of time necessary to change worn out tools. This time can be calculated either directly or indirectly. Direct calculation of tool changing time involves knowledge of tool life for the particular set of cutting conditions. It can also be considered a part of allowances. It is obvious that the operator makes the decision to change the cutting tool in most cases.

4. *Allowances*: Allowances are defined as the amount of additional time necessary to account for items such as inspection, fatigue, etc. This time is not included in the normal operation cycle. Allowances are expressed as a percentage of the normal operation time.

4.3 Primary Variables in a Metal Cutting Operation

Any metal cutting operation such as turning, milling, shaping or drilling is characterized by the following variables when computing machining time:

1. Depth of cut
2. Feed rate
3. Cutting speed

When all three variables are considered together, the result is known as the metal removal rate. This is defined as the rate at which the metal is removed from the work piece. It is obvious that a higher metal removal rate is preferred because it shortens the amount of machining time. However, the downside to a high metal removal rate is apparent in high tool costs and high tool changing costs. In any machining operation, the operator selects the machining parameters in the following order: 1. Depth of cut, 2. Feed rate, and 3. Cutting speed. All three variables depend upon the following factors:

- Material of the cutting tool
- Material of the work piece
- Presence (or absence) of cutting fluids
- Type of machining operation
- Surface finish desired
- Capabilities of the machine

A brief description of each parameter is presented below.

1. *Depth of Cut:* The depth of cut is defined as the chip thickness or the thickness of the material removed from the previous work piece surface to the new intermediate work piece surface. It is a function of part dimensions and machine power requirements. It is expressed in terms of millimeters or thousandths of an inch.
2. *Feed Rate*: Feed rate is defined as the displacement of the tool along the work piece. It occurs in the direction of feed motion. It is expressed in terms of a unit of length per stroke or per revolution

of the work piece or tool depending on the case. Feed rate is usually expressed in terms of thousandths of an inch per stroke or per revolution.

3. *Cutting Speed*: Cutting speed is defined as the rate at which the work piece surface traverses past the cutting tool. It is measured in terms of surface feet per minute.

Three approaches are primarily used to calculate machining time. They are as follows:

1. *Feed based method of calculating machining time*: This method requires detailed knowledge of various machining operations and the specifics of individual processes.
2. *Method of calculating machining time based on length of cutting path and cutting velocity*: This method enables optimization of cutting costs. The variable of interest in this method is the cutting speed.
3. *Feature based method*: This method can be characterized as a hybrid between the feed based method and the cutting path method.

The first two methods will be explained in detail in this chapter.

4.3.1 Feed Based Method of Calculating Machining Time

The feed based method relies upon the feed rate of the operation as well as the length of cut. Machining time for the feed based method is computed using Equation 1.

$$T = \frac{L}{F}. \qquad \text{(Equation 1)}$$

wherein:

T: Machining time or cutting time, expressed in minutes.

L: Length of cut in the direction of feed. This is expressed in inches or millimeters.

F: Feed rate, expressed in inches/minute or millimeters/minute.

L = Length of the work piece to be machined + length of pretravel. (Equation 2)

The feed rate is calculated by multiplying the feed and the rate of rotation of the tool or the work piece. This is expressed by Equation 3 as follows:

$$F = f \times N. \qquad \text{(Equation 3)}$$

wherein:

F: Feed rate, expressed in inches per minute or millimeters per minute.

f: Feed rate, expressed in inches per revolution and inches per stoke. It can also be expressed in millimeters per revolution and millimeters per stroke.

N: Rate of rotation, expressed in terms of revolutions per minute (rpm) and strokes per minute.

In case of the milling operation, the feed rate is computed using Equation 4.

$$f = f_t \times n_t. \quad \text{(Equation 4)}$$

wherein:

f_t: Feed per tooth, expressed in inches per tooth or millimeters per tooth

n_t: Number of teeth per revolution of the milling cutter

In the case of a rotating tool (this is the case with most milling operations) or rotating work piece (this is the case with most turning operations), the cutting speed is expressed by Equation 5.

$$V = \Pi \times D \times N/12. \quad \text{(Equation 5)}$$

wherein:

V: Cutting speed, expressed in feet per minute

D: Diameter of the tool or the average diameter of the work piece, expressed in inches

N: Number of revolutions per minute

Equation 5 morphs into Equation 6 if metric units are used.

$$V = \Pi \times D \times N/1000. \quad \text{(Equation 6)}$$

wherein:

V: Cutting speed, expressed in meters per minute

D: Diameter of the tool or the average diameter of the work piece, expressed in millimeters

N: Number of revolutions per minute

Let us consider a few machining problems in order to better understand the significance and applicability of Equations 1–6. First, we will consider an example involving a turning operation. Next, we will consider the applicability of the aforementioned equations to a milling operation, a shaping operation, and finally a drilling operation.

Cutting speed and feed rate data for turning various materials is depicted in Table 4.1. A list of personal allowances and allowances for delay/interruptions is presented in Table 4.2.

TABLE 4.1

Cutting Speed, Feed Rate, and Depth of Cut for Turning Different Selected Materials

	Cutting Speed Tool Materials					
	High Speed Steel (HSS)		Carbide		Range of Feed Rate	
Material	**Rough**	**Finish**	**Rough**	**Finish**	**Rough**	**Finish**
Mild steel	130	200	300	600	0.025–0.08	0.005–0.03
	(40)	(60)	(90)	(180)	(0.65–2)	(0.125–0.75)
Cast steel	50	80	150	325	0.02–0.05	0.005–0.02
	(15)	(24)	(45)	(100)	(0.5–1.25)	(0.125–0.5)
Stainless steel	50	60	90	150	0.02–0.04	0.003–0.008
	(15)	(18)	(27)	(45)	(0.5–1.0)	(0.075–0.175)
Gray cast iron	60	90	200	325	0.015–0.100	0.0075–0.04
	(18)	(27)	(60)	(100)	(0.4–2.5)	(0.2–1.0)
Aluminum	300	500	800	1200	0004–0.02	0.003–0.01
	(90)	(150)	(240)	(360)	(0.5–1.0)	(0.075–0.25)
Brass	250	325	600	900	0.015–0.08	0.003–0.05
	(75)	(100)	(180)	(270)	(0.375–2.0)	(0.02–1.25)
Bronze	60	120	400	600	0.015–0.03	0.005–0.02
	(18)	(36)	(120)	(180)	(0.375–0.75)	(0.125–0.5)
					Type of Cut	
Units for Depth of Cut					**Rough Cut**	**Finished Cut**
Inches					0.150	0.025
Millimeters					4.0	0.60

Source: Adithan, M and Pabla, B. 1989. *Production Engineering, Estimating and Costing*, Konark Publishers, New Delhi, pp. 149–177 [1].

TABLE 4.2

List of Allowances for Machining Operations

	Manufacturing Operation Control	
Type of Allowance	**Machine (%)**	**Operator (%)**
Personal allowance	5	5
Fatigue allowance	–	6
Delays and interruptions	4	4
Total	9	15

Loading, unloading, and setup times for various operations are depicted in Table 4.3.

In the following pages, we will apply the aforementioned principles of calculating machining time to various machining processes such as turning, milling, shaping, and drilling.

TABLE 4.3

Loading, Unloading, and Setup Times for Various Machining Operations

	Loading and Unloading Times/Piece for Different Work Piece Weights, Expressed in Minutes			
Work Holding Device	**0–10 lbs**	**10–30 lbs**		
Between centers using no dog	0.30	0.40		
Between centers with dog	0.67	0.96		
Universal chuck	0.39	0.53		
Independent chuck	0.69	0.83		
V- Block	0.50	0.59		
Vice	0.30	0.40		
Set up Times for different Machining Operations				
	Work-holding Device	**Set up Time (minutes)**		
Operation		**1 Tool**		**2 Tools**
Turning using NC lathe	Collet	20		28
	Chuck	29		34
	Fixture	31		38
Milling				
Plane surface	Vice	69		–
	Collet or chuck	75		–
Profiles, saw slab	Vice	69		–
	Collet or chuck	75		–
Slot-profile	Vice	87		–
	Collet or chuck	93		–
Peripheral slot	Vice	99		–
	Collet or chuck	103		–
Drilling	**Number of Spindles**			
	1	**2**		**3**
Sensitive	Table or vice	10	15	20
	Parallels, V Block	11	16	21
Radial	Table or vice	17		
	Fixture	24		
		1 Tool		
Shaping	Parallels	11		

Source: Boothroyd, G and Knight, W. 1989. *Fundamentals of Machining and Machine Tools*, Marcel Dekker, pp. 145–204; 399–465 [3].

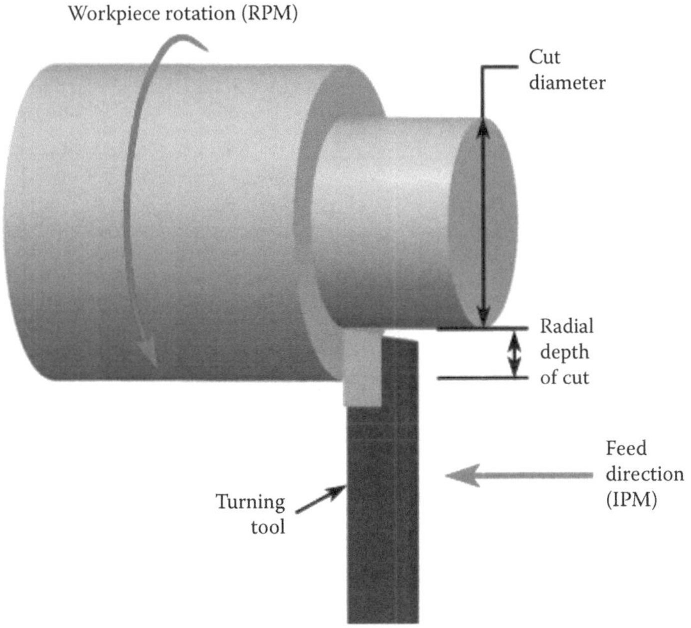

FIGURE 4.1
A typical turning operation. (Custompartnet.com.)

4.3.1.1 Turning

Turning is defined as a machining process in which a work piece is machined on a lathe. The work piece is typically held in a rotating work holding device, referred to as the chuck. The metal removal can take place using a single point cutting tool or more recently, high speed steel (HSS) or carbide inserts. A pictorial representation of a typical turning operation is presented in Figure 4.1.

Consider Example 4.1 in order to better understand the applicability of the feed based method to a turning operation.

EXAMPLE 4.1

Consider the work piece presented in Figure 4.2. Compute the machining time to turn the dimensions shown in the figure for a mild steel bar. The starting bar size is 6 inches in diameter. This is reduced to 5 inches and finally to 3 inches. The length of each step is depicted in the figure. Use the recommended speed, depth of cut, and feed rate for the turning operation from Table 4.2. Assume a lot size of 100 units.

Solution

Referring to Figure 4.1, the first step involves determination of the number of passes required to shrink the diameter down from 6 inches to 5 inches. The second step involves determining the number of passes

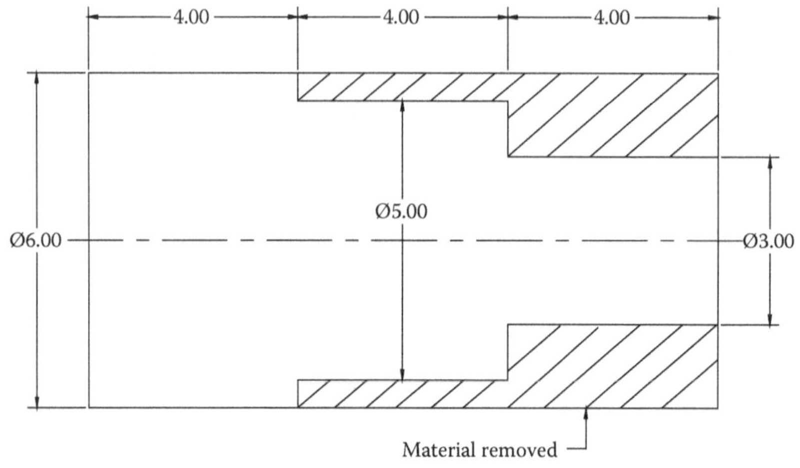

FIGURE 4.2
Turning setup for Example 4.1.

required to shrink the diameter down from 5 inches to 3 inches. This is accomplished as follows.

$$\text{Number of passes} = \frac{\text{Depth of material to be removed}}{\text{Depth of Cut}}$$

In view of the part symmetry and the basic nature of the turning operation, we have the following:

$$\text{Depth of material to be removed} = (\text{large diameter} - \text{small diameter})/2.$$

Step 1: Shrinking the diameter from 6 inches to 5 inches:

$$\text{Depth of material to be removed} = (6 \text{ inches} - 5 \text{ inches})/2 = 0.5 \text{ inches}.$$

From the problem statement, we know that the material being machined is mild steel.

From Table 4.1, the maximum depth of cut allowed for a roughing operation for this material is 0.15 inches.

Substituting these values into the equation, we obtain the following:

$$\text{Number of passes} = 0.5 \text{ inches}/0.150 = 3.33.$$

Since 3.33 is a fraction, we round it up to the next highest integer, namely, four. This is because three passes would be insufficient.

Since four passes are required to accomplish the turning operation, the total depth of material to be removed will be split equally between these four passes.

$$\text{Thus, the depth of cut per pass} = 0.5 \text{ inches}/4 = 0.125 \text{ inches}.$$

In the case of machining operations, much depends on the cutting tool materials as well as the type of cut (such as finish cut or rough cut). In this case, let use assume that the tool is made out of high speed steel (HSS).

In the present case, referring to Table 4.1, the cutting speed for the mild steel/HSS combination for a rough cut corresponds to 130 ft/min. The recommended range of feed for the same work piece/tool combination for a rough cut is 0.025–0.080 inches/revolution. If the operation was a finish turning operation, this range would be 0.005–0.030 inches/revolution. It will be appreciated that there is quite a bit of overlap between the two ranges. Assuming the value of feed rate to be 0.03 inches/revolution, we can proceed with the computation of rpm as follows:

From Equation 5, we know the following relationship:

$$V = \Pi \times D \times N/12. \qquad \text{(Equation 5)}$$

Rewriting this equation to solve for the value of "N", we obtain the following expression:

$$N = 12 \times V/\Pi \times D.$$

Let us compute the rpm requirements for each pass as follows:

RPM Requirements for Pass 1

For the first pass, the average diameter is computed as follows:

Average diameter = outside diameter − depth of cut for pass 1.
Thus, average diameter = 6 inches − 0.125 inches = 5.875 inches.

Substituting this value into the equation for RPM, we obtain the following:

$N = 12 \times 130/3.1415 \times 5.875$
Thus, $N = 1560/18.456 = 84.52 = 85$ revolutions per minute.

RPM Requirements for Pass 2

For the second pass, the average diameter is computed as follows:

Outside diameter = 6 inches − (2 × 0.125 inches) = 5.75 inches.
Thus, average diameter = outside diameter − depth of cut for pass 2.
Thus, average diameter = 5.75 inches − 0.125 inches = 5.625 inches.

Substituting this value into the equation for RPM, we obtain the following:

$N = 12 \times 130/3.1415 \times 5.625$
Thus, $N = 1560/17.67 = 88.32 = 90$ revolutions per minute.

RPM Requirements for Pass 3

For the third pass, the average diameter is computed as follows:

Outside diameter = 5.75 inches − (2 × 0.125 inches) = 5.50 inches.
Thus, average diameter = outside diameter − depth of cut for pass 2.
Thus, average diameter = 5.50 inches − 0.125 inches = 5.375 inches.

Substituting this value into the equation for RPM, we obtain the following:

$N = 12 \times 130/3.1415 \times 5.375$
Thus, $N = 1560/16.88 = 92.386 = 95$ revolutions per minute.

RPM Requirements for Pass 4

For the fourth pass, the average diameter is computed as follows:

Outside diameter = 5.5 inches = (2 × 0.125 inches) = 5.25 inches.
Thus, average diameter = outside diameter = depth of cut for pass 3.
Thus, average diameter = 5.25 inches = 0.125 inches = 5.125 inches.

Substituting this value in the equation for RPM, we obtain the following:

$N = 12 \times 130/3.1415 \times 5.125$
Thus, $N = 1560/16.100 = 96.89 = 100$ revolutions per minute.

Let us assume the value of RPM for all passes to be equal to 95 rpm. The cutting time can be computed using Equation 1 as follows:

$$T = \frac{L}{F}. \qquad \text{(Equation 1)}$$

In this case, the length of cut will be equal to the sum total of the four passes plus any length of pretravel and over-travel. The length of pretravel and over-travel should be as small as possible and just enough to enable a smooth turning operation. Since the operation ends at a point inside the object, there is no over-travel. We assume the length of pretravel to be 0.25 inches.

Referring to Figure 4.2, the total distance traveled for each pass = (4 + 4 + 0.25) inches = 8.25 inches.

This is because the diameter is reduced from 6 inches to 5 inches over a step length of 8 inches during the first step. The second step would involve reducing 4 inches of this first step down to an outside diameter of 3 inches.

From Equation 3, we know the following:

$$F = f \times N. \qquad \text{(Equation 3)}$$

Thus, F = 0.030 × 95 rpm = 2.85.

Substituting the respective value into Equation 3, we obtain the following:

T = 8.25 inches/2.85 = 2.89 minutes per pass.

Thus, the amount of time necessary for all four passes = 4 passes × 2.89 minutes per pass = 11.56 minutes.

Step 2: Reduce the outside diameter from 5 inches to 3 inches.

Using the same logic that we used for step 1, we obtain the following.

Depth of material to be removed = (5 inches − 3 inches)/2 = 1 inch.
Number of passes = 1 inch/0.150 = 6.667.

Since 6.667 is a fraction, we round it up to the next highest integer, namely, seven. This is because six passes would be insufficient.

This also implies that the depth of cut per pass would be equal to 1 inch/7 passes = 0.143 inches/pass.

Let us compute the rpm requirements for each pass as follows:

RPM Requirements for Passes 1 and 2

Average diameter = outside diameter—depth of cut for passes 1 and 2.

Thus, average diameter = 5 inches – 0.143 inches = 4.857 inches.

Substituting this value in the equation for RPM, we obtain the following:

N = 12 × 130/3.1415 × 4.857
Thus, N = 1560/15.258 = 102.23 = 105 revolutions per minute.

Machining time for first two passes = L/F.

Referring to the part geometry in Figure 4.1 and substituting respective values, we obtain the following:

Machining time = (4 + 0.25)/(102.23 × 0.030)
Thus, machining time = 4.25/3.067 = 1.3856 minutes per pass.

This equates to a total of 1.3856 × 2 = 2.77 minutes for the first two passes.

RPM Requirements for Passes 3 and 4

Average diameter = outside diameter − depth of cut for passes 3 and 4.

Thus, average diameter = 4.428 inches − 0.143inches = 4.285 inches.

Substituting this value into the equation for RPM, we obtain the following:

N = 12 × 130/3.1415 × 4.285
Thus, N = 1560/13.461 = 115.88 = 120 revolutions per minute.

Machining time for third and fourth passes = L/F.
Referring to the part geometry in Figure 4.1 and substituting the respective values, we obtain the following:

Machining time = (4 + 0.25)/(115.88 × 0.030)
Thus, machining time = 4.25/3.4766 = 1.222 minutes per pass.

This equates to a total of 1.222 × 2 = 2.445 minutes for passes three and four.

RPM Requirements for Passes 5 and 6

Average diameter = outside diameter − depth of cut for passes 5 and 6.
Thus, average diameter = 3.856 inches − 0.143 inches = 3.713 inches.

Substituting this value into the equation for RPM, we obtain the following:

N = 12 × 130/3.1415 × 3.713
Thus, N = 1560/11.664 = 133.74 = 135 revolutions per minute.

Machining time for third and fourth passes = L/F.

Machining time = (4 + 0.25)/(133.74 × 0.030)
Thus, machining time = 4.25/4.0122 = 1.059 minutes per pass.

This equates to a total of 1.059 × 2 = 2.118 minutes for passes five and six.

RPM Requirements for Pass 7

Average diameter = outside diameter − depth of cut for pass 7.
Thus, average diameter = 3.284 inches − 0.143 inches = 3.141 inches.

Substituting this value into the equation for RPM, we obtain the following:

N = 12 × 130/3.1415 × 3.141
Thus, N = 1560/9.8674 = 158.09 = 160 revolutions per minute.

Machining time for the seventh pass = L/F.

Machining time = (4 + 0.25)/(158.09 × 0.030)
Thus, machining time = 4.25/4.7428 = 0.896 minutes for the seventh pass.

Total cutting time in order to accomplish the entire operation is obtained by adding cutting times for all the passes as follows:

Total machining time = machining time for diameter 5 inch step
+ machining time for diameter 3 inch step.

Thus, total machining time = 11.56 minutes + 2.77 minutes + 2.445 minutes + 2.118 minutes + 0.896 minutes = 19.789 minutes.

This implies that the time for the cutting cycle = 19.789 minutes. This is exclusive of setup time and handling time. Assuming that the machine in question is a lathe with a universal chuck, the unit handling time for such a machine is 0.39 minutes.

In this case, unit piece time = 19.789 minutes + 0.69 minutes = 20.179 minutes.

Consideration of Allowances

Special consideration needs to be given to issues such as personal fatigue, personal needs, inspection, etc. Each of the aforementioned activities is time consuming. As such they need to be accounted for. Accounting for miscellaneous activities is provided by incorporating allowances such as personal allowance, fatigue allowance, and allowances for delays and interruptions.

Referring to Table 4.2, an allowance of 9% of machining time and 15% of personal time needs to be included in the total time consideration.

Thus, total piece time = (19.789 minutes × 1.09) + (0.39 minutes × 1.15).
Total piece time = 21.57 minutes + 0.4485 minutes = 22.0185 minutes.

Next, we will consider the amount of time required to set up the manufacturing operation.

Consideration of Setup Time

Generally, for turning operations, total set up time is about 150% of basic time. In this particular case, the operation being considered is a turning operation. This involves holding the work piece in a chuck. As is evident from Table 4.3, the amount of time for such an operation involving one tool is about 29 minutes. For a lot size of 100 units, the total setup time computation is presented below.

Total Time = (batch size) × (piece time) + setup time.

Substituting the respective values into this equation, we obtain the following:

Total time = (100) × (22.0185minutes) + (29 minutes × 1.50).
Thus, total time = (2201.85 minutes) + (43.5 minutes) = 2,245.35 minutes.

This implies that the time per piece = 2,245.35 minutes/100 = 22.4535 minutes. **Solution**

4.3.1.2 Milling

Milling can be defined as a machining process wherein rotary cutters are used to remove material. The work piece is held in a stationary work holding device such as a vice. Compare this to the process of turning wherein the work piece is held in a rotating work holding device (chuck). In milling, material is removed from a work piece by feeding it in a direction that is at an angle with the tool. Figure 4.2 provides a pictorial representation of the milling process (Figure 4.3).

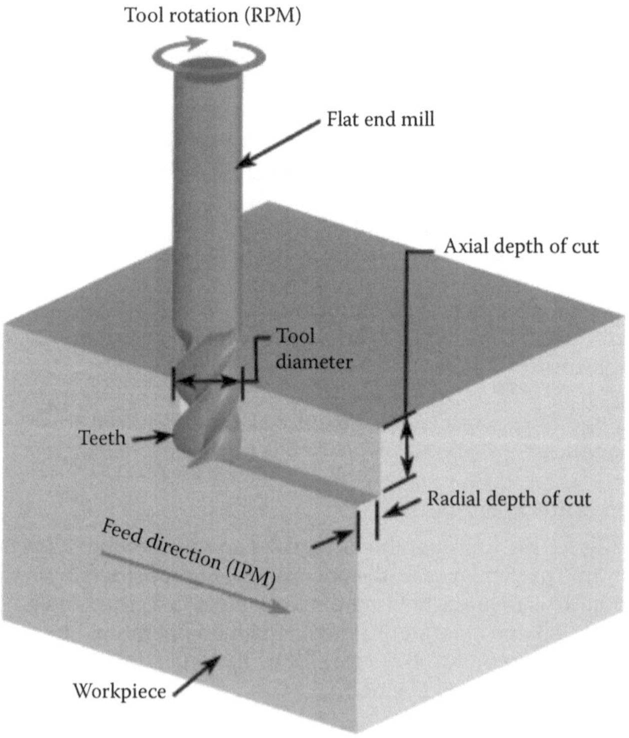

FIGURE 4.3
A typical milling operation. (Custompartnet.com.)

TABLE 4.4

Feeds and Cutting Speeds for a Milling Operation for Various Materials

Work Piece Material	Feed per Tooth		Average Value of Cutting Speed	
	Inch/tooth	Mm/tooth	Ft/min HSS	(m/min) Carbide
Carbon steel	0.003–0.015	0.075–0.381	150 (45)	600 (180)
Cast iron	0.005–0.030	0.130–0.750	125 (40)	300 (90)
Stainless steel	0.002–0.030	0.050–0.750	100 (30)	550 (170)
Aluminum alloys	0.003–0.030	0.075–0.750	750 (230)	2500 (760)
Copper alloys	0.005–0.030	0.130–0.750	225 (70)	800 (250)

Source: Production Processes.

Feeds and cutting speeds for a selection of materials for a milling operation are presented in Table 4.4.

Consider Example 4.2 in order to better understand the applicability of the feed based method to a milling operation.

EXAMPLE 4.2

A carbon steel plate is being face milled. A high speed steel cutter is being used for the machining operation. The cutter is 37.5 millimeters in diameter and has 8 teeth. The plate is 100 millimeters long and 25 millimeters wide. The production engineer plans to use a depth of cut equal to 2.5 millimeters. The machining setup is illustrated in Figure 4.4. Compute the time to machine 100 pieces.

Solution

From Table 4.3, we observe that it takes about 69 minutes to set up the milling machine for plane surface milling. Including a 150% adjustment, the total setup time is computed as follows:

$$\text{Setup time} = 69 \text{ minutes} \times 1.50 = 103.50 \text{ minutes.}$$

From the problem description, we know that the material being processed is carbon steel. We also know that a high speed steel (HSS) cutter will be used for the milling operation.

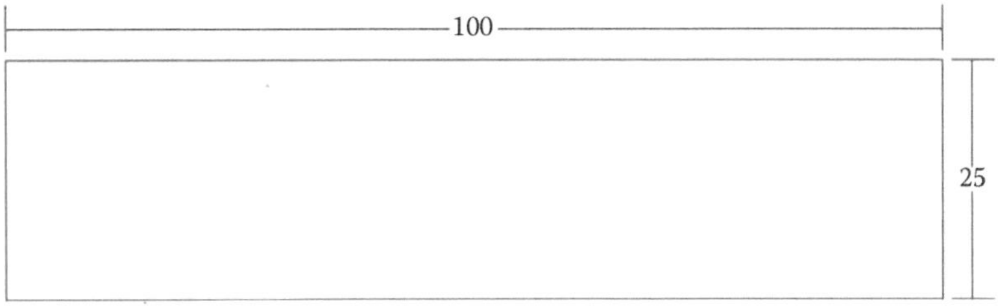

FIGURE 4.4
Milling setup for Example 4.2.

From Table 4.4, a cutting speed of 45 meters/minute is recommended for this particular work piece/toll combination. The feed value would be in the 0.075–0.381 mm/tooth range. Let us assume this value to be 0.25 mm/tooth. Assuming that this is a rough cut operation, the lead is calculated as follows:

$$\text{Lead} = D/2 - 0.5 \times \sqrt{D^2 - W^2}$$

wherein:
D: Diameter of the cutter = 37.5 mm
W: Width of work piece = 25 mm

Substituting the values into the above equation, we obtain the value of the lead as follows:

$$\text{Lead} = 37.5/2 - 0.5 \times \sqrt{(37.5)^2 - (25)^2}$$

Thus, $\text{Lead} = 18.75 - 0.5 \times \sqrt{1406.25 - 625}$
Thus, Lead = 18.75 − 0.5 × (27.95) = 4.774 mm = 4.8 mm.

The total length of the cut is computed as follows:

L = Length of work piece being machined + lead + length of pretravel and over-travel.
We assume that the total amount of pretravel and over-travel = 6 mm.

Substituting the respective values into the above equation, we obtain the following:

L = 100 mm + 4.8 mm + 6 mm.
Thus, L = 110.8 mm.

Next, we compute the value of RPM using Equation 6 as follows.

$$V = \Pi \times D \times N/1000. \qquad \text{(Equation 6)}$$

Rewriting this equation to solve for the unknown value of "N," we obtain the following expression:

$$N = 1000 \times V/\Pi \times D.$$

V = 45 m/min.
D = Diameter of the cutter = 37.5 mm.

Substituting these values into the above equation, we obtain the value of RPM as follows:

$$N = 1000 \times 45/3.1415 \times 37.5$$

Thus, $N = 45{,}000/117.806 = 381.98$ rpm $= 382$ rpm.

The cutting time per piece is calculated by using Equation 1 as follows:

$$T = \frac{L}{F}. \qquad \text{(Equation 1)}$$

The feed "F" is calculated using Equations 3 and 4 as follows:

$$F = f \times N. \qquad \text{(Equation 3)}$$

$$f = f_t \times n_t. \qquad \text{(Equation 4)}$$

Substituting Equation 4 into Equation 3, we obtain the modified form of Equation 3 as follows:

$$F = f_t \times n_t \times N.$$

Substituting the respective values into the above equation, we get the value of feed "F" as follows.

$$F = 0.25 \text{ mm/tooth} \times 8 \text{ teeth} \times 382 \text{ rpm}$$

Thus, $F = 764$ mm/minute.

Using Equation 1, we compute cutting time by substituting the respective values as follows:

$$T = 110.8 \text{ mm}/764 \text{ mm/minute} = 0.145 \text{ minutes}.$$

Next, we compute the loading and unloading time per work piece. We assume that a vice was used as the work holding device for this example. From Table 4.3, this time would be equal to 0.30 minutes/piece.

We also incorporate a 9% machine allowance and a 15% operator allowance. Incorporating these factors, the total time per piece is obtained as follows:

Time per piece = (1.09 × 0.145 minutes) + (1.15 × 0.30 minutes)
Thus, time per piece = (0.15805 minutes) + (0.345 minutes)
Thus, time per piece = 0.50305 minutes

Next, we compute the setup time for milling a plane surface. From Table 4.3, we observe that if a vice was used as a work holding device, this time is 69 minutes. Using an adjustment factor of 150%, the total setup time is equal to 69 minutes × 1.50 = 103.50 minutes.

The total amount of time required for a batch of 100 units is computed as follows:

Total time = setup time + (batch size) × (time/piece).

Substituting the respective values into the above equation, we obtain the total time as follows.

Total time = 103.50 minutes + (100) × (0.50305 minutes).
Thus, total time = 103.50 minutes + 50.305 minutes.

This implies that total time to machine a batch of 100 units using a milling machine under the given constraints = 153.805 minutes.
Total time per piece = 153.805 minutes/100 pieces = 1.538 minutes.
Solution

4.3.1.3 Shaping

Shaping can be defined as a machining process that incorporates material removal by reciprocating the tool across a stationary work piece. Figure 4.5 depicts an example of the shaping process.

The various cutting speeds and feeds for shaping and planing operations for a range of materials are presented in Table 4.5.

Consider Example 4.3 in order to better understand the applicability of the feed based method to a shaping operation.

EXAMPLE 4.3

A carbon steel plate is being machined using a shaper. A high speed steel single point tool is being used for the machining operation. The plate is 100 millimeters long and 25 millimeters wide. The machining setup is illustrated in Figure 4.6. Compute the time to machine 100 pieces.

Solution

From the metallurgical point of view, carbon steel is equivalent to soft steel. From Table 4.5, we can observe that the cutting speed for soft steel is

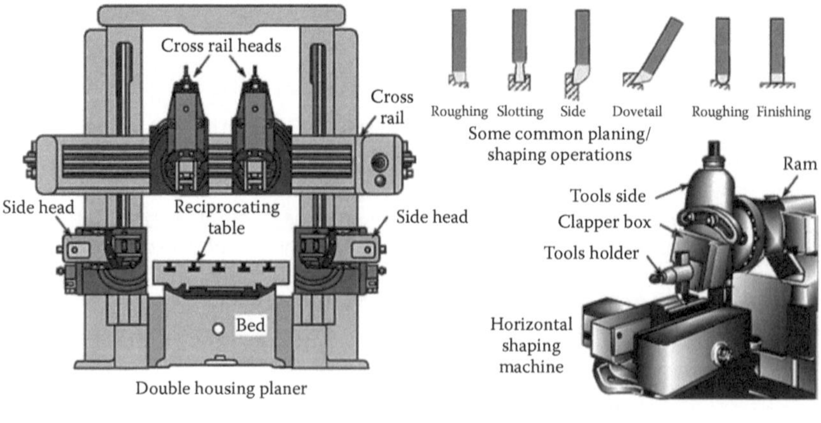

FIGURE 4.5
A typical shaping/planing operation. (Mechanical Technical Solutions.)

TABLE 4.5

Cutting Speeds and Feeds for a Range of Materials for Shaping and Planing Operations

Work Piece Material	Tool Material							
	High Speed Steel (HSS)				Carbide			
	Speed		Feed/Stroke		Speed		Feed/Stroke	
	Ft/min	m/min	Inches	Mm	Ft/min	m/min	Inches	Mm
Hard steel	25	8	0.04	1.00	140	40	0.035	0.90
Medium steel	65	20	0.03	0.75	210	65	0.05	1.25
Soft steel	80	25	0.06	1.50	235	70	0.05	1.25
Cast steel	40	12	0.05	1.25	140	40	0.04	1.00
Hard cast iron	40	12	0.06	1.50	150	45	0.05	1.25
Soft cast iron	65	20	0.12	3.00	165	50	0.05	1.25
Malleable iron	70	21	0.09	2.25	200	60	0.04	1.00
Brass	200	60	0.05	1.25				
Bronze	45	15	0.08	2.00	210	65	0.05	1.25
Aluminum	250	75	0.04	1.00				

Typical depth of cut for a rough cut = 0.15 inches = 4 mm

Typical depth of cut for a finish cut = 0.002 inches = 0.05 mm

25 m/min. The feed rate is 1.5 mm/stroke. We assume that a mechanical shaper with an R value of 1.6 will be used for this machining operation.

First, we compute the amount of travel that the tool must make with respect to the work piece. This is known as the length of cut. It is computed as follows.

Length of cut, L = length of work piece + pretravel + over-travel.

For most shaping operations, the length of pre travel is about 5 mm and the length of over-travel is about 15 mm. These values must be added to the length of the work piece being machined. We know this value to be 100 mm.

Thus, the length of cut, L = 100 mm + 5 mm + 15 mm = 120 mm.

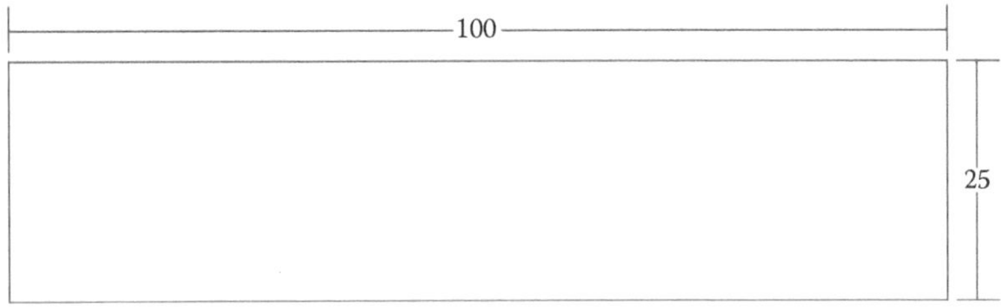

FIGURE 4.6
Shaping setup for Example 4.3.

The feed rate for a shaping operation is defined as the product of feed per stroke and number of strokes per minute. We know that the feed per stroke is equal to 1.5 mm/stroke.

Thus, feed rate, $F = n_s \times f_s$.

The value of number of strokes per minute (n_s) is computed using the following formula:

$$n_s = (1000 \times V)/\{ L \times (1 + 1/R)\}.$$

We know from the problem statement that the value of $R = 1.6$. We also computed the total length of cut, $L = 120$ mm. The cutting speed, is $V = 25$ m/min.

The number of strokes per minute can be computed by substituting these values into the above equation as follows:

$$n_s = (1000 \times V)/\{L \times (1 + 1/R)\}.$$

Thus, $n_s = (1000 \times 25 \text{ m/min})/\{120 \text{ mm} \times (1 + 1/1.6)\}$.

Thus, $n_s = 25{,}000/\{120 \times (1.625)\}$

This implies that $n_s = 25{,}000/195 = 128.20$ strokes/minute $= 129$ strokes per minute.

Substituting this value into the equation for feed rate, we obtain the following expression:

Feed rate, $F = 129$ strokes/minute $\times$ 1.5 mm/stroke.

Thus, Feed rate, $F = 193.5$ mm/minute.

The cutting time for a shaping operation is computed by using the following equation:

$$T = W/F$$

wherein:

W: total width of work piece + pretravel + over-travel

In this case, we add 5 mm in pretravel and another 5 mm in over-travel to the width of the work piece. Thus, the value of $W = 25 \text{ mm} + 5 \text{ mm} + 5 \text{ mm} = 35 \text{ mm}$.

Substituting these values into the cutting time equation, we obtain the following:

$$T = \frac{35\,\text{mm}}{193.5\,\text{mm/minute}} = 0.18\,\text{minutes/piece}.$$

Next, we compute the amount of loading and unloading time from Table 4.3. Notice that this time for a shaping operation using a vice is 0.30 minutes. Once again, incorporating a machine allowance of 9% and an operator allowance of 15%, we obtain the total amount of time per piece as follows:

Time per piece = (1.09 × 0.18 minutes) + (1.15 × 0.30 minutes).
Thus, time per piece = 0.1962 minutes + 0.345 minutes = 0.5412 minutes per piece.

Next, we compute the total time for a batch of 100 pieces as follows.

Total time for the entire batch = setup time + (batch size) × (time per piece).

For a shaping operation, the setup time is assumed to be equal to the setup time of a drill press with parallels. From Table 4.3, we observe that this amounts to 11 minutes. Applying an adjustment of 150%, we obtain the value of setup time as 11 minutes × 1.50 = 16.50 minutes.

Thus total time for the batch = 16.5 minutes + (100) × (0.5412 minutes).
This implies that total time for the entire batch = 16.5 minutes + 54.12 minutes = 70.62 minutes.

The total time per work piece = 70.62 minutes/100 = 0.7062 minutes.
Solution

4.3.1.4 Drilling

Drilling can be defined as a machining process wherein a cutting tool such as a drill bit is used to cut a hole of circular cross section in a solid object. The work piece is generally held stationary with the help of a work holding device such as a vice. The drill is a rotary, multipoint cutting tool that is used to cut and remove material. The drill usually rotates at a high rpm and is pressed against the material of the solid object in order to cut the hole with the required dimensions. Figure 4.6 depicts a graphical representation of the drilling process (Figure 4.7).

Table 4.6 represents the different speeds and feeds for drilling various materials.

Consider Example 4.4 in order to better understand the applicability of the feed based method to a drilling operation.

EXAMPLE 4.4

Five holes are to be drilled in an aluminum plate using a high-speed steel drill. The geometry of the work piece is depicted in Figure 4.8. In order to accomplish the task, a sensitive drill press will be used with two spindles. Calculate the machining time to process a batch of 100 pieces.

Solution

The first step in solving this problem is to determine the setup time for the drilling operation. From the part geometry, it is obvious that two different hole sizes will need to be drilled. Using a two-spindle drill press will facilitate this process.

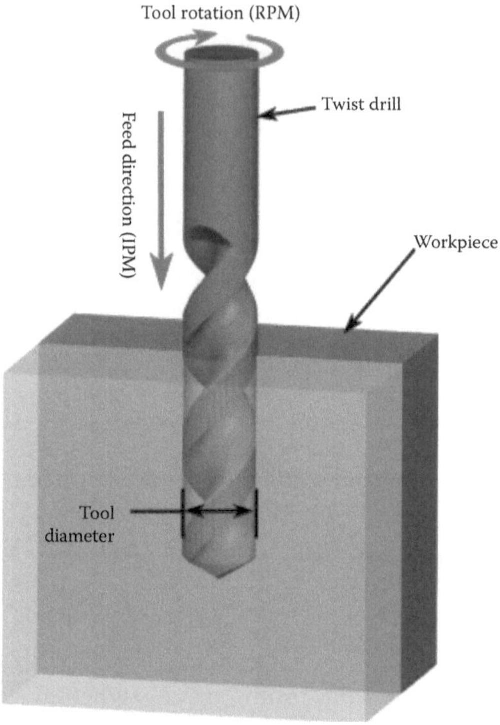

FIGURE 4.7
A typical drilling operation. (Custompartnet.com.)

From Table 4.3, we know that the setup time for a sensitive drill press with two spindles is 15 minutes. Applying the 150% adjustment, we get a total setup time = 22.5 minutes.

From the part geometry, notice that all holes are through holes. Thus, the length of cut will be at least equal to the thickness of the plate. The actual length of cut is calculated as follows:

$$\text{Length of cut} = \text{length of pretravel} + \text{thickness of plate} + \text{length of over-travel} + \text{lead.}$$

We assume that the length of pretravel = length of over-travel = 0.125 inches.

Also, we will use a drill with an included angle of 118 degrees for the machining operation. The lead for the drill is computed using the following equation:

$$\text{Lead} = (D/2) \times \tan(31) = (D/2) \times 0.60 = 0.3 \times D.$$

wherein:

D: Diameter of the drill

TABLE 4.6

Speeds and Feeds for Drilling Different Materials

	Cutting Speed		
Work Material	**Ft/min**	**m/min**	**Feed Class**
Soft cast iron	150	45	1
Medium cast iron	80	25	2
Mild steel	90	27	2
Alloy steel	60	18	3
Tool steel	50	15	3
Brass and bronze	200	60	1
Copper	150	45	1
Aluminum and magnesium alloys	350	105	1

Feed Class	**Material Description**	**Feed Rate**
1	Free Machining Materials	0.02 D
2	Tough or hard materials	0.01 D
3	Very hard materials	0.005 D

Source: Schey, J. 1987. *Introduction to Manufacturing Processes*, McGraw Hill, pp. 505 [16]; Shaw, M. 1984. *Metal Cutting Principles*, Oxford University Press, England, pp. 474 [17].

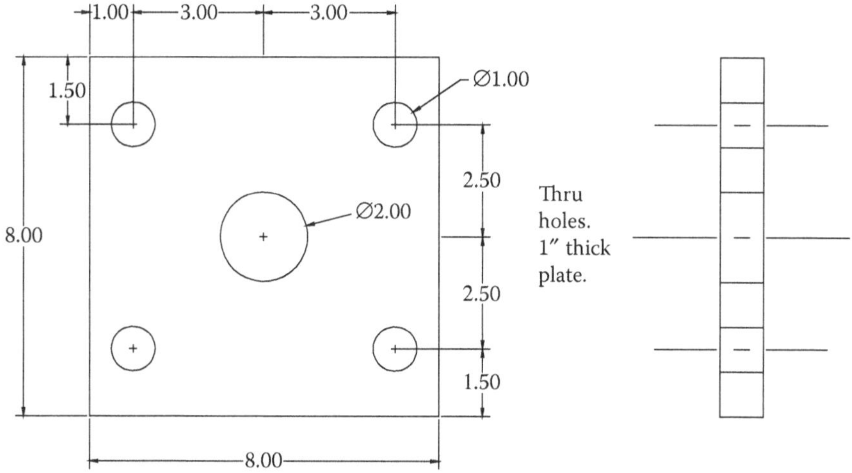

FIGURE 4.8
Drilling setup for Example 4.4.

From the part geometry, we know that two hole sizes need to be drilled, namely four holes each of diameter 1 inch and one hole of diameter 2 inches.

Substituting the respective values into the equation for length of cut, we have the following:

$$\text{Length of cut (diameter 1'' hole)} = 0.125'' + 1'' + 0.125'' + 0.3'' = 1.55 \text{ inches.}$$

$$\text{Length of cut (diameter 2'' hole)} = 0.125'' + 1'' + 0.125'' + 0.6'' = 1.85 \text{ inches.}$$

Next, we compute the RPM for each hole size as follows:

Computation of RPM for 1-inch Diameter Hole

Using Equation 5, we know the following:

$$V = \Pi \times D \times N/12. \qquad \text{(Equation 5)}$$

Solving for the value of N, the above equation is rewritten as follows:

$$N = 12 \times V/\Pi \times D.$$

From the problem statement, we know that the plate is made of aluminum. From Table 4.7, the cutting speed (V) for aluminum is 350 ft/min.

Substituting the respective values into the above equation (D = 1 inch), we obtain the following:

$N = 12 \times (350)/3.1415 \times 1.$

Thus, $N = 4{,}200/3.1415 = 1337$ rpm.

Computation of RPM for 2-inch Diameter Hole

Once again, we use the following equation to determine the RPM for this particular hole size.

$$N = 12 \times V/\Pi \times D.$$

The material of the plate is unchanged. For aluminum, the cutting speed (V) is 350 ft/min.

TABLE 4.7

Different Types of Casting Process

Type of Casting Process	Type of Mold Used	Type of Pattern Used
Sand casting	Single use	Multiple use
Permanent mold casting	Multiple use	
Investment casting	Single use	Single use
Die casting	Multiple use	

Substituting the respective values into the above equation (D = 1 inch), we obtain the following:

$N = 12 \times (350)/3.1415 \times 2.$
Thus, $N = 4{,}200/6.283 = 669$ rpm.

Next we compute the rate for each hole size as follows:

Computation of Feed Rate for 1-inch Diameter Hole

We know that feed rate is the product of feed per revolution and the rpm.
Feed per revolution is computed using the following equation:

$$f_r = 0.02 \times D.$$

Thus, for one-inch diameter hole, this value is 0.02 inches.

$$\text{Feed rate } F = fr \times N.$$

Substituting the respective values into the above equation, we obtain the following.

$$F = 0.02 \times 1337 \text{ rpm} = 26.74 \text{ inches/minute.}$$

Computation of Feed Rate for 2-inch Diameter Hole

$$f_r = 0.02 \times D.$$

Thus, for a two-inch diameter hole, this value is 0.04 inches.

$$\text{Feed rate } F = fr \times N.$$

Substituting the respective values into the above equation, we obtain the following.

$$F = 0.04 \times 669 \text{ rpm} = 26.76 \text{ inches/minute.}$$

Next, we compute the drilling times for each hole size.

Drilling Time for 1-inch Diameter Holes

The drilling time for each hole is obtained using Equation 1 as follows.

$$T = \frac{L}{F}. \qquad \text{(Equation 1)}$$

We know that the length of cut for the one inch hole = 1.55 inches.
We also know the value of "F" for this hole size = 26.74 inches/minute.
Substituting these values into Equation 1, we obtain the cutting time as follows:

$$T = 1.55 \text{ inches}/26.74 \text{ inches/minute} = 0.05796 \text{ minutes.}$$

Since there are four holes of this size, the total time required to drill all four holes = 4 × 0.05796 minutes = 0.2318 minutes.

Drilling Time for 2-inch Diameter Hole

We know that the length of cut for the two inch hole = 1.85 inches.

We also know the value of "F" for this hole size = 26.76 inches/minute.

Substituting these values into Equation 1, we obtain the cutting time as follows:

T = 1.85 inches/26.76 inches/minute = 0.06913 minutes.

The total time to drill all four holes = 0.2318 minutes + 0.06913 minutes = 0.30093 minutes.

Next we will compute the total machining time by including the handling time and allowances.

Computation of Total Machining Time per Piece

From Table 4.3, we know the handling time for this operation is 0.30 minutes. Incorporating an allowance of 9% for machining and 15% for material handling, we obtain the total machining time as follows:

Total machining time per piece = (1.09) × (0.30093 minutes) + (1.15) × (0.30 minutes) = 0.32801 minutes + 0.345 minutes = 0.6730 minutes.

Computation of Total Time

Total time = setup time + (batch size) × (total machining time per piece).

Thus, total time = 22.5 minutes + (100) × (0.6730 minutes) = 22.5 minutes + 67.30 minutes.

Thus, total time for a batch of 100 pieces = 89.801 minutes.

Total time per piece = 89.801 minutes/100 = 0.898 minutes/piece.

Solution

4.3.2 Method of Calculating Machining Time Based on Length of Cutting Path and Cutting Velocity

An alternative is available to the feed based method to compute machining time. It uses the length of the cutting path as well as the cutting velocity to estimate total time. The formula for length of cutting path and cutting velocity is presented in Equation 7.

$$\text{Time} = \frac{\text{Length of the cutting path}}{\text{Cutting Velocity}}. \qquad \text{(Equation 7)}$$

Consider Example 4.5 to better understand this concept.

EXAMPLE 4.5

Consider Example 4.4 that we just solved. Let us solve this example using the cutting path and cutting velocity method and compare the answers.

Solution

The equation to compute the cutting path length for drilling is presented below.

$$\text{Cutting path length} = \Pi \times D_c \times L/12 \times f_r$$

wherein:

D_c: Diameter of the cutter
f_r: Feed rate

The length of cut for the 1-inch diameter hole = 1.55 inches.
The feed for this hole size = 0.02 inch/revolution.
The length of cut for the 2-inch diameter hole = 1.85 inches.
The feed for this hole size = 0.04 inch/revolution.

Substituting these values into the above equation, we obtain the values of cutting path length for each hole size as follows.

Cutting path length (1-inch diameter hole) = (3.1415 × 1 × 1.55)/(12 × 0.020).
Thus, the cutting path length = (4.8693)/(0.24) = 20.28 ft.

Cutting path length (2-inch diameter hole) = (3.1415 × 2 × 1.85)/(12 × 0.040).
Thus, the cutting path length = (11.6235)/(0.48) = 24.21 ft.

Assuming a cutting velocity of 350 ft/min for the entire process, the cutting time for each hole is obtained as follows.

Cutting time (one inch diameter hole) = 20.28 ft/350 ft/min= 0.0579 minutes.
There are four such holes. Therefore, total cutting time for all four holes = 0.0579 × 4 = 0.2317 minutes.

Cutting time (2-inch diameter hole) = 24.21 ft/350 ft/min= 0.069 minutes.
Thus, cutting time for all five holes = 0.2317 minutes + 0.069 minutes = 0.3007 minutes.

Notice that this is almost exactly the same value we obtained using the feed based method. **Solution**

In the next section, we will examine costing of metal shaping operations.

4.4 Metal Shaping Operations

In this section, we will analyze shaping operations from the perspective of process costing. Specifically, we will consider two operations, namely: casting and forging.

4.4.1 Costing of Casting Operations

Casting can be defined as a process that involves pouring liquid metal into a mold. The mold contains a hollow cavity that corresponds to the desired shape of the object being cast. Once poured, the liquid metal is allowed to cool until it turns into a solid block. This block is referred to as a casting. The final step of the operation involves breaking out the casting from the mold. The process of casting is used to make complex shapes that would be too difficult or too expensive to make if other processes were used. The following basic steps are involved in making a casting.

1. *Making the pattern:* This is the first step in the casting process. It involves making a "pattern," which is a replica of the object to be cast.
2. *Molding:* Molding is the process of preparing the mold in order to receive the molten metal.
3. *Melting and pouring the liquid metal:* during this step, both ferrous as well as nonferrous metals are loaded to charge the furnace. Molten metal is poured into the mold by employing a system of gates and risers. When the metal cools and solidifies, it adopts the shape of the mold.
4. *Shakeout and cleaning:* During this process, the solidified casting is removed from the mold by a process referred to as "shakeout".
5. *Heat treatment and inspection:* heat treatment is a process that incorporates both heating as well as chilling in order to reduce stress in a casting. A casting can also be heat treated in order to modify the physical properties of the metal.

Figure 4.9 presents the different parts of a sand mold when it is open and when it is closed.

Figure 4.10 shows an example of a typical sand cast part.

The different types of casting processes as well as the types of mold and pattern used for each process are presented in Table 4.7. Different casting processes use different types of molds. For instance, sand casting uses a single use mold. This is a type of mold that can only be used once. The pattern, in the case of sand casting, however, can be used several times. It should be noted that in the case of processes which use multiple use molds, patterns are not used for each casting. This is because the mold itself is the pattern in such cases. From the costing perspective, the variable production costs are

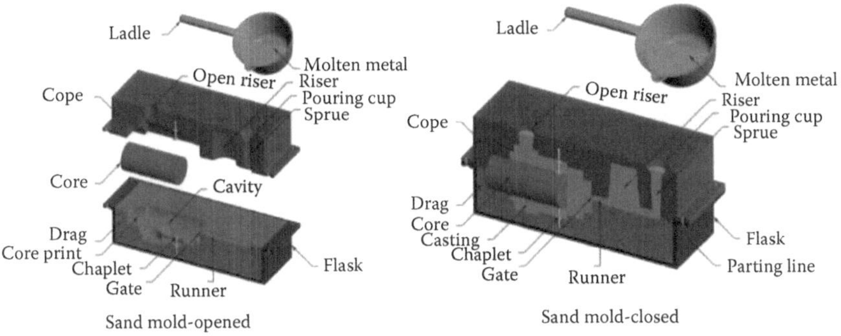

FIGURE 4.9
The process of sand casting. (Custompartnet.com.)

FIGURE 4.10
An example of a sand cast part.

high for single use molds. On the other hand, the mold costs are very high for multiple use molds.

In this chapter, we shall investigate the economics of a sand casting operation. The following are the major elements in the estimation of foundry costs.

- Material cost
- Labor cost
- Direct foundry expenses
- Overhead expenses
- Environmental costs

The environmental costs include costs for disposal of slag, sludge, and sand as well as costs involved in water treatment and cleansing of air. We will consider each of the aforementioned costs in the following discussion.

1. *Material Cost*: In a sand casting operation, the material cost is constituted of metal costs, core costs, and the cost of the sand used for molding. The volume of the flask is equal to the volume of the metal, volume of the core, and that of the molding sand when all of them are considered together.

   a. *Metal Cost*: In order to compute the metal cost, it is essential to know the total amount of metal poured, the return metal in the form of gates, runners, and risers, and the amount of metal loss as well as the casting scrap. The total weight of the metal poured is expressed by Equation 8 as follows.

$$TWP = CW \times \{1/(1\text{-}S\,R)\} \times \{1/Y\} \times \{1/(1\text{-}ML)\} \times \{1/(1\text{-}FSR)\} \quad \text{(Equation 8)}$$

wherein:

TWP: Total weight of metal poured, expressed in terms of lbs (pounds)
CW: Weight of the casting, expressed in terms of lbs
SR: Scrap rate of the casting, expressed as a fraction
Y: Yield of the process, expressed as a fraction
ML: Melt loss, expressed as a fraction
FSR: Finishing and cleaning scrap rate, expressed as a fraction

The yield is a function of the casting weight and poured weight. It is expressed by Equation 9.

$$Y = CW/PW. \quad \text{(Equation 9)}$$

wherein:

CW: Weight of the casting.
PW: Poured weight. This is the sum of casting weight and process returns.

In order to compute poured weight, it is essential to add the casting weight to the weight of metal in the sprue, runner, and gates. The casting scrap per casting can be expressed by Equation 10 as follows.

$$CS = CW \times SR/\{1\text{-}SR\} \times \{1/(1\text{-}FSR)\}. \quad \text{(Equation 10)}$$

wherein:

CS: Weight of casting scrap per casting, expressed in lbs
CW: Weight of the casting, expressed in lbs
SR: Scrap rate of the casting, expressed as a fraction
FSR: Finishing and cleaning scrap rate, expressed as a fraction

It is important to note that the finishing and cleaning scrap rate is a percentage of completed castings that are scrapped during the cleaning and finishing operations. The amount of finishing and cleaning scrap is computed using Equation 11 as follows.

$$CSFC = CW \times FSR/\{1\text{-}FSR\}. \qquad \text{(Equation 11)}$$

wherein:

CSFC: Casting scrap weight from finishing and cleaning operations, expressed in lbs
CW: Weight of casting, expressed in lbs
FSR: Finishing and cleaning scrap rate, expressed as a fraction

The return weight per casting is expressed by Equation 12 as follows.

$$RW = CW \times \{(1\text{-}Y)/Y\} \times \{1/(1\text{-}SR)\} \times \{1/(1\text{-}FSR)\}. \quad \text{(Equation 12)}$$

wherein:

RW = Return weight, expressed in lbs
CW: Weight of the casting, expressed in lbs
Y: Process yield, expressed as a fraction
SR: Scrap rate, expressed as a fraction
FSR: Finishing and cleaning scrap rate, expressed as a fraction

The melt loss is expressed by Equation 13 as follows.

$$MLW = CW \times \{ML/(1\text{-}ML)\} \times (1/Y) \times \{1/(1\text{-}SR)\} \times \{1/(1\text{-}FSR)\}. \quad \text{(Equation 13)}$$

wherein:

MLW: Melt loss weight, expressed in lbs
CW: Weight of casting, expressed in lbs
ML: Melt loss, expressed as a fraction
Y: Process yield, expressed as a fraction
SR: Scrap rate, expressed as a fraction
FSR: Finishing and cleaning scrap rate, expressed as a fraction

The total weight poured is computed by adding equations that are expressed by Equation 14 as follows. This is analogous to Equation 8 that we discussed before.

$$TWP = CW + CS + MLW + CSFC + RW. \qquad \text{(Equation 14)}$$

In order to better understand the concepts presented in the Equations 8–14, let us consider an example.

EXAMPLE 4.6

Compute the total pouring weight, melting weight loss, casting scrap weight, finishing and cleaning scrap weight, and return weight for a casting weighing 30 lbs., yield of 70%, casting scrap rate of 10%, melt loss of 2%, and a finishing and cleaning scrap rate of 3%.

Solution

From the problem statement, we have the following data:

CW: Casting weight = 30 lbs.
SR: Casting scrap rate = 10%.
Y: Process yield = 70%.
ML: Melt loss = 2%.
FSR: Finishing and cleaning scrap rate = 3%.

In order to compute the total weight poured, we use Equation 6 as follows.

$$\text{TWP} = \text{CW} \times \{1/(1\text{-S R})\} \times \{1/\text{Y}\} \times \{1/(1\text{-ML})\} \times \{1/(1\text{-FSR})\} \quad \text{(Equation 6)}$$

Substituting the respective variables into Equation 6, we obtain the following:

TWP = 30lbs × {1/(1 − 0.10)} × {1/0.70} × {1/(1 − 0.02)} × {1/(1 − 0.03)}.
Thus, TWP = 30 lbs × {1/0.9} × (1.42857) × {1/0.98} × {1/0.97}.
Thus, TWP = 30 lbs × (1.111) × (1.42857} × (1.0204) × (1.0309).
Thus, TWP = 49.992 lbs.

The weight of the melt loss can be computed from Equation 13 as follows.

$$\text{MLW} = \text{CW} \times \{\text{ML}/(1\text{-ML})\} \times (1/\text{Y}) \times \{1/(1\text{-SR})\} \times \{1/(1\text{-FSR})\}. \quad \text{(Equation 13)}$$

Substituting the respective variables into Equation 13, we obtain the following:

MLW = 30 lbs × {0.02/(1 − 0.02)} × (1/0.70) × {1/(1 − 0.10)} × {1/(1 − 0.03)}.
Thus, MLW = 30 lbs × (0.0204) × (1.42857) × (1.1111) × (1.0309).
MLW = 1 lb.

The casting scrap weight is determined from Equation 10 as follows.

$$\text{CS} = \text{CW} \times \text{SR}/\{1\text{-SR}\} \times \{1/(1\text{-FSR})\}. \quad \text{(Equation 10)}$$

Substituting the respective values into Equation 10, we obtain the following.

$$CS = 30 \text{ lbs.} \times \{0.10/(1 - 0.10)\} \times \{1/(1 - 0.03)\}.$$
$$\text{Thus, } CS = 30 \text{ lbs.} \times (0.111) \times (1.03092)$$
$$\text{Thus, } CS = 3.432 \text{ lbs.}$$

The finishing and cleaning scrap weight is determined from Equation 11 as follows.

$$CSFC = CW \times FSR/\{1\text{-}FSR\}. \qquad \text{(Equation 11)}$$

Substituting the respective values into Equation 11, we obtain the following.

$$CSFC = 30 \text{ lbs} \times \{0.03/0.97\} = 0.9278 \text{ lbs.}$$

The return weight is computed from Equation 12 as follows.

$$RW = CW \times \{(1\text{-}Y)/Y\} \times \{1/(1\text{-}SR)\} \times \{1/(1\text{-}FSR)\}. \qquad \text{(Equation 12)}$$

Substituting the respective values into Equation 12, we obtain the following.

$$RW = 30 \text{ lbs} \times \{(1 - 0.7)/0.70\} \times \{1/(1 - 0.10)\} \times \{1/(1 - 0.03)\}.$$
$$\text{Thus, } RW = 30 \text{ lbs} \times (0.42857) \times \{1.1111\} \times \{1/0.97\}.$$
$$\text{Thus, } RW = 30 \text{ lbs} \times (0.42857) \times (1.14546).$$

Thus, RW = 14.72 lbs. **Solution**

b. *Core Costs:* The core costs include items such as cost of the cores used in the casting process, the cost of cores that were broken, and the cost of cores in the scrapped castings as well as the core disposal costs. The total core weight is computed using Equation 13 as follows.

$$TCW = CCW \times \{1/(1\text{-}CSR)\} \times \{1/(1\text{-}SR)\} \times \{1/(1\text{-}FSR)\}. \qquad \text{(Equation 13)}$$

wherein:
TCW: Total weight of the core, expressed in lbs
CCW: Casting core weight, expressed in lbs
CSR: Core and molding scrap rate, expressed as a fraction
SR: Scrap rate, expressed as a fraction
FSR: Finishing and cleaning scrap rate, expressed as a fraction

In order to understand the practical applicability of the concept presented in Equation 13, consider Example 4.7.

EXAMPLE 4.7

The core weight of a particular casting is 20 lbs. The core and molding room core scrap rate is 20%. The casting scrap rate is 15%. The finishing and cleaning scrap rate is 5%. Compute the core weight to be used in each casting.

Solution

From the data given in the example, we already know the following values:

CCW: Casting core weight = 20 lbs.
CSR: Core and molding scrap rate = 20%.
SR: Scrap rate = 15%.
FSR: Finishing and cleaning scrap rate = 5%.

Substituting these values into Equation 13, we obtain the following.

$$TCW = CCW \times \{1/(1\text{-}CSR)\} \times \{1/(1\text{-}SR)\} \times \{1/(1\text{-}FSR)\} \quad \text{(Equation 13)}$$

Thus, $TCW = 20 \text{ lbs} \times \{1/(1 - 0.2)\} \times \{1/(1 - 0.15)\} \times \{1/(1 - 0.05)\}$.
Thus, $TCW = 20 \text{ lbs} \times \{1.25\} \times \{1.1765\} \times \{1.05263\}$.
Thus, $TCW = 30.96$ lbs. **Solution**

c. *Cost of Molding Sand:* Core sand is generally more expensive than molding sand. However, in view of the sheer amount of molding sand used in sand casting, the cost of molding sand constitutes the largest element of cost in any sand molding operation. Molding sand can be reclaimed after having been used once. There is some amount of molding sand lost due to factors such as dust, fines, burn-on, etc. This necessitates the addition of a fresh amount of molding sand in order to compensate for the loss. Also, disposal costs for molding sand are high due to environmental concerns. The total cost of molding sand per casting is expressed in Equation 14.

$$TSW = SMW \times \{1/(1\text{-}SML)\} \times \{1/(1\text{-}SR)\} \times \{1/(1\text{-}FSR)\}. \quad \text{(Equation 14)}$$

wherein:

TSW: Total weight of molding sand per casting, expressed in terms of lbs
SMW: Weight of sand mold per casting, expressed in lbs
SML: Sand mold losses, expressed as a fraction
SR: Scrap rate of casting, expressed as a fraction
FSR: Finishing and cleaning scrap rate, expressed as a fraction

Cores can suffer breakage in the core room or during transportation as well as during insertion into the mold. Thus, the sand mold losses as expressed in Equation 14 are generally smaller than the core scrap rate and the molding scrap rate. Sand mold losses usually occur during molding time.

New system sand (NSS) is a percentage of the total weight of the molding sand. It is expressed by Equation 15.

$$NSS = A \times TSW. \qquad \text{(Equation 15)}$$

wherein:

NSS: New system sand, expressed in lbs
A: Fraction of new system sand to total molding sand
TSW: Total weight of molding sand per casting, expressed in terms of lbs

The waste sand for disposal is expressed by Equation 16 as follows.

$$WSD = B \times TSW. \qquad \text{(Equation 16)}$$

wherein:

WSD: Waste sand for disposal, expressed in lbs
B: Fraction of waste sand to total molding sand
TSW: Total weight of molding sand per casting, expressed in terms of lbs

Consider Example 4.8 in order to better understand the aforementioned concepts.

EXAMPLE 4.8

The sand mold weight for a casting is 200 lbs. The sand mold losses are 7% and the casting scrap rate is 10%. The finishing and cleaning scrap rate is 5% and the new system sand is 10%. The waste sand fraction is 5%. The molding sand costs $0.03/lb and the new sand costs $0.05/lb. The disposal costs are $0.08/lb. Compute the molding sand costs.

Solution

From the data presented in the example, we know the values of the following variables:

SMW: Weight of sand mold per casting, expressed in lbs. = 200 lbs.
SML: Sand mold losses, expressed as a fraction = 7%.
SR: Scrap rate of casting, expressed as a fraction = 10%.
FSR: Finishing and cleaning scrap rate, expressed as a fraction = 5%.

Substituting the above values into Equation 14, we obtain the following.

$$TSW = SMW \times \{1/(1\text{-}SML)\} \times \{1/(1\text{-}SR)\} \times \{1/(1\text{-}FSR)\}. \quad \text{(Equation 14)}$$

Thus, $TSW = 200 \text{ lbs} \times \{1/(1-0.07)\} \times \{1/(1-0.10)\} \times \{1/(1-0.05)\}$.
Thus, $TSW = 200 \text{ lbs} \times \{1.07527\} \times \{1.1111\} \times \{1.05263\}$.
Thus, TSW = 251.52 lbs.

The amount of new system sand can be computed from Equation 15 as follows:

$$NSS = A \times TSW. \quad \text{(Equation 15)}$$

From the data presented in the example, we know the value of "A" to be 10%. Substituting this value into Equation 15, we obtain the following.

$$NSS = 0.10 \times 251.52 \text{ lbs} = 25.152 \text{ lbs}.$$

Cost of molding sand = (251.52 lbs − 25.152 lbs) × ($0.03/lb)
Thus, cost of molding sand = (226.368 lbs) × ($0.03/lb) = $6.79.
Cost of new system sand = 25.152 lbs × $0.05/lb = $1.2576.

The waste sand disposal is obtained from Equation 16 as follows.

$$WSD = B \times TSW. \quad \text{(Equation 16)}$$

We know the value of "B" to be 5%. Substituting this value into Equation 16, we obtain the following.

$$WSD = 0.05 \times 251.52 \text{ lbs} = 12.576 \text{ lbs}.$$

Waste and sand disposal cost = 12.576 lbs × ($0.08/lb) = $1.006.

The total molding sand cost is computed as the sum of the molding sand cost, new system sand cost, and the waste sand disposal costs.
Thus, total cost of molding sand = $6.79 + $1.2576 + $1.006 = $9.0536.
Solution

4.4.1.1 Labor Cost in Sand Casting

We have already seen earlier in this chapter that a series of steps needs to be undertaken in order to create a sand casting. A brief discussion of labor costs associated with each step is presented in the following section.

- *Pattern making:* The costs of pattern making are treated as a fixed cost. This is similar to the setup cost in machining.
- *Core making:* Cost associated with making the core is considered to be variable cost. This cost is adjusted for losses that occur due to the core making process itself and also for the scrap castings.
- *Molding:* The labor cost for molding is determined for each mold and is computed in a manner similar to the core making cost. Once again, the labor cost in molding should be adjusted for losses that occur during the process.
- *Melting and pouring:* The labor costs for melting and pouring are included in the metal cost. There is no need for separate consideration of labor costs associated with this part of the process.
- *Cleaning and finishing:* Cleaning and finishing costs are determined for each casting. The amount of cleaning and finishing required per casting is a direct function of the complexity of the casting.

The labor cost in making the core is expressed in Equation 17.

$$\text{CMUC} = \text{CMUT} \times \text{CMUL} \times \{1/(1\text{-CSR})\} \times \{1/(1\text{-SR})\} \times \{1/(1\text{-FSR})\}. \qquad \text{(Equation 17)}$$

wherein:

CMUC: Unit labor cost involved in making the core ($/casting)
CMUT: Time required per unit in making the core (min/casting)
CMUL: Labor rate in making the core ($/min)
CSR: Core and molding scrap rate, expressed as a fraction
SR: Scrap rate, expressed as a fraction
FSR: Finishing and cleaning scrap rate, expressed as a fraction

The labor cost in making the molding is expressed in Equation 18.

$$\text{MULC} = \text{MULT} \times \text{MLR} \times \{1/(1\text{-SML})\} \times \{1/(1\text{-SR})\} \times \{1/(1\text{-FSR})\}. \qquad \text{(Equation 18)}$$

wherein:

MULC: Unit labor cost involved in making the molding ($/casting)
MULT: Time required per unit in making the molding (min/casting)
MLR: Labor rate in making the molding ($/min)
SML: Sand mold loses, expressed as a fraction
SR: Scrap rate, expressed as a fraction
FSR: Finishing and cleaning scrap rate, expressed as a fraction

The labor costs involved in cleaning and finishing the casting are expressed in Equation 19.

$$\text{CULC} = \text{CULT} \times \text{CLR} \times \{1/(1\text{-FSR})\}. \qquad \text{(Equation 19)}$$

wherein:

CULC: Unit labor cost involved in cleaning and finishing operations ($/casting)

CULT: Time required per unit in cleaning and finishing (min/casting)

CLR: Unit labor rate for cleaning and finishing operations ($/min)

FSR: Finishing and cleaning scrap rate, expressed as a fraction

4.4.1.2 Other Direct Expenses

During the process of sand casting, additional expenditure has to be incurred on processes such as making the patterns and the core boxes. These are treated as fixed costs akin to setup costs in machining operations. By their nature, they have no relationship to the number of castings produced. These costs must be recovered regardless of the number of castings made. It is obvious that in the case of large production quantities, the fixed cost can be spread over the expected life of the pattern. However, in case of quick changes in design, the pattern life is somewhat short and recovering the fixed cost can become quite a problem.

4.4.1.3 Overhead Expenses

In a sand casting operation, overhead expenses have to be borne in the form of expenses for equipment usage, salaries paid to supervisory staff and inspection staff, administrative overhead, etc. The overhead rates are assigned at about 150%–200% of direct labor costs (involved in core making, molding, and cleaning operations).

In the next subsection, we will analyze the costs of a forging operation.

4.4.2 Costing of Forging Operations

Forging is a type of metal forming process. Metal forming constitutes a plastic deformation process in which a basic shape is transformed into another shape. The advantage of metal forming lies in the fact that such processes have much higher yields. The parts obtained by metal forming processes also exhibit better mechanical properties than machined parts.

Shapes such as structural beams, pipes, and rails are generally manufactured using metal forming processes. In terms of production cost, metal forming is characterized by high tooling cost. As a consequence, it is best suited to high production volumes. In this section, we will consider the costing of forging processes. An example of some products manufactured using forging is depicted in Figure 4.11.

The general principle of the forging process is depicted in Figure 4.12.

Process costs for the forging operation can be determined using two methods as follows.

- *Basic standard time costing:* In this approach, the cost of the part is determined directly.

FIGURE 4.11
Example of parts manufactured using forging.

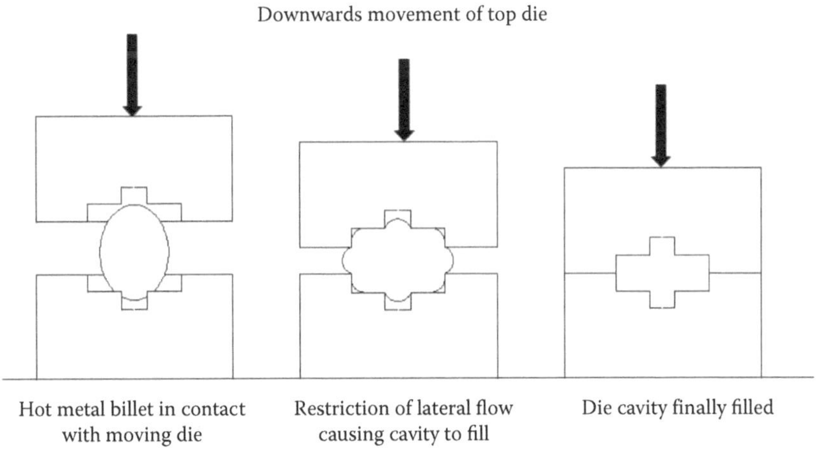

FIGURE 4.12
Process of forging.

- *Relative costing:* In this approach, a quantity referred to as the relative cost is determined. The relative cost is defined as the ratio of the cost of the component under consideration to that of a reference component. This method allows for evaluating different designs quickly on a relative basis. It can also be easily computerized. One of the drawbacks of this approach is that a variety of relative factors to the reference part first needs to be developed in order to use this approach.

In the following section, we will deal with each of the aforementioned approaches separately.

1. *Basic standard time costing method:* This approach computes the cost of the forged part directly. In forging operations, the major cost items are the following:
   a. Material cost
   b. Labor cost
   c. Direct expenses
   d. Overhead expenses

   Observe that these costs are the same as for any processing operation. For instance, as we have seen previously in this chapter, machining has the same overall cost structure. When compared to casting, though, the material costs for forging are a bit higher. This is due to the fact that the initial material required for forging is of an overall higher quality than casting. However, the labor costs for forging are lower than the labor costs for casting. This is because the processing times are shorter for forging than for casting. Direct expenses for forging are higher than direct expenses for casting due to the fact that tooling for forging processes requires better mechanical properties. It is also subjected to more adverse temperatures than casting. Next, we will take a closer look at each of the aforementioned elements of cost for the basic standard time costing method.

   a. *Material Cost:* Material costs in forging are a direct function of material loss. It is obvious that a higher degree of material loss implies higher material costs overall and vice versa. The following material losses are encountered in forging operations: Shear loss, scale loss, sprue loss, flash loss, and tong-hold loss. These losses account for approximately 60% of the net weight of the product. Each type of material loss is defined in more detail as follows:
      i. *Shear Loss*: Shear loss accounts for about 5% of the net weight of the product. It occurs when a blank is cut from bars or billets. Shear loss constitutes the material removed during sawing as well as the small bits left at the end.
      ii. *Scale Loss*: Surface scaling occurs due to reaction of the metal with an oxygen rich atmosphere. Scaling is the direct result of surface oxidation. Every time the hammer hits the surface of the metal, the scale falls off and is replaced by another layer of scale. Scale loss accounts for about 6% of the net weight. It is also a function of the material being forged. Thus, a material that can oxidize easily is bound to have a greater amount of scale loss and vice versa.

iii. *Sprue Loss*: During the forging process, the tong-hold and metal in the die are connected by a portion of metal. This is referred to as the sprue. The sprue is removed when the forging operation is complete. It accounts for about 7% of the net weight of the part.

iv. *Flash Loss*: Dies are used during the forging operation in order to shape the piece of metal in the shape of the die. During the forging operation, some metal overflows at the parting line of the top half and bottom half of the die. Some flash is actually beneficial since it prevents damage to the die. Flash volume depends upon the surface area of the product at the parting line and is usually the largest loss item in a forging operation.

v. *Tong-hold Loss*: In the case of drop forging operations, the stock is held at one end with tongs. In order to enable this, a small length of about 1 inch or the diameter of the stock (whichever is greater) is added to the length of the stock. Tong-hold loss is defined as the cross sectional area multiplied by the length of the tong-hold.

In the case of forging operations, the gross amount of metal required for the operation is denoted by Equation 20.

$$\text{Gross amount of metal} = \text{net amount of metal} \times (1 + \text{losses}). \quad \text{(Equation 20)}$$

This can be further elaborated using Equation 21 as follows.

$$GA = NA \times \{1 + (SL + TL + SCL + FL + SRL)\}. \quad \text{(Equation 21)}$$

wherein:

GA: Gross amount, expressed in lbs or kgs
NA: Net amount, expressed in lbs or kgs
SL: Shear loss, expressed as a fraction
TL: Tong-hold loss, expressed as a fraction
SCL: Scale loss, expressed as a fraction
FL: Flash loss, expressed as a fraction
SRL: Sprue loss, expressed as a fraction

In order to better understand the concepts presented in the preceding paragraphs, consider Example 4.9.

EXAMPLE 4.9

Compute the gross weight of a low carbon steel part with a density of 0.25 lb/cu.in and a volume of 10 cubic inches. The flash volume is 1.2 cubic inches, the sawing loss is 6%, tong-hold loss is 6%, scale loss is 5%, and sprue loss is 7%. Also, if the scrap rate was 5%, compute the total gross amount. Compute the total material cost if low carbon steel costs $5/lb. and scrap costs $0.35/lb.

Solution

In order to solve this problem, we first need to compute the net weight of the part. We know the following equation:

$$\text{Density} = \frac{\text{Mass}}{\text{Volume}}$$

Thus, Mass = Density × Volume.

From the data given in the problem, we know that density of low carbon steel = 0.25 lb/cu in

Also, the volume of the component = 10 cubic inches.

Substituting these values into the above equation, we obtain the mass of the part as follows.

$$\text{Mass} = 0.25\ \text{lb/cu in} \times 10\ \text{cu in} = 2.5\ \text{lbs.}$$

The flash fractional loss is calculated as follows;

$$\text{FL} = \frac{\text{Flash Volume}}{\text{Net Volume}}$$

From the data given in the problem, we know that flash volume = 1.2 cubic inches.

Net volume of the part = 10 cubic inches.
Thus, flash volume = 1.2 cu in/10 cu in = 0.12 cubic inches.

The gross amount is now computed by using Equation 21 as follows.

$$\text{GA} = \text{NA} \times \{1 + (\text{SL} + \text{TL} + \text{SCL} + \text{FL} + \text{SRL})\}. \qquad \text{(Equation 21)}$$

Substituting the respective values into the above equation, we obtain the value of gross amount as follows:

GA = 2.5 lbs × (1 + 0.06 + 0.06 + 0.05 + 0.12 + 0.07)
Thus, GA = 2.5 lbs × (1.36) = 3.4 lbs.

$$\text{Total gross amount} = \text{gross amount}/(1\text{-SR}).$$

wherein:

SR: Scrap rate.

Thus, for our present example, we can compute the total gross amount as follows:

$$\text{TGA} = 3.4\ \text{lbs.}/(1\text{–}0.05).$$

Thus, TGA = 3.4 lbs./0.95 = 3.579 lbs.

We compute return scrap as follows:

Return scrap = net amount × {(TL + FL + SRL)}/(1-SR)} + {SR/(1-SR)}.

Thus, return scrap = 2.5 lbs × {(0.06 + 0.12 + 0.07)}/(0.95)} + {0.05/0.95}.
Thus, return scrap = 2.5 lbs × {(0.25 + 0.05)/0.95} = 2.5 lbs × {(0.30)/0.95}.
Thus, return scrap = 0.789 lbs.

The remaining amount of material lost is waste material, which has no value. The return scrap can be sold in the market for $0.35/lb.

Thus, total material cost = {(3.4 lbs) × ($5/lb)} – {(0.789 lbs × $0.35/lb)}.
This implies that total material cost = $17 – $0.276 = $16.72.
Solution

Next, we will consider labor cost for forging operations.

b. *Labor Cost*: Table 4.8 depicts some basic data for forging operations in terms of setup time and crew time.

In order to better understand the computation of labor cost for forging operations, consider Example 4.10.

EXAMPLE 4.10

A 3000 lb forging press is used to make two parts with a gross weight of 1.5 lb per forging. They are made from a bar that is 1 inch in diameter and 13 inches long. It weighs 3.7 lbs. The operation calls for a crew of 2 workers. Compute the amount of time it takes to produce a batch of 300 units. Also compute the labor cost and the machine cost if the labor rate was $35/hour and the machine rate was $52.50/hour.

Solution

In order to solve this problem, we first compute the amount of time it takes to process the part by referring to Table 4.9.

TABLE 4.8

Set up Times and Crew Times for Different Ratings of Forge Presses

Rating of Forge Press (lbs)	Crew Size	Set up Time (minutes)	Crew Minutes/Piece Processed
1500	2	60	2.67+ (0.70 × D × L)
3000	2	90	2.91+ (0.100 × D × L)
5000	3	120	4.95 + (0.0586 × D × L) + 0.0205 × GA
12000	4	180	5.19 + (0.1066 × D × L) + 0.0355 × GA

Source: Ostwald, P. 1982. *American Machinist Manufacturing Cost Estimating Guide*, McGraw Hill, pp. 382 [12].

TABLE 4.9

Material Data for Relative Costing for Selected Materials

Material Data for Relative Costing						
Material Class	**Material Description**	**Mlf (kg-m/sq mm)**	**Mdlf**	**Dmf**	**Dr**	**Cr**
0	*Aluminum alloy*					
	Group A	0.054	1.00	1.00	0.35	4.70
	Group B	0.087	1.00	1.00	0.35	4.94
2	*Carbon and alloy steels*					
	Group A	0.065	1.00	1.00	1.00	1.00
	Group B	0.076	0.800	1.00	1.00	1.29
4	*High alloy steels*					
	Stainless-A	0.130	0.300	1.200	1.01	7.37
	Nickel alloy	0.130	0.300	1.200	1.020	16.160

Source: Knight, W and Poli, C. 1958. 'A Systematic Approach to Forging Design', *Machine Design*, pp. 94–99 [9].

Note: Mlf: material load factor; Mdlf: material die life factor; Dmf: die material cost factor; Dr: relative material density; Cr: Relative material cost.

For a press rated at 3000 lbs and a crew of two workers, the equation for crew minutes/piece processed is given as follows:

$$T = 2.91 + (0.100 \times D \times L).$$

wherein:

D: Diameter of the bar = 1 inch
L: Length of the bar = 13 inches

Thus, $T = 2.91 + (0.10 \times 1 \times 13) = 4.21$ min/bar.

Dividing this time equally into two parts, we get time/worker = 4.21/2 = 2.105 min/worker.

A batch of 300 units implies processing 150 bars. The total labor time for 150 bars is computed as the labor time plus the setup time. From Table 4.9, we know that the setup time for 3000 lb press is 90 minutes.

Thus, total labor time = (150 × 4.21 min) + (90 minutes).
Total labor time = 631.5 minutes + 90 minutes = 721.5 minutes = 12.025 hours.

The total operating time is given as:

total operating time = 150 bars × 4.21 crew minutes/bar/2 crew members.
Thus, total operating time = 315.75 minutes.

Additionally, the press is occupied for a certain amount of time. This is computed by adding the total operating time to the setup time.

Thus, the machine will be occupied for a total of (315.75 minutes + (90/2) minutes) = 360.75 minutes = 6.0125 hours.

From the data given in the problem, we know that labor rate = \$35/hour.

Also, total labor time = 12.025 hours.

Thus, labor cost = total labor time × labor rate.

This implies that labor cost = 12.025 hours × \$35/hour = \$420.875.

From the data given in the problem, we know that machine rate = \$52.50/hour.

We also know that the machine will be occupied for 360.75 minutes = 6.0125 hours.

Thus, machine cost = machine rate × machine occupancy.

This implies that machine cost = \$52.50/hour × 6.0125 hours = \$315.65. **Solution**

c. Other Direct Expenses in Forging Operations

The cost to produce the die for the forgings constitutes the largest element of cost in most forging operations. It is usually higher than the material cost and the operating cost of the forging press. This is especially the case when complex shapes have to be forged. In such cases, due to the complexity of the shape to be produced, the life of the die is reduced. Thus, not only is it expensive to produce the die for a complex shape, it is also necessary to produce it often because the life of the die is shorter owing to the complexity of the shape being produced. Similarly, the nature of the material being forged has a direct impact on die costs. For instance, alloys that can only be worked under high temperatures further reduce the life of the die, thus leading to increased die costs.

As we have seen in the preceding discussion, die costs for forging operations depend on a variety of factors. Thus, die costs span an entire range of costs. They can start at as little as \$550 on the lower end of the price range all the way to about \$400,000 at the upper limit of that range. The price of the die is largely a function of parameters such as part surface area, volume of metal to be forged, material of the part, complexity of the part, the degree of precision required in the forged part, and the die material. The cost of a die for a forging operation can be computed using Equation 22.

$$\text{Die cost} = 350 - 2.38X + 6.08X^2. \qquad \text{(Equation 22)}$$

wherein:

Die cost is expressed in \$

X: Length of the part expressed in inches

EXAMPLE 4.11

Compute the cost of the forging die for a part that is 15 inches long.

Solution

We can estimate the cost of making the die by substituting the value of $X = 15$ inches into Equation 22.

Thus, die cost $= 350 - 2.38 \times (15) + 6.08 \times (15)^2$.
This implies that the cost of the die $= 350 - 35.7 + 1368$.
This gives us the cost of the die = \$1682.3. **Solution**

d. Overhead Expenses for Basic Standard Time Costing

As for all operations, overhead expenses play an important role in computing the total cost of parts produced by forging processes. They can be determined by using the methods discussed in the chapter on overhead costs. In general, though, overhead expenses in forging operations range from 150% to 300% of the direct labor rate or machine hour rate. In the next section, we will discuss the costing of forging operations using the relative costing approach.

4.4.3 Relative Costing Approach

The relative costing approach, as its name suggests, compares the costs of different forging designs. The different costs are compared to a reference forging. The reference forging has a reference cost ratio for several variables such as material costs, tooling costs, and operating costs as well as total costs. Generally, the following reference ratio is used for each of the aforementioned variables:

- Reference cost ratio is 50% for material costs
- Reference cost ratio is 30% for operating costs
- Reference cost ratio is 20% for tooling costs

Combining the above values we get Equation 23, which expresses the different ratios for a specific part.

$$CF/CFR = (0.5 \times Cmr) + (0.30 \times Cor) + (0.20 \times Ctr). \quad \text{(Equation 23)}$$

wherein:

CF: Cost of the forging for the part being considered
CFR: Cost of reference forging
Cmr: Relative cost of material
Cor: Relative cost of operation
Ctr: Relative cost of tooling (relative die cost)

Next, we determine how to compute the values of each of Cmr, Cor, and Ctr.

First, we start with the equation for relative cost of material. The relative cost of material, Cmr, can be determined using Equation 24 as follows.

$$\text{Cmr} = \text{Dr} \times \text{Cr} \times \text{Kf} \times \text{V}/196{,}000. \qquad \text{(Equation 24)}$$

wherein:

Cmr: Relative cost of material.
Dr: Ratio of material density to density of reference material.
Cr: Ratio of material price to price of reference material, both expressed as ($/lb).
Kf: Ratio of gross to net weight for the part and reference material.
V: Part volume, expressed in cubic millimeters.
196,000: Volume of reference material in cubic millimeters. This can be converted to cubic inches as 11.96.

Next, we compute the relative cost of operation, Cor, as follows.

$$\text{Cor} = \text{Co} \times \text{N} \qquad \text{(Equation 25)}$$

wherein:

Cor: Relative cost of operation
Co: Relative cost of operation per operation
N: Number of operations

The relative operation cost is determined as a function of the energy capacity of the equipment. This is computed using Equation 26.

$$\text{E} = \text{Ap} \times \text{Mlf} \times \text{Slf}. \qquad \text{(Equation 26)}$$

wherein:

E: Energy capacity for the equipment, expressed in ft-lbs
Ap: Plane area of the part being processed, expressed in square inches
Mlf: Load factor of the material, expressed in ft-lb/square inch
Slf: Shape load factor

Figure 4.13 depicts the relationship between relative operating cost and energy capacity.

The relative die cost can be obtained by using Equation 27.

$$\text{Ctr} = \text{Ap} \times \text{Cdc} \times \text{N} \times \{(\text{Dmf} \times \text{R}) + (\text{Sdmf} \times (1\text{-R}))\}/\{\text{Mdlf} \times \text{Sdlf} \times \text{K}\}. \qquad \text{(Equation 27)}$$

wherein:

Ctr: Relative cost of die.
Ap: Plane area of the part, expressed in square inches.

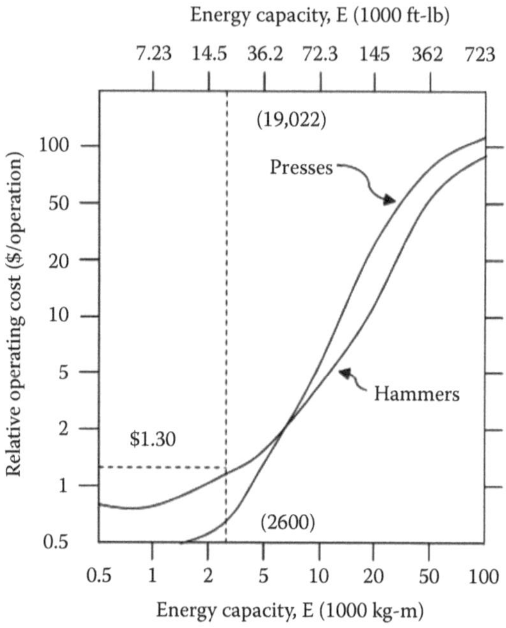

FIGURE 4.13
Relationship between relative operating cost and energy capacity. (Adapted from Knight, W and Poli, C. 1958. 'A Systematic Approach to Forging Design', *Machine Design*, pp. 94–99 [9].)

Cdc: Relative cost of die per unit area.
N: Number of operations.
Dmf: Material cost factor for the die.
R: Material cost proportion for the die.
Sdmf: Shape die machining factor.
Mdlf: Material die life factor.
Sdlf: Shape die life factor.
K: Reference die surface area. This is taken as 27.39 square inches.

Figure 4.14 illustrates the relationship between material cost proportion for the die (R) and plane area including the flash.

Table 4.9 presents material data for the relative costing method for different materials. It presents values of factors such as Mlf, Mdlf, Dmf, Dr, Cr, etc., for different materials. Note that the values presented in Table 4.9 are not exhaustive.

Table 4.10 presents shape data for relative costing for selected materials. It contains information on several factors such as Kf, Slf, Sdlf, etc. Once again, this list is not exhaustive.

In Table 4.10, a score of "0" for shape complexity implies a shape that is very simple to manufacture. On the other hand, a complexity score of 6 implies a shape that is complex and involves a greater level of difficulty. This is borne

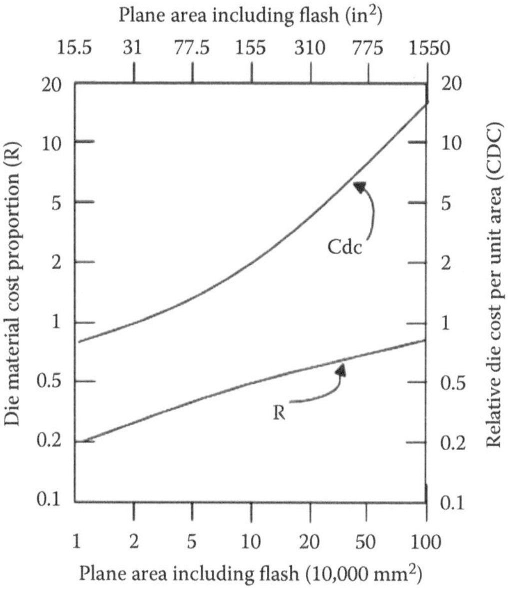

FIGURE 4.14
Relationship between material cost proportion for the die and plane area including the flash. (Adapted from Knight, W and Poli, C. 1958. 'A Systematic Approach to Forging Design', *Machine Design*, pp. 94–99 [9]).

TABLE 4.10
Shape Data for Relative Costing for Selected Materials

Shape Data for Relative Costing						
Shape Code	**Shape Complexity**	**Kf**	**Slf**	**Sdlf**	**Sdmf**	**N**
1-0	0	1.00	1.00	1.00	1.00	1.00
1-1	5	1.06	1.15	0.80	1.10	3.00
1-1	6	1.09	1.35	0.60	1.10	4.00
1-2	6	1.10	1.40	0.55	1.15	4.00

Source: Knight, W and Poli, C. 1958. 'A Systematic Approach to Forging Design', *Machine Design*, pp. 94–99 [9].
Note: Kf: ratio of gross weight to net weight; Slf: shape load factor; Sdlf: shape die life factor; Sdmf: shape die machining factor.

out by the number of operations required to produce the desired shape. For a shape complexity score of "0", only one operation will suffice. However, for more complex shapes, notice that more operations are required to achieve the desired result. Thus, four operations are required to produce a shape with a complexity score of "6".

In order to better understand the utility value of the relative costing approach, consider Example 4.12.

EXAMPLE 4.12

Compute the relative cost of a nickel alloy steel part with a shape configuration of 1-1 and a complexity score of 5. The plane area of the part is 30 square inches and has a volume of 15.74 cubic inches. It is formed using a hammer forging operation.

Solution

The relative material cost for the part can be obtained by using Equation 24 as follows.

$$Cmr = Dr \times Cr \times Kf \times V/11.96. \qquad \text{(Equation 24)}$$

The values of each of the factors such as Dr, Cr, and Kf can be obtained from Tables 4.10 and 4.11.

Thus, for a nickel alloy part, we have the following material data values:

Dr: 1.02
Cr: 16.16

From Table 4.11, the value of Kf for a shape code of 1-1 and a part complexity of 5 is:

Kf: 1.06
V: Volume of the part = 15.74 cubic inches

Substituting these values into Equation 24, we obtain the following:

$Cmr = 1.02 \times 16.16 \times 1.06 \times 15.74/11.96.$
Thus, $Cmr = 275.012/11.96 = \$22.994.$

TABLE 4.11

Setup Times for Selected Welding Operations

Welding Operation	Level of Automation	Minimum Setup Time in Minutes	
		Mean	Range
Shielded metal arc welding	Manual	18	–
Flux cored arc welding	Semiautomatic	24	–
Flux cored arc welding	Manual	180	120–240
Submerged arc welding	Semiautomatic	30	–
Submerged arc welding	Manual	180	120–240
Gas metal arc welding	Semiautomatic	24	–
Gas tungsten arc welding	Manual	24	–

Source: Ostwald, P. 1982. *American Machinist Manufacturing Cost Estimating Guide*, McGraw Hill, pp. 382 [12].

We can compute the energy capacity of the equipment using Equation 26 as follows.

$$E = Ap \times Mlf \times Slf. \qquad \text{(Equation 26)}$$

From the problem statement, we know the following values:

Ap: 30 square inches
Mlf: 0.130 for nickel alloy steel
Slf: 1.15 for a 1-1 part and complexity of 5

Substituting these values into Equation 26, we obtain the energy capacity of the equipment as follows.

$$E = 30 \times 0.130 \times 1.15 \times 4{,}662 \text{ ft-lb/sq in} = 4.485 \times 4{,}662 = 20{,}909 \text{ ft-lbs.}$$

In the above equation, we multiply by 4662 ft-lb/sq in in order to convert kg-m/sq mm into ft-lb/sq in. Note that Mlf values are expressed in kg-m/sq mm.

The operation cost for the hammer is determined using Equation 25 as follows.

$$Cor = Co \times N \qquad \text{(Equation 25)}$$

We use Figure 4.13 in order to determine the relative operating cost for a hammer with energy capacity of 20,909 ft-lbs. This corresponds to about 1.35. We also from Table 4.11 see that it takes 3 operations to produce the desired shape. Thus, we have the following:

Co: 1.35
N: 3

Substituting the values into Equation 25, we obtain the following.

$$Cor = \$1.35 \times 3 = \$4.05.$$

Next, we use Equation 27 to determine the relative die cost as follows.

$$Ctr = Ap \times Cdc \times N \times \{(Dmf \times R) + (Sdmf \times (1\text{-}R))\}/\{Mdlf \times Sdlf \times K\}. \qquad \text{(Equation 27)}$$

We already know the following values:

Ap: 30 square inches
R: From Figure 4.14, the die material cost proportion for a plane area of 30 square inches is about 0.4
Cdc: Relative die cost per unit area is $ 1.00
N: 3
Dmf: 1.2 for nickel alloy steel
Sdmf: 1.10 from Table 4.11
Mdlf: 0.3 for nickel alloy steel
Sdlf: 0.8 from Table 4.11
K: 27.39 square inches

It should be noted that the data in Figures 4.13 and 4.14 is subject to continual change and needs to be periodically updated.

Substituting these values into Equation 27, we obtain the following.

$$Ctr = 30 \times \$1.00 \times 3 \times \{(1.2 \times 0.4) + (1.1 \times (1 - 0.4)\}/(0.3 \times 0.8 \times 27.39).$$

Thus, $Ctr = 90 \times \{0.48 + 0.66\}/6.5736$.
Thus, $Ctr = 90 \times 1.14/6.5736$.
This implies that $Ctr = \$15.60$.

The cost ratio is obtained by using Equation 23 as follows:

$$CF/CFR = (0.5 \times Cmr) + (0.30 \times Cor) + (0.20 \times Ctr). \qquad \text{(Equation 23)}$$

We know from before the values of each of the variables in the above equation as follows.

Cmr: $22.994.
Cor: $4.05.
Ctr: $15.60.

Substituting each value into Equation 23, we obtain the following:

$CF/CFR = (0.5 \times 22.994) + (0.30 \times 4.05) + (0.20 \times 15.60)$.
Thus, $CF/CFR = (11.497) + (1.215) + (3.12)$.
Thus, $CF/CFR = 15.832$.

This implies that the cost of the nickel steel part is 15.832 times the cost of the reference part. **Solution**

In the next section of this chapter, we will examine the nature of welding operations and the associated costs involved with welding.

4.5 Welding Operations

Since most products are an assembly of individual components, it is necessary to incorporate efficient and effective fastening processes in order to join one component to another. Welding is a type of fastening method. Other fastening methods include adhesive bonding, soldering, brazing, and using fasteners such as screws and bolts. Fastening using screws and bolts is termed as temporary fastening. Processes such as welding, soldering, brazing, etc., are regarded as permanent joining methods. In this chapter we shall examine the cost structure of welding operations in particular.

4.5.1 Types of Welding Operations

Welding involves the joining of two materials together using heat and filler material. There are three main types of welding used in industry. They are as follows.

- *Gas Metal Arc Welding (GMAW)*: This type of welding is commonly referred to as MIG welding. It is the most common welding operation and is used to fuse mild steel, stainless steel, and aluminum. It is especially useful for joining thin sheet metal.
- *Tungsten Inert Gas or GTAW Welding*: This type of welding is also referred to as TIG welding. It is used to perform high quality fusion when the application calls for a better standard of finish. This type of welding does not require much post processing in the form of cleaning, sanding, etc.
- *Arc Welding or SMAW*: This type of welding is also referred to as stick or arc welding. This is the most common of all forms of welding. It is commonly used for manufacturing and construction applications especially with respect to thicker cross sections (4 mm and above).

The general principle of the welding process is depicted in Figure 4.15.

Some of the more commonly used terminology in the arc welding process is described in the following section. The symbol used to depict each variable is indicated in parentheses.

- *Travel Speed* (S): Travel speed refers to the speed at which the electrode travels during the welding operation. The travel speed can range from as low as 5 inches/minute to as high as 1000 inches/minute for

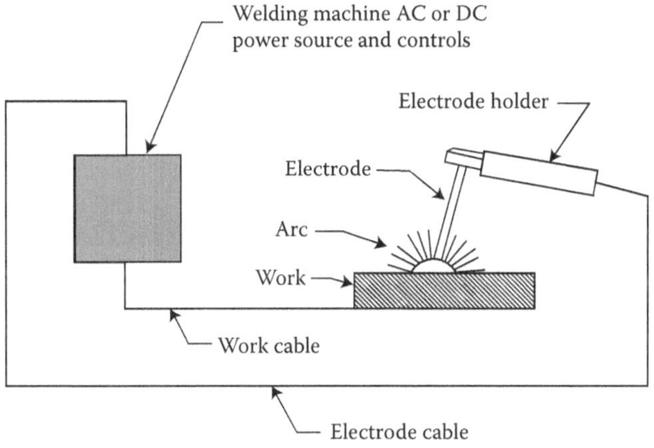

FIGURE 4.15
Basic principle of arc welding. (Adapted from *The Procedure Handbook of Arc Welding*, 1973. The Lincoln Electric Company, Cleveland, Ohio [6].)

automatic pipe welding applications. Travel speed is an important consideration in determining process cost. It is also an indication of the rate at which a manual welder should move the electrode. It is measured in inches/minute.

- *Welding Current* (I): Welding current has a direct impact on factors such as penetration of the weld and the deposition rate. If the energy input were to remain the same, welding current has a greater impact on weld penetration than travel speed or voltage. It is measured in amperes.
- *Welding Voltage* (V): Like welding current, welding voltage is a primary variable in determining the quality of the weld. Welding voltage is directly proportional to the width of the weld bead in particular. This implies a higher welding voltage produces a wider weld bead. It is measured in volts.
- *Operator Factor* (OF): Operator factor is defined as the ratio of arc time to the total cycle time for the welding operation. The total cycle time for the welding operation includes time taken to position and tack the pieces, arc time, cleaning and brushing time, removal of slag, and miscellaneous elements that have to be incorporated into the welding cycle. Setup time for the weld job is not included in the cycle time.
- *Wire Feed Rate* (WFR): The wire feed rate is used in automatic welding processes. It is defined as the rate at which the welding filler wire is fed into the arc of the weld and melts. The welder filler wire also acts as the electrode. It is measured in inches per minute.
- *Electrode Weight per Unit Length* (EWL): In the case of automatic and semiautomatic welding processes, EWL is defined as the weight per unit length of the welder filler wire. It depends upon the diameter of the wire and its material. It is obtained by dividing the weight of a piece of wire by its length. It is measured in pounds per inch.
- *Deposition Rate* (DR): The deposition rate is defined as the amount of weld metal that can be deposited in one hour. It is a function of the particular type of welding process, welding current, yield of filler metal, and the size of the electrode.
- *Duty Cycle* (DC): In any ten minute time interval, the power supply can be run at the rated current and voltage value without overheating. This is referred to as the duty cycle. In welding operations, the power supply can only dissipate a limited amount of heat as expressed by the duty cycle.
- *Electrode Metal Yield* (EMY): Electrode metal yield is defined as the fraction of the electrode weight that is transformed into weld metal. The electrode metal yield varies for different welding processes. It is also referred to as the deposition efficiency ratio.

- *Machine Efficiency* (M): Machine efficiency for welding operations refers to the efficiency of the power supply. In most welding operations, the power supply suffers a loss of about 10%. This implies that the machine efficiency for most cases is about 90%. It is measured as a percentage.
- *Gas Flow Rate* (GFR): The gas flow rate measures the rate of flow of the shielding gas in cubic feet/hour. The shielding gas protects the weld metal in the case of gas welding. It is measured in cubic feet per hour.
- *Flux Consumption Rate* (FCR): A granular flux is used to shield the weld metal from the atmosphere in submerged arc welding. Some of this flux crystallizes and thus cannot be reused. The amount of crystallized flux depends on the amount of weld metal deposited. The flux consumption rate is defined as the amount of flux that has either crystallized or is consumed per unit of weld metal deposited.
- *Energy Consumed per Unit of Weld Length* (H): Energy consumed per unit of weld length is a metric that is often used to compare different welding process. It is dependent upon the three primary variables of welding, namely: welding voltage, welding current, and the travel speed. The equation for (H) is expressed as follows:

$$H = 60 \times \frac{V \times I}{S}. \qquad \text{(Equation 28)}$$

- *Weight of the Weld Metal Deposited* (WMD): This measures the amount of weld metal deposited per unit length of the weld. It is the product of the cross sectional area of the weld metal and the density of the weld metal. Thus, we have the following equation:

$$\text{WMD (lb/in)} = \text{cross sectional area of weld (sq inches)} \times \text{density of metal (lb/cu in)}. \qquad \text{(Equation 29)}$$

- *Length of Weld* (L): The length of the weld is a major determinant in computing the welding time and related costs. The length of the weld is measured in inches or feet (US units).
- *Welding Time* (WT): Welding time is defined as the amount of time allowed for the entire welding cycle. It includes both the arc time as well as nonarc time per piece. The welding time is expressed as a product of length of weld (L), travel speed (S), and Operator factor (OF). This is depicted in Equation 30.

$$\text{WT (minutes)} = \{L \times 12 \text{ inches/ft}\}/\{S \times OF\}. \qquad \text{(Equation 30)}$$

4.5.2 Cost Components for Welding Operations

The two main components of cost of welding operations are fixed cost and variable cost. Notice that this is the same structure for machining as well as forming operations. This section will analyze each component separately.

- *Fixed Cost*: For welding processes, fixed costs consist of tooling cost and setup cost.
  - *Tooling Cost*: Tooling is an essential component of most mass manufacturing operations. Welding is no exception. Tooling in the form of appropriate jigs and fixtures is required in the case of high volume automatic welding operations. The jigs and fixtures specially designed for welding serve to hold the work piece and/or welding electrode. This enables the welding operation to be performed rapidly.
  - *Setup Costs*: Setup costs vary from one welding process to another. They also depend on the type of work being performed. Manual and semi-automatic welding operations have comparatively less complex welding setups that automatic welding operations. The setup times for selected welding operations are presented in Table 4.11.
- *Variable Cost*: For welding operations, variable costs are classified as labor cost, material (electrode) cost, shielding (gas or flux) cost, power cost, and overhead cost. Variable cost is measured in terms of $/ft or $/lb of weld metal deposited. Each of the five variable costs is expressed through mathematical equations as follows.
  - *Labor Cost*: The labor cost for welding operations is a function of the travel speed, labor rate, and operator factor. It is expressed in Equation 31 as follows.

$$LC(\$/\text{ft}) = \frac{LR \times 12\ \text{in/ft}}{\left\{60\dfrac{\text{min}}{\text{hr}} \times S \times OF\right\}}. \qquad \text{(Equation 31)}$$

wherein:
LC: Labor cost
LR: Labor rate expressed in $/hour
S: Travel speed, expressed in inches/minute
OF: Operator factor

Travel speed depends on the specific type of welding process being used, material, and weld thickness. The operator factors for general and specific welding processes are presented in

TABLE 4.12

Operator Factor for Different Welding Processes

	Operator Factor	
Welding Process Used	**Range**	**Typical Value**
Manual welding processes	0.05–0.40	0.30
Semiautomatic welding processes	0.10–0.60	0.40
Automatic welding processes	0.40–1.00	0.60
Shielded metal arc welding		0.30
Gas metal arc welding	0.41–0.66	0.50
Flux cored arc welding		0.50
Submerged arc welding	0.34–0.63	0.50

Table 4.12. Travel speeds for arc welding processes are presented in Table 4.13. Labor cost can also be expressed as a function of deposition rate as presented in Equation 32.

$$LC(\$/ft) = \frac{LR \times 12\frac{in}{ft} \times WMD}{\{DR \times OF\}}. \qquad \text{(Equation 32)}$$

TABLE 4.13

Travel Speeds for Different Arc Welding Processes

Travel Speeds for Arc Welding Processes for 3/8 inch Butt Weld		
Welding Process	**in/min**	**mm/sec**
Shielded metal arc welding (SMAW)	4.25	1.80
Flux cored arc welding	10.00	4.20
Semi-automatic submerged arc welding	15.00	6.30
Automatic submerged arc welding	18.00	7.60
Range of Travel Speeds for Arc Welding on Steel		
SMAW < 0.14 inches	15–25	6–11
SMAW > 0.14 inches	5–15	2–6
Submerged arc welding < 0.25 inches	30–100	13–42
Submerged arc welding > 0.25 inches	8–30	3–13
Flux core arc welding < 0.14 inches		
Fully automatic	100–200	42–85
Semi-automatic	10–25	4–11
Gas metal arc welding	10–25	4–11
Gas tungsten arc welding	10–25	4–8

Source: *The Procedure Handbook of Arc Welding*, 1973. The Lincoln Electric Company, Cleveland, Ohio [6].

wherein:
LC: Labor cost
LR: Labor rate expressed in \$/hour
WMD: Weld metal deposited, expressed in lb./inches
DR: Deposition rate, expressed in lb./hr
OF: Operator factor

Travel speed can be estimated using Equation 33 as follows:

$$S = DR/(WMD \times 60 \text{ min/hour}) \qquad \text{(Equation 33)}$$

wherein:
S: Travel speed, expressed in inches/minute
WMD: Weld metal deposited, expressed in lb/inches
DR: Deposition rate, expressed in lb/hr

The expression for deposition rate is given in Equations 34 and 35 for automatic welding and manual welding, respectively.

For automatic welding,

$$DR = 0.033 \times I. \qquad \text{(Equation 34)}$$

wherein:
DR: Deposition rate expressed in lb./hr
I: Welding current, expressed in amperes

For manual welding,

$$DR = 0.033 \times I\text{-}1.7. \qquad \text{(Equation 35)}$$

- *Material (Electrode) Cost*: From the perspective of welding operations, the material cost refers to the cost of the electrode. Bear in mind that this has nothing to do with the material cost of the parts being welded. The material cost is expressed by the symbol: VEC. This is short for variable electrode cost. The VEC for automatic and semiautomatic welding is expressed by Equation 36 as follows.

$$VEC = \frac{WFR \times EWL \times EC \times 12 \text{inches/ft}}{S \times EMY}. \qquad \text{(Equation 36)}$$

wherein:
VEC: Variable electrode cost, expressed in \$/ft
WFR: Wire fee rate, expressed in inches/minute
EC: Electrode cost, expressed in \$/lb
S: Travel speed, expressed in inches/minute

EMY: Electrode metal yield
EWL: Electrode weight per unit length, expressed in lb/inch

The VEC for manual welding is computed using Equation 37 as follows.

$$VEC = WMD \times EC \times 12 \text{ inches/ft/EMY}. \qquad \text{(Equation 37)}$$

wherein:
VEC: Variable electrode cost, expressed in $/ft
WMD: Weld metal deposited, expressed in lb/in
EC: Electrode cost, expressed in $/lb
EMY: Electrode metal yield

Electrode metal yield factors for different welding processes are presented in Table 4.14.

- *Shielding Gas/Flux Cost*: In all welding processes, shielding gases and fluxes are used to "shield" or protect the weld metal from the atmosphere. The cost associated with procuring the shielding gases and fluxes is referred to as the shielding gas/flux cost. In the case of shielded metal arc welding, this cost is included in the cost of the electrode material. In the case of flux core arc welding, it is included in the cost of flux costs.

  In the casees of gas metal arc welding, gas tungsten arc welding, and in flux core arc welding (depending on whether it is used in this process), the shielding gas costs must be determined separately. In the case of submerged arc welding, the flux is used to shield the weld from exposure to the atmosphere. As a consequence, the cost of the flux must be included in this process.

TABLE 4.14

Electrode Metal Yield Factors for Different Welding Processes

	Electrode Metal Yield	
Welding Process	**Normal Range**	**Typical Value**
Shielded metal arc welding		
Average		0.60
14-inch electrode	0.55–0.65	
18-inch electrode	0.60–0.70	
28-inch electrode	0.65–0.75	
Submerged arc welding	0.95–1.00	1.00
Gas metal arc welding		
Using carbon dioxide	0.90–0.95	0.92
Using other gases	0.95–1.00	1.00
Flux core arc welding using carbon dioxide	0.80–0.85	0.85

Source: Cary, H. 1989. *Modern Welding Technology*, Prentice Hall, pp. 616 [4].

In general, the cost of shielding gas can be computed using Equation 38.

$$VGC = \frac{GFR \times GC \times 12\,in/ft}{\{S \times 60\,min/hr\}}. \qquad \text{(Equation 38)}$$

wherein:
VGC: Variable gas cost, expressed in \$/ft
GFR: Gas flow rate, expressed in cubic feet/hour
GC: Gas cost, expressed in \$/cubic feet
S: Travel speed, expressed in inches/minute

In the case of submerged arc welding, the flux cost is determined using Equation 39.

$$VFC = WMD \times FCR \times FC \times 12\text{inches/ft} \qquad \text{(Equation 39)}$$

wherein:
VFC: Variable flux cost, expressed in \$/ft.
WMD: Weld metal deposited, expressed in lb/inch.
FCR: Flux consumption rate, expressed in pounds of flux consumed per pound of metal. This value generally ranges from 1.0 to 1.5. A value of 1.0 is frequently used.
FC: Cost of flux, expressed in \$/lb.

- *Power Cost*: Variable power cost is a function of utility rates. It can be computed using Equation 40 as follows.

$$VPC = \frac{(I \times V \times PC/1000) \times 12\,in/ft}{\{S \times M \times 60\,min/hr\}}. \qquad \text{(Equation 40)}$$

wherein:
VPC: Variable power cost, expressed in \$/ft
I: Welding current, expressed in amperes
V: Welding voltage, expressed in volts
PC: Power cost, expressed in \$/KW-hr
S: Travel speed, expressed in inches/minute
M: Efficiency of machine, expressed as a fraction

Variable power cost is also computed as a function of the deposition rate and the weight of weld metal deposited. This is expressed by Equation 41.

$$VPC = \frac{(I \times V \times PC/1000) \times WMD \times 12\,in/ft}{\{DR \times M\}}. \qquad \text{(Equation 41)}$$

wherein:
VPC: Variable power cost, expressed in \$/ft
I: Welding current, expressed in amperes
V: Welding voltage, expressed in volts
PC: Power cost, expressed in \$/KW-hr
WMD: Weight of weld metal deposited, expressed in lb/in
DR: Deposition rate, expressed in lb/hr
M: Efficiency of machine, expressed as a fraction

- *Overhead Cost*: Overhead cost computation for welding operations is similar to the computation of labor cost. Instead of labor rate, the overhead rate is used. The corresponding equations are presented below.

$$OC(\$/ft) = \frac{OR \times 12\,in/ft}{\left\{60\frac{min}{hr} \times S \times OF\right\}}. \qquad \text{(Equation 42)}$$

wherein:
OC: Labor cost
OR: Overhead rate expressed in \$/hour
S: Travel speed, expressed in inches/minute
OF: Operator factor

Overhead cost can also be expressed as a function of deposition rate as presented in Equation 43.

$$OC(\$/ft) = \frac{OR \times 12\frac{in}{ft} \times WMD}{\{DR \times OF\}} \qquad \text{(Equation 43)}$$

wherein:
LC: Overhead cost
LR: Labor cate expressed in \$/hr
WMD: Weld metal deposited, expressed in lb/in
DR: Deposition rate, expressed in lb/hr
OF: Operator factor

It should be borne in mind that labor costs and overhead costs constitute the largest component of variable welding costs and can amount to as much as 60% to 90% of the total variable welding cost.

4.5.3 Welding in Multiple Passes

In most welding applications, it is essential to perform more than one pass in order to complete the welding process. It should be noted that in such cases, the travel speed of each pass can be different than the other passes. In such cases, the overall travel speed is determined using Equation 44 as follows.

$$S = \frac{1}{\frac{1}{S1} + \frac{1}{S2} + \frac{1}{S3} + \frac{1}{S4} + \cdots + \frac{1}{Sn}}. \qquad \text{(Equation 44)}$$

In Equation 44, it should be noted that "S" represents the overall travel speed. The travel speed of the first pass is denoted by "S1", that of the second pass is denoted by "S2" and so on. Consider Example 4.13 in order to better understand the concept of variable travel speeds in multiple pass welding.

EXAMPLE 4.13

A welding operation involves three passes. The first pass is made at a travel speed of 15 inches per minute. The travel speed of the second pass was 20 inches per minute and that for the third pass was 25 inches per minute. Compute the overall travel speed for the entire operation.

Solution

The following data is given in the problem statement:

S1 = 15 inches/minute. Thus, 1/S1 = 0.06667 min/inch
S2 = 20 inches per minute; 1/S2 = 0.05 min/inch
S3 = 25 inches per minute. 1/S3 = 0.04 min/inch

The overall travel speed for the entire operation is computed using Equation 44 as follows.

$$S = \frac{1}{\frac{1}{S1} + \frac{1}{S2} + \frac{1}{S3} + \frac{1}{S4} + \cdots + \frac{1}{Sn}}. \qquad \text{(Equation 44)}$$

Substituting the respective values into the above equation, we obtain the value for overall travel speed as follows.

$$S = \frac{1}{0.06667 + 0.05 + 0.04} = \frac{1}{0.1566667}.$$

This implies that the overall travel speed for the entire operation is 6.383-inches/minute. **Solution**

We will now solve a few examples to illustrate practical applications of the various equations that were presented for different variables affecting welding processes. Consider Example 4.14.

EXAMPLE 4.14

Consider the welding setup presented in Figure 4.16. Two pieces of metal are to be welded together as depicted in the figure. Both pieces are composed of low carbon steel and are 15 inches long. They are welded using a 20% duty cycle. The current rating is 220 A. The welding process requires 160 A, 20 V, and travel speed of 15 inches per minute. The operator factor is 0.35. Compute the amount of time necessary to weld the two pieces together. This should exclude setup time. Also compute the energy generated per unit length during the welding procedure.

Solution

The primary objective of solving this problem is to determine the welding time. The equation for welding time is given as follows:

$$\text{WT (minutes)} = \{L \times 12\text{ inches/ft}\}/\{S \times OF\}.$$

From the data given in the problem, we know the following:

L: length of the weld measured in feet = 15 inches/12 = 1.25 feet.

From the figure, notice that there are two locations that need to be welded. Thus, total length to be welded = 2.50 feet.

S: Travel speed = 15 inches/minute.
OF: 0.35.

Substituting each of these values into the equation for welding time, we obtain the final value as follows.

WT (minutes) = {2.50 ft × 12 inches/ft}/{15in/min × 0.35}.
Thus, welding time = 5.71 minutes.

Next, we compute the energy generated per unit length of the weld. This can be accomplished using Equation 28 as follows.

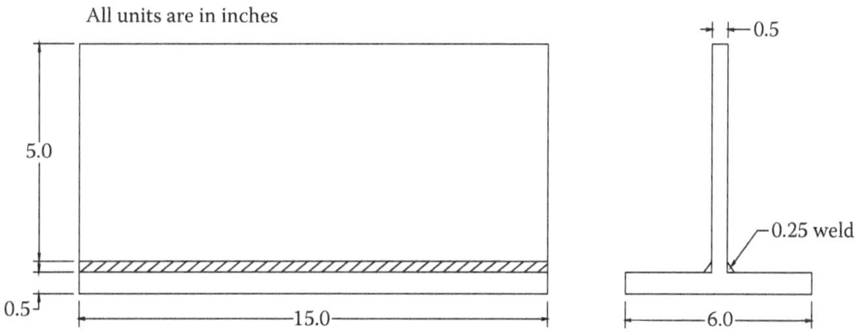

FIGURE 4.16
Welding setup for Example 4.14.

$$H = 60 \times \frac{V \times I}{S}. \qquad \text{(Equation 28)}$$

From the problem statement, we know the following values:

V: 20 volts
I: 160 amps
S: 15 inches/minute

Substituting each value into Equation 28, we obtain the following:

$$H = 60 \times \frac{20\,V \times 160\,\text{amps}}{15\,\text{in/min}}$$

Thus, H = 12,800 J/inch. **Solution**

EXAMPLE 4.15

Consider the metal box represented in Figure 4.17. As shown in the figure, the box has a basal area of 15inches × 9 inches. The wall height is 6 inches. The box is made out of low carbon steel. The objective is to place 0.25 inch fillet welds at the inside corners using shielded metal arc welding. The wall thickness of the material is 0.2 inches. Compute the cost to produce a batch of 60 boxes excluding material cost. Process data is presented below.

Travel speed: 10 inches/minute
Welding current: 200 Amps
Welding voltage: 20 V
Efficiency of machine: 85%
Duty cycle: 60%

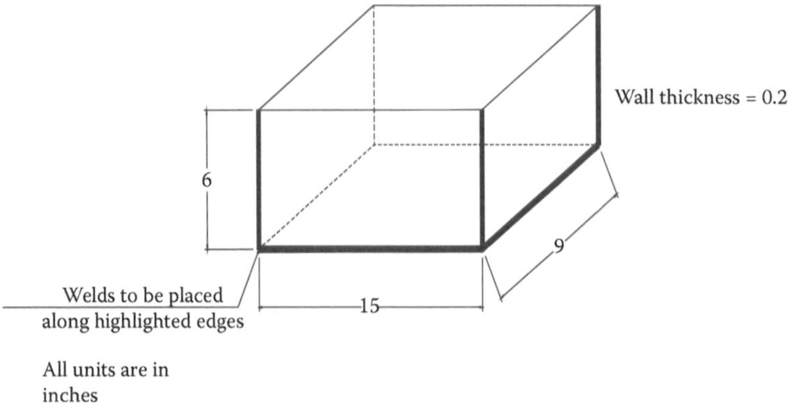

FIGURE 4.17
Welding setup - low carbon steel metal box for Example 4.15.

Operator factor: 0.30
Electrode metal yield for 18-inch electrode: 0.65
Labor rate: \$20/hr
Overhead rate: \$30/hr
Electrode cost: \$0.30/lb
Power cost: \$0.15/KW-hr

Solution

The first step in solving this problem is to ascertain the weight of the weld metal deposited per unit length. This can be computed by multiplying the cross sectional area and density of the weld metal. The density of steel is 0.28-lb/cubic inch.

A fillet weld is triangular in cross section. Its area is computed using the basic area equation for a triangle as follows:

$$\text{Cross sectional area of fillet weld} = 1/2 \times \text{base} \times \text{height.}$$

From the problem statement, we know that the inside corners of the box are to be 0.25 inch fillet welds.

Substituting this value as the value for base and height into the above equation, we obtain the following.

$$\begin{aligned}\text{Cross sectional area of fillet weld} &= 0.5 \times 0.25 \times 0.25\\ &= 0.03125 \text{ square inches.}\end{aligned}$$

Next, we compute weld metal deposition (WMD) using Equation 29 as follows:

$$\begin{aligned}\text{WMD (lb/in)} &= \text{cross sectional area of weld (sq. inches)}\\ &\times \text{density of metal (lb/cu.in).} \qquad \text{(Equation 29)}\end{aligned}$$

We know that the density of steel is 0.28 lb/cubic inch.
Thus, WMD = 0.03125 sq inches × 0.28 lb/cu inch = 0.00875 lb/inch.

Next, we compute the length of the weld as follows.

The assignment calls for welds to be placed along the inside corners of the base and the four vertical corners. Thus, welds will be placed along the inside corners of the 15 inch × 9 inch base as well as the four vertical inside corners (each weld will be 6 inches long).

Thus, the total length of the weld is computed as follows:

Along the length of the base of the box: 15 inches + 15 inches = 30 inches.
Along the width of the base of the box: 9 inches + 9 inches = 18 inches.
Along the four vertical corners of the box: 6 inches + 6 inches + 6 inches + 6 inches = 24 inches.

Thus, total length of weld = 30 inches + 18 inches + 24 inches = 72 inches = 6 ft.

The labor cost per ft is obtained from Equation 31 as follows:

$$LC(\$/ft) = \frac{LR \times 12\,in/ft}{\left\{60\frac{min}{hr} \times S \times OF\right\}}. \qquad \text{(Equation 31)}$$

From the problem statement, we know the following:

LR: Labor rate = \$20/hr
S: Travel speed = 10 inches/minute
OF: Operator factor = 0.30

Substituting these values into Equation 31, we obtain the following:

$$LC(\$/ft) = \frac{\$20/hr \times 12\,in/ft}{\left\{60\frac{min}{hr} \times 10\,in/min \times 0.30\right\}}. \qquad \text{(Equation 31)}$$

Thus, Labor cost, LC = \$1.33/ft.
Next, we compute the overhead cost per ft using Equation 42 as follows.

$$OC(\$/ft) = \frac{OR \times 12\,in/ft}{\left\{60\frac{min}{hr} \times S \times OF\right\}}. \qquad \text{(Equation 42)}$$

Once again, we refer to the problem statement to obtain the value of overhead rate. This is given as \$30/hr. We already know the remaining values for Equation 42. Substituting these values into Equation 42, we obtain the following:

$$OC(\$/ft) = \frac{\$30/hr \times 12\,in/ft}{\left\{60\frac{min}{hr} \times 10\,in/min \times 0.30\right\}}$$

Thus, Overhead cost = \$2/ft.
Next, we compute the variable electrode cost using Equation 37 as follows.

$$VEC = WMD \times EC \times 12\ \text{inches/ft/EMY}. \qquad \text{(Equation 37)}$$

We already computed the value of weld metal deposited as 0.00875 lb/inch.
From the problem statement,

EC: Electrode cost = \$0.30/lb.

Also, EMY: Electrode metal yield = 0.65.

Substituting each value into Equation 37, we obtain the variable electrode cost as follows:

VEC = 0.00875 lb/inch × \$0.30/lb × 12 in/ft/0.65
Thus, VEC = \$ 0.0485/ft.

In order to compute the power cost, we use Equation 40 as follows.

$$\text{VPC} = \frac{(\text{I} \times \text{V} \times \text{PC}/1000) \times 12\,\text{in/ft}}{\{\text{S} \times \text{M} \times 60\,\text{min/hr}\}}. \qquad \text{(Equation 40)}$$

From the problem statement, we know the following:

I: Welding current = 200 Amps
V: Welding voltage = 20 V
PC: Power cost = \$0.15/KW-hr
S: Travel speed = 10 in/min
M: Machine efficiency = 85% = 0.85

Substituting each value into Equation 40, we obtain the following:

$$\text{VPC} = \frac{\left(200\text{A} \times 20\,\text{V} \times \frac{\$0.15}{\text{KW}} - \text{hr}/1000\right) \times 12\,\text{in/ft}}{\{10\,\text{in/min} \times 0.85 \times 60\,\text{min/hr}\}}$$

$$\text{Thus, VPC} = \frac{\$0.6 \times 12/\text{ft}}{510} = \$0.014/\text{ft}.$$

The total variable cost is computed as the sum of labor cost, overhead cost, variable electrode cost, and variable power cost.

Thus, Total variable cost = LC + OC + VEC + VPC.

Substituting each value into the above equation, we obtain the following:

Total variable cost = \$1.33/ft + \$2/ft + \$0.0485/ft + \$0.014/ft
Thus, Total variable cost = \$3.3925/ft.

We have already computed the total weld length per part to be 6 ft.
Therefore, the variable cost per part = variable cost/ft × number of ft to be welded.

Thus, variable cost per part = \$3.3925/ft × 6 ft
This gives us the variable cost per part = \$20.355 per part.
Thus, variable cost for welding 60 parts = \$20.355/box × 60 boxes = \$1221.30.

In order to compute setup time, we refer to Table 4.12. Notice that the setup time for a shielded metal arc welding process is given to be 18 minutes.

Total setup cost = (setup time) × (labor rate + overhead rate)/60 min/hour.
Thus, Total setup cost = (18 minutes) × ($20/hr + $30/hr)/60 min/hour.
This implies that total setup cost = $15.

Remember that this time represents the setup time for the entire batch of 60 boxes.

Total cost for welding 60 boxes = setup cost + total variable cost for welding 60 boxes.
Thus, Total cost for welding 60 boxes = $15 + $1221.30 = $1236.30. **Solution**

EXAMPLE 4.16

Consider the metal box represented in Figure 4.17. As before, the objective is to place 0.25-inch fillet welds at the inside corners using gas metal arc welding. The wall thickness of the material is 0.2 inches. The setup time for this operation is 24 minutes. Compute the cost to produce a batch of 60 boxes excluding material cost. Process data is presented below.

Travel speed: 20 inches/minute
Welding current: 400 Amps
Welding voltage: 27 V
Efficiency of machine: 85%
Wire feed rate: 87 inches/min
Gas flow rate: 40 cu ft/hr
Electrode weight per length: 0.00193 lb/in
Duty cycle: 60%
Operator factor: 0.50
Electrode metal yield for 18-inch electrode: 0.95
Labor rate: $20/hr
Overhead rate: $40/hr
Electrode cost: $0.30/lb
Gas cost: $0.40/cu ft
Power cost: $0.15/KW-hr

Solution

The dimensions of the box to be welded remain the same as in Example 4.15. This implies that the weld metal deposited will remain the same as in the previous example. The total length of weld will also remain the same. Thus, we start solving this problem beginning with labor cost estimation.

$$LC\,(\$/ft) = \frac{LR \times 12\,in/ft}{\left\{60\frac{min}{hr} \times S \times OF\right\}}. \qquad \text{(Equation 31)}$$

From the problem statement, we know the following:

LR: Labor rate = \$20/hr
S: Travel speed = 20 inches/minute
OF: Operator factor = 0.50

Substituting these values into Equation 31, we obtain the following:

$$LC(\$/ft) = \frac{\$20/hr \times 12\,in/ft}{\left\{60\frac{min}{hr} \times 20\,in/min \times 0.50\right\}}. \qquad \text{(Equation 31)}$$

Thus, Labor cost, LC = \$0.4/ft.

Next, we compute the overhead cost per ft using Equation 42 as follows.

$$OC(\$/ft) = \frac{OR \times 12\,in/ft}{\left\{60\frac{min}{hr} \times S \times OF\right\}}. \qquad \text{(Equation 42)}$$

Once again, we refer to the problem statement to obtain the value of overhead rate. This is given as \$40/hr. We already know the remaining values for Equation 42. Substituting these values into Equation 42, we obtain the following:

$$OC(\$/ft) = \frac{\$40/hr \times 12\,in/ft}{\left\{60\frac{min}{hr} \times 20\,in/min \times 0.50\right\}}$$

Thus, Overhead cost = \$0.80/ft.

Next, we compute the variable electrode cost using Equation 36 as follows.

$$VEC = \frac{WFR \times EWL \times EC \times 12\,inches/ft}{S \times EMY}. \qquad \text{(Equation 36)}$$

From the problem statement, the value of WFR: Wire feed rate = 87 inches/min.

Also, EC: Electrode cost = \$0.30/lb.
Electrode weight per length, EWL = 0.00193 lb/in.
Travel speed, S = 20 in/min.
Also, EMY: Electrode metal yield = 0.65.

Substituting each value into Equation 36, we obtain the variable electrode cost as follows:

$$\text{VEC} = 87\ \text{in/min} \times 0.00193\ \text{lb/in} \times \$0.30/\text{lb} \times 12\ \text{in/ft}/20\ \text{in/min} \times 0.65.$$

Thus, VEC = 0.604476/13 = \$ 0.0465/ft.

Since this example deals with gas metal arc welding, we compute the variable gas cost next using Equation 38.

$$\text{VGC} = \frac{\text{GFR} \times \text{GC} \times 12\,\text{in/ft}}{\{\text{S} \times 60\,\text{min/hr}\}}. \qquad \text{(Equation 38)}$$

From the problem statement, we know the value of the gas flow rate, GFR = 40 cu ft/hr.

Also, Gas cost, GC = \$0.40/cu ft.

Substituting these values into Equation 38, we obtain the following value for variable gas cost.

$$\text{VGC} = \frac{40\,\text{cu}\dfrac{\text{ft}}{\text{hr}} \times \dfrac{\$0.40}{\text{cu}}\text{ft} \times 12\,\text{in/ft}}{\left\{20\,\text{in/min} \times 60\,\text{min/hr}\right\}} = \$0.16/\text{ft}.$$

In order to compute the power cost, we use Equation 40 as follows.

$$\text{VPC} = \frac{(\text{I} \times \text{V} \times \text{PC}/1000) \times 12\,\text{in/ft}}{\{\text{S} \times \text{M} \times 60\,\text{min/hr}\}}. \qquad \text{(Equation 40)}$$

From the problem statement, we know the following:

I: Welding current = 400 Amps
V: Welding voltage = 27 V
PC: Power cost = \$0.15/KW-hr
S: Travel speed = 20 in/min
M: Machine efficiency = 85% = 0.85

Substituting each value into Equation 40, we obtain the following:

$$\text{VPC} = \frac{\left(400\,\text{A} \times 27\,\text{V} \times \dfrac{\$0.15}{\text{KW}} - \text{hr}/1000\right) \times 12\,\text{in/ft}}{\{20\,\text{in/min} \times 0.85 \times 60\,\text{min/hr}\}}.$$

$$\text{Thus, VPC} = \frac{\$1.62 \times 12/\text{ft}}{1020} = \$0.019/\text{ft}.$$

The total variable cost is computed as the sum of labor cost, overhead cost, variable electrode cost, and variable power cost.

Thus, Total variable cost = LC + OC + VEC + VPC.

Substituting each value into the above equation, we obtain the following:

Total variable cost = \$0.40/ft + \$0.80/ft + \$0.0465/ft + \$0.019/ft.
Thus, Total variable cost = \$1.2655/ft.

We have already computed the total weld length per part to be 6 ft.
Therefore, the variable cost per part = variable cost/ft × number of ft to be welded.

Thus, variable cost per part = \$1.2655/ft × 6 ft.
This gives us the variable cost per part = \$7.593 per part.
Thus, variable cost for welding 60 parts = \$7.593/box × 60 boxes = \$455.58.

Next, we compute the setup cost as follows.

Total setup cost = (setup time) × (labor rate + overhead rate)/60 min/hour.
Thus, Total setup cost = (24 minutes) × (\$20/hr + \$40/hr)/60 min/hour.
This implies that total setup cost = \$24.

Remember that this time represents the setup time for the entire batch of 60 boxes.

Total cost for welding 60 boxes = setup cost + total variable cost for welding 60 boxes.
Thus, Total cost for welding 60 boxes = \$24 + \$455.58 = \$479.58.
Solution

In the next section, we will examine the cost structure of parts manufactured using additive manufacturing techniques.

4.6 Additive Manufacturing

Additive **M**anufacturing (**AM**) is a term used to describe the technologies that are used to build three-dimensional objects (3D objects) by *adding* layer-upon-layer of material. Additive manufacturing can work with a variety of materials including plastic, metal, concrete, etc. This can be compared with traditional machining processes wherein material is removed from the formed part in order to produce the finished component (also referred to as subtractive manufacturing). Most additive manufacturing methods make extensive use of computer equipment, 3D modeling Computer Aided Design (CAD) software, machine equipment, and layering material. The following sequential steps are used to produce a component using additive manufacturing:

- The first step involves the production of a CAD sketch using CAD software.
- The second step involves reading the data by the additive manufacturing equipment from the CAD file.
- The third step involves laying down and adding successive layers of material (in the form of liquid, sheet metal or powder) in order to fabricate the final object. The material is laid down in a layered format with one layer on top of the other.

One of the principal benefits of AM is the wide variety and range of shapes that can be produced using this technology. Designs that cannot be manufactured in one entire piece with traditional means can easily be fabricated using AM technology. For instance, shapes that incorporate a scooped out or hollow center can be manufactured with ease as a single piece. There is no need to weld or attach individual components together if the part is being produced using additive manufacturing. This process renders the part stronger. It does not involve any weak spots, which can be compromised or stressed during operation.

Additive manufacturing imparts a great deal of speed to the manufacturing operation as well. CAD software can be used to make any changes quickly. All it usually takes is a click of the mouse. The fabrication speed afforded by AM provides companies with a greater degree of flexibility, while simultaneously reducing costs. The three most common techniques of additive manufacturing are: selective laser sintering (SLS), direct metal laser sintering (DMLS), and selective laser melting (SLM).

- *Selective Laser Sintering (SLS)*: This additive manufacturing technique utilizes a laser to sinter powdered material. The material is usually in the form of nylon or a polyamide. The laser is aimed at specific points in space that have been defined by the 3D model. This enables binding the material together in order to create a solid structure.
- *Direct Metal Laser Sintering (DMLS)*: Direct metal laser sintering is very similar to SLS. It uses a precise high-powered laser to micro weld powdered metals and alloys in order to form fully functional components based on the CAD design. DMLS parts can be made from a wide range of materials such as aluminum, stainless steel, and

titanium. DMLS is able to create durable as well as intricate metal parts that were beyond the scope of traditional manufacturing processes.

- *Selective Laser Melting (SLM):* Like SLS and DMLS, selective laser melting uses a laser to melt successive layers of metallic powder. The laser heats particles at predetermined points in space on a bed of metallic powder until the particles have melted completely. A successive layer of powder is added by the AM machine above the melted layer in order to enable fusion. This process is repeated over and over until the entire part has been fabricated.

The ASTM F2792 Standard categorizes additive manufacturing technologies more broadly along the following lines:

- *Binder Jetting*: Binder jetting utilizes a liquid bonding agent that is deposited using an inkjet printer head. It serves the purpose of joining powder materials in a powder bed.
- *Powder Bed Fusion*: Thermal energy from a laser or electron beam is used to selectively fuse powder in a powder bed in the process of powder bed fusion.
- *Directed Energy Deposition*: Directed energy deposition utilizes thermal energy generated using a laser in order to fuse materials by melting them as they are deposited.
- *Sheet Lamination*: Sheets of material are bonded to form a three-dimensional object in the process of sheet lamination.
- *Vat Photo Polymerization*: A liquid photopolymer is selectively cured in a vat using light in the process of vat photo polymerization.
- *Material Extrusion:* Material extrusion machines push a thermoplastic filament through a nozzle onto a platform. The platform is capable of moving in horizontal and vertical directions.
- *Material Jetting*: Material jetting uses a moving inkjet print head in order to deposit material across a specified build area.

Figure 4.18 depicts the layout of the DMLS process.

Figures 4.19 and 4.20 depict examples of parts fabricated using additive manufacturing.

Additive manufacturing is performed using so called 3D printers. An example of a 3D printer is depicted in Figure 4.21.

Additive manufacturing has some distinct advantages vis-à-vis traditional manufacturing processes. These advantages are listed below.

- AM exhibits a strong potential to match traditional manufacturing methods on a direct cost basis for production applications. As far as direct costs are concerned, the drivers are substantially different for the two technologies.

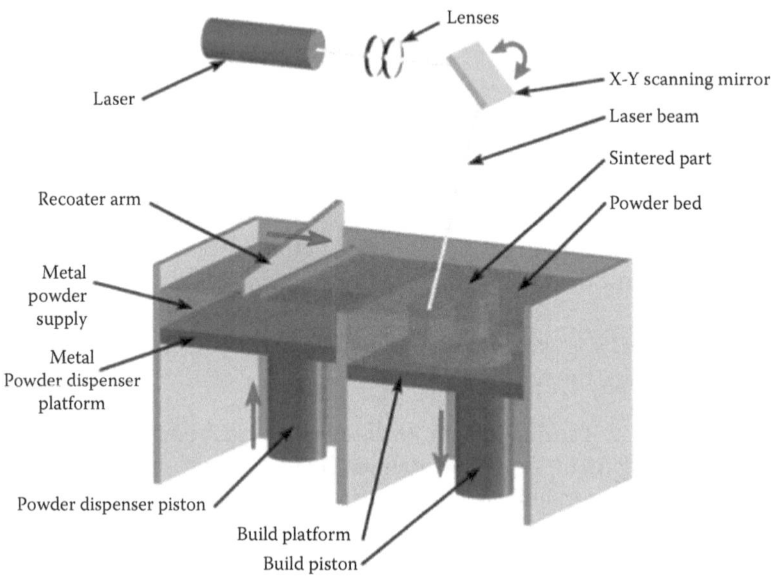

FIGURE 4.18
Layout of the DMLS process. (custompartnet.com.)

- AM technologies have a unique ability to transform supply chains, products, and business models. This enables companies that adopt AM to differentiate themselves by creating unique market offerings and positions.

From a cost perspective, adoption of AM technologies reduces the amount of capital required to achieve economies of scale, while simultaneously

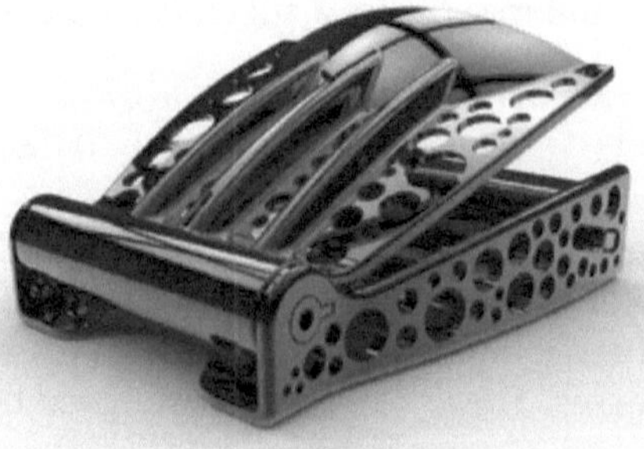

FIGURE 4.19
Example of a part fabricated using additive manufacturing. (Image Credit: Crucible Industrial Design. Design Products and Applications. www.dpaonthenet.net.)

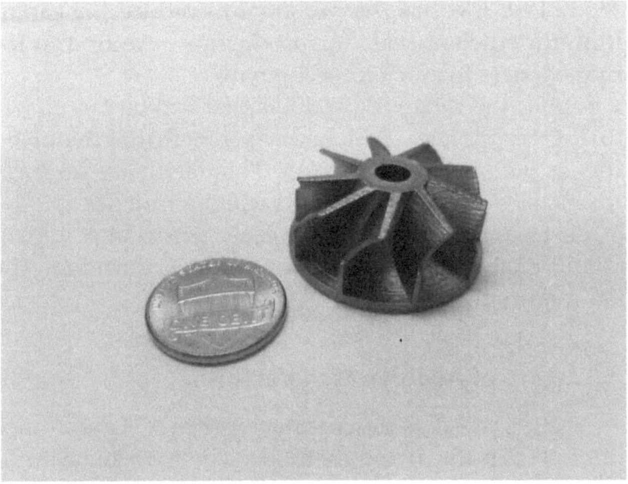

FIGURE 4.20
Example of a part fabricated using additive manufacturing. (rapidsol.org.)

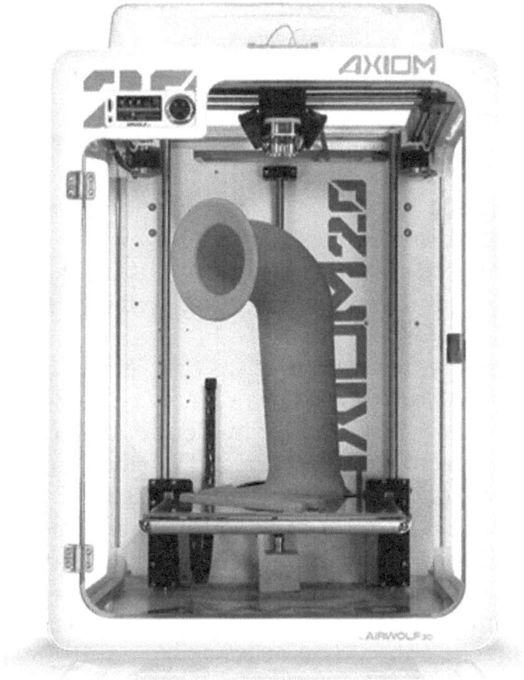

FIGURE 4.21
An example of a 3D printer. (airwolf3d.com.)

increasing flexibility. AM has the potential to reduce the capital required to reach minimum efficient scale for production resulting in lowering the manufacturing barriers to entry for a company.

Flexibility implies the range of products that can be manufactured using a manufacturing technology. AM imparts a high degree of flexibility to manufacturers as we have already seen. This facilitates an increase in the variety of products that can be produced using a unit of capital. Additional flexibility leads to a reduction in the costs associated with production changeovers and customization, which ultimately results in lowering the overall capital requirement.

4.6.1 Cost Structure of Additive Manufacturing

The following section presents the cost structure of AM and compares it with that of traditional manufacturing methods. The four key differentiators in this comparison are:

- *Tooling Costs*: A key attribute of AM is its ability to improve upon or entirely eliminated costs associated with tooling. This is significant in light of the fact that beyond actual production, it is essential to maintain, store, and track tooling over extended periods of time.
- *Machine Costs*: Machine cost is one of the most significant costs involved in additive manufacturing. The average selling price of an industrial additive manufacturing system in 2011 was about $73,000. Between 2001 and 2011, the price for an AM machine has decreased almost 51% after adjusting for inflation. The overall trend in machine costs has been generally downward. However, large differences persist between the costs for polymer-based systems and metal-based systems. Also, the huge growth in sales of low cost, polymer-based systems has had a strong impact on the average selling price of additive manufacturing systems, thereby pushing prices lower.

  Machine costs represent as much as 60%–70% of total direct costs associated with AM. AM systems capable of new production can require hundreds of thousands or millions of dollars in investment. Build volume, machine utilization, and depreciation play a crucial role in terms of comparisons with traditional manufacturing. Build volume is a critical factor in deciding whether or not to adopt AM technology. This is because AM technologies often focus on the fabrication of a single object. Other issues include expected machine life and maintenance requirements on AM machines. The cost difference between the different types of additive manufacturing machinery is quite significant, ranging between $100,000 for polymer systems and up to $1.0 million for metal systems.

The size of the build envelope and the utilization of this envelope both have an impact on the cost of products manufactured using AM. The size of the build envelope has a twofold impact on the cost of the final product. First, based on AM principles, products can only be built to the size of the build envelope. This implies that it might not be possible to build some products using additive manufacturing technologies without enlarging the build envelope. The second impact of the build envelope is related to utilizing the total amount of build capacity. A significant efficiency factor lies in the ability to exhaust the available build space.

Build time is a crucial component in estimating the cost of additive manufacturing. Numerous software packages are available for estimating this metric. Build time estimations are specific to the system and material being used to fabricate an individual part.

Last, energy consumption is an important factor in determining the cost of parts fabricated using additive manufacturing. It should be noted that although this is a significant element of the cost, it amounts to only about one percent of the final cost. In metal additive manufacturing, energy consumption is driven by the solid-to-cavity volume ratio. At low ratios, the additive manufacturing process of direct metal deposition minimizes energy, while at high ratios, computer numeric controlled milling minimizes energy consumption. Other studies tend to focus on accurately predicting energy consumption and minimizing energy consumption for additive manufacturing. Envelope utilization and build orientation are among the issues for reducing energy consumption (Morrow et al, 2007).

- *Materials Costs*: Material costs represent approximately 30% of the cost per unit of parts manufactured using additive manufacturing processes. Compare this with about 0.2%–2.7% for conventional manufacturing processes such as injection molding. The extreme difference in materials cost can be attributed to the price differential that exists between AM materials and material used for more traditional manufacturing processes. For instance, thermoplastics and photopolymers used for AM applications can range in cost from $175/kg to as much as $250/kg. The steel powders used for AM applications are about 100 times as expensive as those used for commercial grade applications. Also, counterintuitive as it may seem, AM applications are not actually zero waste. Rates of material recycling vary by process, system, and application. However, they are still favorable when compared to traditional subtractive manufacturing processes.

  In light of the geometric freedom afforded by it, additive manufacturing is capable of producing products using less material while not compromising on performance. Products can be produced at the requisite level of performance. As discussed previously, metal and plastic are the primary materials used for this technology. Currently, the cost of material for additive manufacturing can be quite high when compared to traditional manufacturing as we have already seen in the preceding paragraph.

Although material costs for additive manufacturing are significant, technologies can be used that are complementary to each other. This involves the adoption of two technologies alongside each other. This allows for the overall benefits to be greater than if each technology were used individually. For instance, consider the example of computer aided design (CAD) and computer aided manufacturing (CAM). This is an example of the complementary use of two technologies to support each other. Each individual technology is required to be utilized for the other to be valuable. Thus, CAD complements CAM and vice versa. Economies of scale can result in reduced cost of raw materials as additive manufacturing becomes more commonplace. The reduced cost of raw material could propagate further adoption of additive manufacturing, thus resulting in a virtuous cycle. Economies of scale in raw material costs can also be achieved if particular materials are adopted more widely than a mélange of different materials.

- *Labor Costs*: Labor comprises only a small portion of AM costs. Activities such as removing the finished product or refilling the raw material can be considered as labor intensive activities. There is no significant difference in labor rates when a comparison is made between AM and traditional manufacturing. However, part simplification (as in reduction in number of parts using the Design for X [DfX] principle) can result in savings of as much as 67% in assembly time. Labor cost can vary from as little as 1% of the final cost to as much as 3% of the final cost of the finished product. In either case, labor cost constitutes only a small portion of the cost of the finished product. Additional labor is also built into other costs. Such costs include material cost and machine cost. It is obvious that production of these items is labor intensive.

4.6.2 Costing Models for Additive Manufacturing for a Single Part

A comprehensive review of literature on the topic of additive manufacturing costing reveals the presence of three models. They are as follows:

- The costing model developed by Hopkinson and Dickens
- The model developed by Ruffo et al.
- The model developed by Allen to compare additive manufacturing to machining for aero-engine parts

Each model is described in detail in the following paragraphs.

4.6.2.1 The Hopkinson and Dickens Model

The model developed by Hopkinson and Dickens in 2003 calculates the cost of parts produced by additive manufacturing. It includes factors such as labor cost, machine cost, and material cost. The average cost per part is computed by dividing the total cost by the total number of parts manufactured annually. The model makes the following additional assumptions:

- Only one type of part is produced by the system per year.
- The AM system utilizes a maximum amount of volume.
- It is assumed that the AM machine operates 90% of the time.

The analysis compared the three methods of AM: laser sintering, stereo lithography, and fused deposition from the point of view of total cost of manufactured part. It was observed that the highest cost allocation for each method was as follows:

- Machine cost contributed the greatest amount to total cost for parts produced by stereo lithography.
- Machine cost contributed the greatest amount to total cost for parts produced by fused deposition.
- Material cost was the greatest contributor to total cost for parts produced by laser sintering.

Based on this model, the machine cost per part is calculated using Equation 45.

$$\text{Machine Cost / Part} = \frac{(\text{Depreciation cost of machine / year} + \text{Machine maintenance cost / year})}{\text{Production Volume per year}}. \qquad \text{(Equation 45)}$$

It is to be noted that the AM machine is depreciated over a span of eight years in this model using the straight line method of depreciation. Depreciation is discussed at length in Chapter 13 of this book.

This analysis results in an annual machine cost that is constant over time. Next, we examine the model developed by Ruffo et al.

4.6.2.2 The Ruffo et al. Model

The model developed by Ruffo et al. in 2006 computes the cost of parts produced by additive manufacturing. The costs are computed using an activity based cost model. As its name implies, each cost is associated with a particular activity in this model. Time and material are the two main variables used in this costing model. Ruffo et al.'s model calculates three different times in order to facilitate its analysis. These times are described as follows:

- The amount of time required to laser scan a section and its border in order to sinter.
- The amount of time required to add layers of powder.
- The amount of time required to heat the bed before scanning as well as that required to cool down slowly after scanning, adding layers of powder or the waiting time to reach the right temperature.

The total cost of a "build" is computed using Equation 46.

$$C = (P_{Material} \times M) + (P_{Indirect} \times T). \qquad \text{(Equation 46)}$$

wherein:

C: Total cost of a build, expressed in euros.
$P_{Material}$: Raw material cost, expressed in euros/kg.
M: Mass of material deposited, expressed in kg.
T: Total build time, expressed in minutes. The value of "T" is the sum of the three varieties of time as described above.
$P_{Indirect}$: Cost rate, expressed in euros/minute.

The total cost of a part is computed using Equation 47.

$$\text{The total cost of a part} = \text{C/number of parts in the build.} \qquad \text{(Equation 47)}$$

4.6.2.3 *The Allen Model*

Allen (2006) developed a model to compute the cost of parts fabricated using AM for use in the aerospace industry. This industry often uses parts made out of expensive raw materials. These materials are known for their high performance and low weight. Such parts are very expensive to machine using traditional manufacturing processes. This is one example where AM can be used advantageously. Equation 48 represents the computation of cost for producing a near net shape product using additive manufacturing.

$$C_a = V \times \rho \times C_d. \qquad \text{(Equation 48)}$$

wherein:

C_a: Cost of producing a near net shape product using additive manufacturing
V: Volume of component
ρ: Density of titanium
C_d: Specific cost of deposited titanium

It was concluded that additive manufacturing is cost effective when the buy/fly ratio is 12 times that of conventional ratios. The buy/fly ratio is computed as the volume of billet divided by the volume of the component. Thus, it represents the amount of material that needs to be machined "away." In general, it was concluded that additive manufacturing should be used in the following instances:

- Components with a high buy/fly ratio
- Components having a complex shape that require significant machining
- Components that have a high material cost and slow machining rates

4.6.3 Costing Models for Additive Manufacturing for a Product Mix

It will be noted that the above models assume a single type of part being manufactured using AM technology. However, one of the greatest assets of AM lies in its flexibility. AM technologies are able to produce a variety of different types of parts simultaneously. Thus it is vital to pick and choose a smart mix of components in the same build in order to reduce cost. As we have already seen, if only one part was being produced using AM, the cost per part for a build would be equal to the total cost divided by the number of parts. If, however, different parts were being built simultaneously, the equation to compute the cost per part is more complicated. The following section presents two methodologies proposed by Ruffo and Hague in 2007 to accomplish this objective.

4.6.3.1 Method I: Based on Parts Volume

In a given product mix, the cost per part can be computed based on parts volume. This is accomplished using Equation 49.

$$\text{Cost}_{\text{Pi}} = \{V_{\text{Pi}}/V_{\text{B}}\} \times \text{Cost}_{\text{B}}. \qquad \text{(Equation 49)}$$

wherein:

Cost Pi: Cost of part "A"
V_A: Volume of part "A"
V_B: Volume of part "B"

$$\text{Cost B: Total Cost of build} = \sum \frac{\text{Indirect Costs}}{\text{Working Time}}(t_{xy} + t_z + t_{HC}) + \left(\frac{\text{Direct Cost}}{\text{Unit Mass}}\right) \times m_B.$$

wherein:

m_B: Mass of the planned production proportional to the volumes of the objects and the time to manufacturing the entire build.

t_{xy}: Amount of time required to laser scan the section and its border to sinter powder.

t_z: Amount of time required to add layers of powder.

t_{HC}: Amount of time required to heat the bed prior to scanning and to cool down after scanning and adding layers of powder.

i: An index, whose value ranges from 1 all the way to the number of parts in the build.

4.6.3.2 Method II: Based on the Cost of Building a Single Part

In a given product mix, the cost per part can be computed based on the cost of a single part. This is accomplished using Equation 50.

$$\text{Cost}_{\text{Pi}} = \frac{\gamma \text{i} \times \text{Cost(B)}}{\text{ni}}. \qquad \text{(Equation 50)}$$

where:

$$\gamma \text{i} = \frac{\text{Cost}^*_{\text{Pi}} + \text{ni}}{\sum(\text{Cost}^*_{\text{Pj}} \times \text{nj})}$$

wherein:

i: Index of parts being calculated

j: Index of all parts manufactured in the same bed

n_i: Number of parts identified by "i"

Cost^*_{pi}: Cost of a single part "i" estimated using Equation 49

4.7 Conclusion

In this chapter, we addressed the all-important issue of process costing. Traditional manufacturing processes were addressed. These include processes such as casting and forging as well as traditional subtractive manufacturing processes such as machining. From the point of view of machining, the costing of processes such as turning, milling, and shaping were determined. Costing of welding processes was also presented as well as emerging nontraditional manufacturing processes such as additive manufacturing. Production processes are a crucial component of any production and manufacturing operation because they add value. The entire manufacturing facility is built to enable the process. It is obvious that the importance of process costing cannot possibly be neglected.

PROBLEMS

1. A 30 lb. casting has a yield of 75%, scrap rate of 7%, and a melt loss of 4%. Compute the following for each casting:
   a. Total poured weight
   b. Scrap weight
   c. Return weight
   d. Melt loss weight
2. The cost of hot metal for the casting in Q1 is $3/lb. The scrap and return credit is $0.60/lb and the environmental cost is $5/lb. Compute the total metal cost for the casting.
3. A 20 lb casting has a yield of 70%, scrap rate of 10%, and a melt loss of 4%. The core used in this process weighs 4 lbs. The core and molding scrap rate is 10%. Compute the total amount of core material weight to be charged per casting.
4. For the process in Q4, the core sand costs $0.15/lb. The core sand disposal costs $0.07/lb. Compute the total material cost for the core.
5. Compute the relative cost ratio for an aluminum alloy group A, for a part of shape configuration code 1-1 and shape complexity code 6. The part volume is 15 cubic inches. It has a plane area of 30 square inches.
6. A multiple pass weld can be completed in three passes. The travel speeds are 12 inches/minute, 15 inches/minute, and 17 inches per minute. Compute the overall travel speed in inches per minute.
7. Define the three primary variables in a metal cutting operation.
8. Describe the productive and nonproductive elements that are used to compute the unit piece time.
9. For the process in Q4, the core sand costs $0.20/lb. The core sand disposal costs $0.11/lb. Compute the total material cost for the core.
10. The cost of hot metal for the casting in Q1 is $4.42/lb. The scrap and return credit is $0.72/lb. and the environmental cost is $6.35/lb. Compute the total metal cost for the casting.
11. An operator controls five automatic machines. They work free of the operator once they are set up. The time to make one piece is 6 seconds or 600 pieces hourly when working at 100% efficiency. The setup time is 0.005 minute per piece. If the efficiency rate is at 81% for the operator controlling five machines and the labor rate is at $.0603 per minute, what is the operation cost per piece?

12. What is the cutting time for a copper shaft 4 × 35 square inches long when there is a surface velocity of 372 surface feet per minute (sfpm) and a feed of 0.011 inches per revolution?
13. Calculate the total pouring weight, melting weight loss, casting scrap weight, finishing and cleaning scrap weight, and return weight for a casting weighing 55 lbs, yield of 70.4%, casting scrap rate of 14%, melt loss of 3.27%, and a finishing and cleaning scrap rate of 4%.
14. An end-facing cut needs a 19-inch diameter work piece. The revolutions (per minute) of the lathe maintain 313 surface feet per minute from the center to the surface. The feed is 0.0006. What is the time for the cut?
15. Find the time for the cut in the scenario above if:
    - You need an 8-inch diameter work piece
    - The revolutions per minute of the lathe were 650 feet per minute
    - The feed is 0.0101
16. A box has a basal area of 23 in. × 16.5 in. and the wall height is 9 in. The box is made out of low carbon steel. The objective is to place 0.42 inch fillet welds at the inside corners using shielded metal arc welding. The wall thickness of the material is 0.33 inches. Compute the cost to produce a batch of 80 boxes excluding material cost. The process data is shown below.
    - Travel speed: 14 inches/minute
    - Welding current: 228 Amps
    - Welding voltage: 40 V
    - Efficiency of machine: 88%
    - Duty cycle: 64%
    - Operator factor: 0.40
    - Electrode metal yield for 18-inch electrode: 0.75
    - Labor rate: $16.25/hr
    - Overhead rate: $37/hr
    - Electrode cost: $0.54/lb
    - Power cost: $0.18/KW-hr
17. Five holes are going to be drilled in an aluminum plate using a high-speed steel drill. In order to accomplish this task, a sensitive drill press with two spindles shall be used. Calculate the machining time to process a batch of 350 pieces. Refer to Figure 4.8 for the geometry of the work piece in question.

5

Overhead Expenses

5.1 What Are Overhead Expenses?

In any manufacturing or production operation, there are certain costs that have to be born that cannot be clearly identified and charged to any specific product. These costs are referred to as overhead expenses. Simply put, overhead expenses can be characterized as expenditure incurred during the production or manufacturing process that cannot be classified as direct material cost, direct labor cost or any other direct expenses such as engineering expenses. In keeping with this definition, overhead expenses can be termed as indirect expenses. Examples of these costs include administrative costs, secretarial costs, supervisor salary, and costs of oil, grease, and lubricants. Despite the fact that such expenses cannot be attributed to any particular product, they still need to be accounted for in order to compute metrics such as selling price, profitability, return on equity, and so on.

As a matter of fact, the magnitude of overhead expenses in a typical production operation makes these costs impossible to ignore. Overhead expenses constitute the single greatest cost item in such operations, larger in magnitude than even direct costs such as direct material and direct labor costs. Overhead expenses can often run as high as 200%–400% of direct labor costs. This implies that if the manufacturer incurred $10 per item in direct labor costs, they would have to pay an additional $20 to $40 in overhead expenses such as selling expenses and administrative expenses. The importance of direct labor costs in their role in producing the end product is obvious. However, overhead expenses play a much smaller role in the actual production process itself. Yet, the magnitude of overhead expenses makes it quite easy to realize the significant impact they can have on the company's profitability.

Overhead expenses are classified as follows:

- Factory overhead expenses
- Administrative overhead expenses
- Selling overhead expenses

The following section will examine each class of overhead expenses in detail.

5.2 Classification of Factory Overhead Expenses

Factory overhead is also termed as manufacturing overhead or work overhead. It is defined as the factory related indirect costs that have to be incurred when a product is manufactured or produced. Thus, factory overhead costs are classified as: indirect labor costs, indirect material costs, indirect engineering costs, and other related indirect expenses. Most overhead expenses are clumped together and are not attributable to any particular product. This makes it difficult to account for indirect costs by distributing them across a family of products. The significance of each element of factory overhead expenses will be examined in this section.

5.2.1 Indirect Material Costs

Every product is made out of some material. This material is termed as the direct material. Thus, indirect material is all material that is not a part of the final product, but is consumed during the production process anyway. The production process can be broadly defined as the process of converting raw material into finished product. Therefore, all material that does not constitute the finished product but is consumed during the production process can be defined as indirect material. Examples of indirect material costs include cost of grinding wheels, cutting fluids in machining, lubricants for forming, cleaning materials, and the cost of lubricants such as oils, grease, and other lubricants that are an essential constituent of the manufacturing process but not a component of the finished product.

5.2.2 Indirect Labor Costs

Direct labor is defined as all labor that is directly responsible for the manufacturing or production operation. Thus, indirect labor constitutes labor that does not play a direct role in the actual production process itself. However, the services of such personnel are required to ensure smooth functioning of the operation. Examples of indirect labor expenses include salaries paid to supervisors, repair and maintenance staff, care operators, janitorial staff, foremen, and miscellaneous factory personnel. Administrative staff, such as secretaries, is an essential component of the overall functioning of the factory. Thus, wages paid to such staff need to be accounted for as well. It should be noted that unlike foremen, maintenance staff, etc., administrative staff are not necessarily present on site where the production process takes place. As a result, salaries paid to such personnel are often considered as an element of administrative overhead as well as sales and distribution overhead.

5.2.3 Indirect Engineering Costs

During a production process, expenses need to be incurred that are outside the realm of direct and indirect material expenses and direct and indirect labor expenses. Examples of such costs include cost to procure computer supplies, computer software, electronic hardware, and any other equipment that is necessary to ensure successful design, development, and production of the finished product. Such costs are defined as indirect engineering costs. They can include expenses incurred on equipment necessary for engineering design, process design, and process control as well as equipment to conduct tests during the product development stage.

5.2.4 Other Indirect Expenses: Burden

Overhead expenses that cannot be classified as indirect material costs, indirect labor costs or indirect engineering costs are defined as other indirect expenses, also referred to as burden. Examples of burden include such items as depreciation charges taken on production equipment and manufacturing machinery, expenses necessary for upkeep and maintenance of the factory building, costs incurred on utilities such as electricity, water, and waste disposal, taxes paid on the factory building and site, and so on.

Salaries paid to corporate staff and research and development staff are listed as administrative overhead. Salaries paid to sales, marketing, and distribution teams are listed as sales and distribution overhead.

It will be appreciated that allocation of some of the aforementioned overhead costs to specific product lines is impractical. It is generally not done as a result.

5.3 How and Why Are Overhead Costs Distributed?

A production operation usually consists of several product lines. It will be clear from the preceding discussion that it is impractical to specifically allocate elements of overhead costs, especially those that constitute indirect labor and burden to particular products. Consider a hypothetical case wherein a company produces just one product. In such a case, it would be very easy to allocate all elements of overhead expenses to just that one product. However, the situation becomes rather complex when product lines are added to the company's portfolio of products. But why bother allocating overhead expenses to any single product in the first place?

The answer to this question lies in the fact that companies are businesses that are driven by the need to be profitable. Improper allocation of overhead expenses can result in wrong decisions with respect to not just product manufacturing, but product selection itself. This implies that the wrong product could be chosen at the outset. Needless to say this error would occur at the expense of a product that would have been more profitable. It should have been selected but was not due to improper allocation of overhead costs. In light of the fact that the overhead costs constitute perhaps the largest element of cost, they should be carefully analyzed and allocated. The remainder of this chapter will present some methods that are useful in allocation of overhead costs. These methods include the following:

- Single Parameter Method
- Multiple Parameter Method
- Unused Capacity Costs

5.4 Single Parameter Method of Allocation

The single parameter method of allocation uses one parameter on the basis of which indirect costs are allocated. Any of the following eight methods can be used to distribute the indirect costs among different products. The choice of a particular method is up to the discretion of the company based on their specific requirements.

- Allocation based on direct cost proportioning
- Allocation based on direct labor cost proportioning
- Allocation based on direct material cost proportioning
- Allocation based on direct labor hour proportioning
- Allocation based on machine hour proportioning
- Allocation based on production quantity proportioning
- Allocation based on production floor area
- Predetermined allocation base

EXAMPLE

Let us solve a numeric problem in order to fully comprehend each method of allocation. Consider the data presented in Tables 5.1–5.3. A company manufactures two products: product A and product B. Details regarding direct material, direct labor, production quantity, machine hours, production floor space, etc., are presented in the tables as well. A breakdown of indirect expenses for both products is provided in Table 5.3. The objective is to determine the allocation of indirect cost to each product based on the data presented in the table. This exercise will serve two objectives. First, it will help us solve the problem. Second, it will help further our understanding of each method of indirect cost allocation based on a single parameter.

TABLE 5.1

Production Data for Overhead Allocation

Production Parameter	Product A	Product B
Production quantity	2000	4000
Production floor space (sq ft)	200	80
Machine hours	120	80
Direct labor hours	600	800

TABLE 5.2

Direct Costs for Two Products for Overhead Cost Allocation

	Labor Rate ($/hr)	Hours of Labor	Material Rate ($/unit)	Amount of Material (units)	Cost ($)
Product A					
Direct material			4.00	2000	8000
Direct labor	20.00	600			12,000
Product B					
Direct material			10.00	4000	40,000
Direct labor	12.00	800			9600

TABLE 5.3

Direct and Indirect Expense Description

			Overhead Expenses ($)			
Cost	Product A	Product B	Factory	Administrative	Sales	Total Cost ($)
Direct labor cost ($)	12,000	9600				21,600
Indirect labor cost ($)			4000			4000
Direct material cost ($)	8000	40,000				48,000
Indirect material cost ($)			8500			8500
Direct engineering cost ($)	1500	2500				4000
Indirect engineering cost ($)			2500			2500
Direct expense ($)	1500	2000				3500
Other factory costs ($)			2500			2500
Administrative expense ($)				9000		9000
Sales and distribution costs						
Direct costs ($)	1500	1200				2700
Indirect costs ($)					4000	4000
Total cost ($)	24,500	55,300	17,500	9000	4000	110,300

5.4.1 Allocation of Overhead Expenses Based on Direct Cost Proportioning Method

In this method, the overhead expenses are allocated based on direct cost proportioning. This is expressed as a ratio of direct cost for product "x" to total direct costs. The total overhead cost is then multiplied by this ratio in order to compute the overhead cost to be allocated to product "x".

Thus, the formula for the direct cost proportioning method can be written as follows:

$$\text{Allocation of overhead to product ``x''} = \frac{\text{Direct cost of product ``x''}}{\text{Total direct costs}} \times \text{Total overhead cost}$$

(Equation 1)

In order to compute the allocation of overhead expenses to product A, we use Equation 1 as follows:

5.4.1.1 Allocation of Overhead Expenses to Product A

$$\frac{\text{Direct cost of product ``A''}}{\text{Total direct costs}} \times \text{Total overhead costs}$$

From Table 5.3, we observe that the total direct cost for Product A is represented by the summation of all figures in the direct expense category for product A. This amount is equal to \$15,500.

In this example, there are two products, A and B. Thus, total direct costs are obtained by adding the direct cost for product A and the direct cost for product B.

We observe that the direct cost for product A = \$24,500

Direct cost for product B = \$55,300

Thus, total direct costs = \$24,500 + \$55,300 = \$79,800

We also observe from Table 5.3 that the total overhead costs amount to \$30,500. This is obtained by adding the following numbers: \$17,500 + \$9000 + \$4000.

Substituting the respective values into Equation 1, we obtain the allocation of overhead expenses to product A as follows:

$$\frac{\text{Direct cost of product ``A''}}{\text{Total direct costs}} \times \text{Total overhead costs}$$

$$= (\$24,500/\$79,800) \times \$30,500 = 0.3070 \times \$30,500 = \$9364.03$$

Thus, total manufacturing cost for product A = Direct cost + Indirect cost

Direct cost for product A = \$24,500

Indirect cost for product A = \$9364.03

Thus, total cost for product A = \$24,500 + \$9364.03 = \$33,864.03

This implies that the unit cost for product A can be obtained by dividing the total manufacturing cost for product A by the number of units of that product.

We can observe from Table 5.1, that 2000 units of product A will be manufactured.

Thus, unit cost of product A = \$33,864.03/2000 = \$16.93/unit **Solution 1**

5.4.1.2 Allocation of Overhead Expenses to Product B

Overhead expenses can be allocated to product B using the same technique as for product A. This is demonstrated as follows.

From Table 5.3, we observe that the total direct cost for Product B is represented by the summation of all figures in the direct expense category for product B. This amount is equal to \$55,300.

Total direct costs = \$24,500 + \$55,300 = \$79,800

Substituting the respective values into Equation 1, we obtain the allocation of overhead expenses to product A as follows:

$$\frac{\text{Direct cost of product "B"}}{\text{Total direct costs}} \times \text{Total overhead costs}$$
$$= (\$55,300/\$79,800) \times \$30,500 = 0.6929 \times \$30,500 = \$21,135.96$$

Thus, total manufacturing cost for product B = Direct cost + Indirect cost

Direct cost for product B = \$55,300

Indirect cost for product B = \$21,135.96

Thus, total cost for product B = \$55,300 + \$21,135.96 = \$76,435.96

We can observe from Table 5.1, that 4000 units of product B will be manufactured.

Thus, unit cost of product B = \$76,435.96/4000 = \$19.11/unit **Solution 2**

In the next subsection, we shall examine how overhead expenses can also be allocated to individual products on the basis of their direct labor cost.

5.4.2 Allocation of Overhead Expenses Based on Direct Labor Cost Proportioning Method

In this method, the overhead expenses are allocated based on direct labor cost proportioning. This is expressed as a ratio of direct labor cost for product "x" to total direct labor costs. The total overhead cost is then multiplied by this ratio in order to compute the overhead cost to be allocated to product "x".

Thus, the formula for the direct cost proportioning method can be written as follows:

$$\text{Allocation of overhead to product "x"} = \frac{\text{Direct labor cost of product "x"}}{\text{Total direct labor costs}} \times \text{Total overhead cost} \qquad \text{(Equation 2)}$$

In order to compute the allocation of overhead expenses to product A, we use Equation 2 as follows:

5.4.2.1 Allocation of Overhead Expenses to Product A

$$\frac{\text{Direct cost of product "A"}}{\text{Total direct costs}} \times \text{Total overhead costs}$$

From Table 5.3, it will be observed that:

Direct labor cost for product A = \$12,000
Direct labor cost for product B = \$9600
Thus, total direct labor cost = \$21,600

Substituting the respective values into Equation 2, we obtain the following:

$$\begin{aligned}\text{Overhead expense allocation to product A} &= (\$12{,}000/\$21{,}600) \times \text{Total overhead costs} \\ &= 0.555 \times \$30{,}500 = \$16{,}944.44\end{aligned}$$

Thus, total cost for manufacturing product A = Direct cost for product A + Indirect cost for product A.

Direct cost for product A = \$24,500
Indirect cost for product A = \$16,944.44
Thus, total cost for product A = \$24,500 + \$16,944.44 = \$41,444

Once again, as before, the unit cost for product A can be obtained by dividing the total cost for product A by the number of units of product A that were manufactured.

We know that 2000 units of product A were manufactured.

Thus, unit cost of product A = \$41,444/2000 = \$20.722 **Solution 1**

5.4.2.2 Allocation of Overhead Expenses to Product B

Overhead expenses can be allocated to product B using the same technique as for product A. This is demonstrated as follows.

We know from before that direct labor cost for product B = \$9600

Also, total direct labor cost = \$21,600

Substituting the respective values into Equation 2, we obtain the following:

$$\text{Overhead expense allocation to product A} = (\$9600/\$21{,}600) \times \text{Total overhead costs} = 0.444 \times \$30{,}500 = \$13{,}555.55$$

Thus, total cost for manufacturing product B = Direct cost for product B + Indirect cost for product B.

Direct cost for product B = \$55,300

Indirect cost for product B = \$13,555.55

Thus, total cost for product B = \$55,300 + \$13,555.55 = \$68,855.55

Once again, as before, the unit cost for product B can be obtained by dividing the total cost for product B by the number of units of product B that were manufactured.

We know that 4000 units of product B were manufactured

Thus, unit cost of product B = \$68,855.55/4000 = \$17.21 **Solution 2**

In the next subsection, we shall examine how overhead expenses can also be allocated to individual products on the basis of their direct material cost.

5.4.3 Allocation of Overhead Expenses Based on Direct Material Cost Proportioning Method

In this method, the overhead expenses are allocated based on direct material cost proportioning. This is expressed as a ratio of direct material cost for product "x" to total direct material costs. The total overhead cost is then multiplied by this ratio in order to compute the overhead cost to be allocated to product "x".

Thus, the formula for the direct cost proportioning method can be written as follows:

$$\text{Allocation of overhead to product "x"} = \frac{\text{Direct material cost of product "x"}}{\text{Total direct material costs}} \times \text{Total overhead cost} \quad \text{(Equation 3)}$$

In our current example, the allocation of overhead cost based on this method is demonstrated as follows.

5.4.3.1 Allocation of Overhead Expenses to Product A

$$\frac{\text{Direct material cost of product "A"}}{\text{Total direct material costs}} \times \text{Total overhead costs}$$

From Table 5.3, it will be observed that:

Direct material cost for product A = \$8000
Direct material cost for product B = \$40,000
Thus, total direct material cost = \$48,000

Substituting the respective values into Equation 3, we obtain the following:

$$\begin{aligned}\text{Overhead expense allocation to product A} &= (\$8000/\$48,000) \\ &\quad \times \text{Total overhead costs} \\ &= 0.1667 \times \$30,500 = \$5083.33\end{aligned}$$

Thus, total cost for manufacturing product A = Direct cost for product A + Indirect cost for product A.

Direct cost for product A = \$24,500
Indirect cost for product A = \$5083
Thus, total cost for product A = \$24,500 + \$5083 = \$29,583.33

Once again, as before, the unit cost for product A can be obtained by dividing the total cost for product A by the number of units of product A that were manufactured.

We know that 2000 units of product A were manufactured.

Thus, unit cost of product A = \$29,583.33/2000 = \$14.79 **Solution 1**

5.4.3.2 Allocation of Overhead Expenses to Product B

Overhead expenses can be allocated to product B using the same technique as for product A. This is demonstrated as follows:

We know from before that direct material cost for product B = \$40,000
Also, total direct material cost = \$48,000

Substituting the respective values into Equation 2, we obtain the following:

$$\text{Overhead expense allocation to product B} = (\$40{,}000/\$48{,}000) \times \text{Total overhead costs} = 0.833 \times \$30{,}500 = \$25{,}416.67$$

Thus, total cost for manufacturing product B = Direct cost for product B + Indirect cost for product B.

Direct cost for Product B = $55,300

Indirect cost for Product B = $25,416.67

Thus, total cost for Product B = $55,300 + $25,416.67 = $80,716.67

Once again, as before, the unit cost for product B can be obtained by dividing the total cost for product B by the number of units of product B that were manufactured.

We know that 4000 units of product B were manufactured

Thus, unit cost of product B = $80,716.67/4000 = $20.18 **Solution 2**

In the next subsection, we shall examine how overhead expenses can also be allocated to individual products on the basis of the number of direct labor hours invested into processing each product.

5.4.4 Allocation of Overhead Expenses Based on Direct Labor Hour Proportioning Method

In this method, the overhead expenses are allocated based on direct labor hour proportioning. This is expressed as a ratio of number of direct labor hours necessary for processing product "x" to total number of direct labor hours. The total overhead cost is then multiplied by this ratio in order to compute the overhead cost to be allocated to product "x".

Thus, the formula for direct cost proportioning method can be written as follows:

$$\text{Allocation of overhead to product ``x''} = \frac{\text{Direct labor hours for product ``x''}}{\text{Total direct labor hours}} \times \text{Total overhead cost}$$

(Equation 4)

In our current example, the allocation of overhead casts based on this method is demonstrated as follows.

5.4.4.1 Allocation of Overhead Expenses to Product A

$$\frac{\text{Direct labor hours for product ``A''}}{\text{Total direct labor hours}} \times \text{Total overhead costs}$$

From Table 5.1, it will be observed that:

Direct labor hours for product A = 600
Direct labor hours for product B = 800
Thus, total direct labor hours = 1400

Substituting the respective values into Equation 4, we obtain the following:

$$\text{Overhead expense allocation to product A} = (600/1400) \times \text{Total overhead costs} = 0.4285 \times \$30,500 = \$13,071.43$$

Thus, total cost for manufacturing product A = Direct cost for product A + Indirect cost for product A.

Direct cost for product A = $24,500
Indirect cost for product A = $13,071.43
Thus, total cost for product A = $24,500 + $13,071.43 = $37,571.43

Once again, as before, the unit cost for product A can be obtained by dividing the total cost for product A by the number of units of product A that were manufactured.

We know that 2000 units of product A were manufactured.

Thus, unit cost of product A = $37,571.43/2000 = $18.79 **Solution 1**

5.4.4.2 Allocation of Overhead Expenses to Product B

Overhead expenses can be allocated to product B using the same technique as for product A. This is demonstrated as follows.

We know from before that direct labor hours for product B = 800.

Also, total direct labor hours = 1400.

Substituting the respective values into Equation 4, we obtain the following:

$$\text{Overhead expense allocation to product B} = (800/1400) \times \text{Total overhead costs} = 0.5714 \times \$30,500 = \$17,428.57$$

Thus, total cost for manufacturing product B = Direct cost for product B + Indirect cost for product B.

Direct cost for product B = $55,300
Indirect cost for product B = $17,428.57
Thus, total cost for product B = $55,300 + $17,428.57 = $72,728.57

Once again, as before, the unit cost for product B can be obtained by dividing the total cost for product B by the number of units of product B that were manufactured.

We know that 4000 units of product B were manufactured

Thus, unit cost of product B = \$72,728.57/4000 = \$18.182 **Solution 2**

In the next subsection, we shall examine how overhead expenses can also be allocated to individual products on the basis of the number of machine hours required to process each product.

5.4.5 Allocation of Overhead Expenses Based on Machine Hour Proportioning Method

In this method, the overhead expenses are allocated based on machine hour proportioning. This is expressed as a ratio of the number of machine hours necessary for processing product "x" to the total number of machine hours. The total overhead cost is then multiplied by this ratio in order to compute the overhead cost to be allocated to product "x".

Thus, the formula for the direct cost proportioning method can be written as follows:

$$\text{Allocation of overhead to product ``x''} = \frac{\text{Machine hours for product ``x''}}{\text{Total machine hours}} \times \text{Total overhead cost}$$

(Equation 5)

In our current example, the allocation of overhead cost based on this method is demonstrated as follows.

5.4.5.1 Allocation of Overhead Expenses to Product A

$$\frac{\text{Machine hours for product ``A''}}{\text{Total machine hours}} \times \text{Total overhead costs}$$

From Table 5.1, it will be observed that:

Direct labor hours for product A = 120

Direct labor hours for product B = 80

Thus, total direct labor hours = 200

Substituting the respective values into Equation 5, we obtain the following:

$$\begin{aligned}\text{Overhead expense allocation to product A} &= (120/200) \\ &\times \text{Total overhead costs} \\ &= 0.60 \times \$30,500 = \$18,300\end{aligned}$$

Thus, total cost for manufacturing product A = Direct cost for product A + Indirect cost for product A.

Direct cost for product A = \$24,500

Indirect cost for product A = \$18,300

Thus, total cost for product A = \$24,500 + \$18,300 = \$42,800

Once again, as before, the unit cost for product A can be obtained by dividing the total cost for product A by the number of units of product A that were manufactured.

We know that 2000 units of product A were manufactured.

Thus, unit cost of product A = \$42,800/2000 = \$21.40 **Solution 1**

5.4.5.2 Allocation of Overhead Expenses to Product B

Overhead expenses can be allocated to product B using the same technique as for product A. This is demonstrated as follows.

We know from before that machine hours for product B = 80

Also, Total machine hours = 200

Substituting the respective values into Equation 5, we obtain the following:

$$\text{Overhead expense allocation to product B} = (80/120) \times \text{Total overhead costs}$$
$$= 0.4 \times \$30,500 = \$12,200$$

Thus, total cost for manufacturing product B = Direct cost for product B + Indirect cost for product B.

Direct cost for product B = \$55,300

Indirect cost for product B = \$12,200

Thus, total cost for product B = \$55,300 + \$12,200 = \$67,500

Once again, as before, the unit cost for product B can be obtained by dividing the total cost for product B by the number of units of product B that were manufactured.

We know that 4000 units of product B were manufactured.

Thus, unit cost of product B = \$67,500/4000 = \$16.875 **Solution 2**

In the next subsection, we shall examine how overhead expenses can also be allocated to individual products on the basis of their production quantities.

5.4.6 Allocation of Overhead Expenses Based on Production Quantity Proportioning Method

In this method, the overhead expenses are allocated based on production quantity proportioning. This is expressed as a ratio of production quantity of product "x" to total production quantity. The total overhead cost is then multiplied by this ratio in order to compute the overhead cost to be allocated to product "x".

Thus, the formula for direct cost proportioning method can be written as follows:

$$\text{Allocation of overhead to product "x"} = \frac{\text{Production quantity of product "x"}}{\text{Total production quantity}} \times \text{Total overhead cost}$$

(Equation 6)

In our current example, the allocation of overhead cost based on this method is demonstrated as follows.

5.4.6.1 Allocation of Overhead Expenses to Product A

$$\frac{\text{Production quantity for product "A"}}{\text{Total production quantity}} \times \text{Total overhead costs}$$

From Table 5.1, it will be observed that:

Production quantity for product A = 2000
Production quantity for product B = 4000
Thus, total production quantity = 6000

Substituting the respective values into Equation 5, we obtain the following:

$$\begin{aligned}\text{Overhead expense allocation to product A} &= (2000/6000) \\ &\quad \times \text{Total overhead costs} \\ &= 0.333 \times \$30,500 = \$10,166.67\end{aligned}$$

Thus, total cost for manufacturing product A = Direct cost for product A + Indirect cost for product A.

Direct cost for product A = $24,500
Indirect cost for product A = $10,166.67
Thus, total cost for product A = $24,500 + $10,166.67 = $34,666.67

Once again, as before, the unit cost for product A can be obtained by dividing the total cost for product A by the number of units of product A that were manufactured.

We know that 2000 units of product A were manufactured.

Thus, unit cost of product A = $34,666.67/2000 = $17.33 **Solution 1**

5.4.6.2 Allocation of Overhead Expenses to Product B

Overhead expenses can be allocated to product B using the same technique as for product A. This is demonstrated as follows.

We know from before that production quantity for product B = 4000

Also, total production quantity = 6000

Substituting the respective values into Equation 5, we obtain the following:

$$\text{Overhead expense allocation to product B} = (4000/6000) \times \text{Total overhead costs} = 0.6667 \times \$30{,}500 = \$20{,}333.33$$

Thus, total cost for manufacturing product B = Direct cost for product B + Indirect cost for product B.

Direct cost for product B = \$55,300

Indirect cost for product B = \$20,333.33

Thus, total cost for product B = \$55,300 + \$12,200 = \$67,500

Once again, as before, the unit cost for product B can be obtained by dividing the total cost for product B by the number of units of product B that were manufactured.

We know that 4000 units of product B were manufactured

Thus, unit cost of product B = \$67,500/4000 = \$16.875 **Solution 2**

In the next subsection, we shall examine how overhead expenses can also be allocated to individual products on the basis of the amount of production floor area dedicated to processing each product.

5.4.7 Allocation of Overhead Expenses Based on Production Floor Area Proportioning Method

In this method, the overhead expenses are allocated based on production floor area proportioning. This is expressed as a ratio of the amount of production floor area dedicated to processing product "x" to total production floor area in the plant. The total overhead cost is then multiplied by this ratio in order to compute the overhead cost to be allocated to product "x".

Thus, the formula for direct cost proportioning method can be written as follows:

$$\text{Allocation of overhead to product "x"} = \frac{\text{Production floor area of product "x"}}{\text{Total production floor area}} \times \text{Total overhead cost}$$

(Equation 7)

In our current example, the allocation of overhead costs based on this method is demonstrated as follows.

5.4.7.1 Allocation of Overhead Expenses to Product A

$$\frac{\text{Production floor area for product "A"}}{\text{Total production floor area}} \times \text{Total overhead costs.}$$

From Table 5.1, it will be observed that:

Production floor area for product A = 200 sq ft
Production floor area for product B = 80 sq ft
Thus, total production floor area = 280 sq ft

Substituting the respective values into Equation 7, we obtain the following:

$$\begin{aligned}\text{Overhead expense allocation to product A} &= (200/280) \times \text{Total overhead costs} \\ &= 0.714 \times \$30,500 = \$21,785.71\end{aligned}$$

Thus, total cost for manufacturing product A = Direct cost for product A + Indirect cost for product A.

Direct cost for product A = $24,500
Indirect cost for product A = $21,785.71
Thus, total cost for product A = $24,500 + $21,785.71 = $46,285.71

Once again, as before, the unit cost for product A can be obtained by dividing the total cost for product A by the number of units of product A that were manufactured.

We know that 2000 units of product A were manufactured.

Thus, unit cost of product A = $46,285.71/2000 = $23.14 **Solution 1**

5.4.7.2 Allocation of Overhead Expenses to Product B

Overhead expenses can be allocated to product B using the same technique as for product A. This is demonstrated as follows.

We know from before that production floor area for product B = 80 sq ft
Also, total production floor area = 280 sq ft

Substituting the respective values into Equation 7, we obtain the following:

$$\text{Overhead expense allocation to product B} = (80/280) \times \text{Total overhead costs}$$
$$= 0.2857 \times \$30,500 = \$8714.29$$

Thus, total cost for manufacturing product B = Direct cost for product B + Indirect cost for product B.

Direct cost for product B = $55,300

Indirect cost for product B = $8714.29

Thus, total cost for product B = $55,300 + $8714.29 = $64,014.29

Once again, as before, the unit cost for product B can be obtained by dividing the total cost for product B by the number of units of product B that were manufactured.

We know that 4000 units of product B were manufactured

Thus, unit cost of product B = $64,014.29/4000 = $16.00 **Solution 2**

We examined a variety of methods to allocate overhead expenses in this section. But the question arises: which method should be used? Generally speaking, if the production numbers were based on a fixed period of time such as a month or a quarter, etc., direct labor hours are used as a base on which to allocate overhead expenses.

Overhead expenses are also allocated on the basis of machine hours. This makes sense because the machine time is being encumbered due to the manufacturing process of the product in question. It will be appreciated from the preceding discussion that when overhead expenses are allocated on the basis of a single parameter, a wide range of solutions are obtained due to the variety of methods on the basis of which allocation can take place. However, allocation can also be achieved based on multiple parameter analysis. This method is presented in the following section.

5.5 Multiple Parameter Method of Allocation

The multiple parameter method of allocation aims to achieve a better distribution of overhead expenses than what is achieved using the single parameter alternative. In this method, factory expense is allocated differently than administrative expense, which in turn is allocated differently. For instance, direct labor hours may be used to allocate factory expense. Prime cost could be used to allocate administrative expenses and production quantity could be used to allocate sales expenses.

The number of items sold is more representative of sales and thus related sales expenses. Production quantity is the one metric that most closely reflects this. A similar argument can be made for the relationship between prime cost and administrative expenses. The relationship between direct labor hours and factory expense is clear enough and does not need to be elaborated upon.

Let us solve the same problem from the point of view of multiple parameters as explained in the preceding paragraph.

5.5.1 Allocation of Overhead Expenses for Product A

5.5.1.1 Factory Overhead

From Table 5.3, we notice that the total factory overhead to be allocated between products A and B = \$17,500.

From Table 5.2, Direct labor hours for product A = 600 hours

Total direct labor hours = 1400 hours

Using direct labor hours to allocate factory overhead, we obtain the following allocation for product A:

$$\text{Factory expense allocation for product A} = (600\ \text{hours}/1400\ \text{hours}) \times \$17{,}500 = \$7500$$

5.5.1.2 Administrative Overhead

Similarly, from Table 5.3, we notice that the total administrative overhead expense to be allocated between products A and B = \$9000.

This will be allocated based on prime cost.

Prime cost is calculated by adding direct elements of cost for each product. In our current example, these elements include direct material, direct labor, direct engineering, and direct expense. The direct costs for sales and distribution are excluded from this calculation.

Thus, using data presented in Table 5.3, prime cost for product A = \$12,000 + \$8000 + \$1500 + \$1500

Thus, prime cost for product A = \$23,000

Using a similar calculation, prime cost for product B = \$9600 + \$40,000 + \$2500 + \$2000 = \$54,100

Thus, total prime cost = \$23,000 + \$54,100 = \$77,100

Using prime cost to allocate administrative overhead, we obtain the following allocation for product A:

$$\text{Administrative overhead expense allocation for product A} \\ = (\$23{,}000/\$77{,}100) \times \$9000 = \$2684.8$$

5.5.1.3 Sales Overhead

Once again, from Table 5.3, we notice that the total sales overhead expense to be allocated between products A and B = \$4000.

We will use production quantity to allocate the sales overhead. From Table 5.1, it will be observed that 2000 units of product A and 4000 units of product B are manufactured.

Thus, total number of units manufactured = 2000 + 4000 = 6000

$$\text{Sales expense overhead allocation for product A} = \frac{2000}{6000} \times \$4000 = \$1333.33$$

5.5.2 Allocation of Overhead Expenses for Product B

Using similar logic for product B, we obtain the allocation of respective overhead expenses as follows:

$$\text{Factory expense allocation for product B} = (800\text{ hours}/1400\text{ hours}) \\ \times \$17{,}500 = \$10{,}000$$

$$\text{Administrative overhead expense allocation for product B} \\ = (\$54{,}100/\$77{,}100) \times \$9000 = \$6315.18$$

$$\text{Sales expense overhead allocation for product B} \\ = (4000/6000) \times \$4000 = \$2666.67$$

5.6 Activity Based Costing Method for Allocation of Overhead Costs

The relentless forward march of automation has ushered in a new age of greater productivity. Large gains in productivity have been enabled by new technologies that are characterized by a greater degree of automation such as computer software, advanced manufacturing technologies, robotics, etc. Greater productivity implies that a fewer number of hours need to be invested in the manufacturing process. The amount of time that can be characterized as direct labor has decreased substantially. There used to be a time when direct labor cost used to account for as much as 40% to 45% of total product cost. This share has decreased almost exponentially in recent years. Direct labor cost now accounts for as little as 10% to 15% of total product cost. This is reflective of modern manufacturing operations that are substantially more productive than their counterparts from a few decades ago.

However, there is a price to pay for this jump in productivity. The share of indirect costs as a percentage of total product costs has increased. Indirect costs now account for as much as 40%–60% of total product costs. This change has necessitated the adoption of different strategies to allocate overhead costs. One such method is referred to as activity based costing.

Activity based costing is a method that has been developed to suit the needs of industries that incur high overhead costs. Activity based costing achieves its objective by identifying cost centers, cost drivers, and activities. Definitions of each term are listed below.

- Cost Center: Cost center, within the purview of activity based costing, is defined as the final product or service rendered by the company. The overhead expense is allocated to cost centers.
- Cost Driver: Cost drivers are responsible for consumption of shared resources. For instance, number of machine setups, number of safety violations, administrative expenses, and sales expenses as described earlier in this chapter. Thus, cost driver is another term for overhead expenses.
- Activities: Support departments such as quality control, purchasing, engineering, maintenance, etc., that are responsible for generating overhead expenses are termed as activities.

Activities based costing for overhead expenses involve the following steps:

- The first step involves identification of each activity. Next, the total cost associated with that activity is determined.
- The second step involves identifying individual cost drivers. A certain usage volume is associated with each driver. Ascertainment of this volume is the next logical step.
- The final step involves computation of the overhead cost rate for each activity. The formula for this rate is presented in Equation 8.

$$\begin{array}{c}\text{Activities based costing}\\ \text{overhead expense rate}\end{array} = \frac{\text{Total cost of activity}}{\text{Total volume of cost driver}} \qquad \text{(Equation 8)}$$

- Overhead expenses are allocated to each cost center for each activity based on the overhead expense rate as determined by the equation above.

5.7 Conclusion

Overhead expenses have to be incurred in order to run a business. However, they need to be controlled since they do not contribute directly to value addition. In this chapter, various methods of allocating overhead expenses were discussed at length. Each method can be used in a specific setting based on the requirements of the situation. It is important to allocate overhead expenses as precisely as possible in order to enhance the accuracy of the final cost estimate.

PROBLEMS

1. Consider the data presented in the following tables. Compute the unit cost of products X and Y using the following single parameter allocation methods:
   a. Direct cost proportioning
   b. Direct labor cost proportioning
   c. Direct material cost proportioning
   d. Direct labor hour proportioning
   e. Machine hour proportioning
   f. Production quantity proportioning

Production Parameter	Product X	Product Y
Production quantity	2000	4000
Production floor space (sq ft)	200	80
Machine hours	120	80
Direct labor hours	600	800

	Labor Rate ($/hr)	Hours of Labor	Material Rate ($/unit)	Amount of Material (units)	Cost ($)
Product X					
Direct material			4.00	2000	8000
Direct labor	20.00	600			12,000
Product Y					
Direct material			10.00	4000	40,000
Direct labor	12.00	800			9600

			Overhead Expenses ($)			
Cost	Product X	Product Y	Factory	Administrative	Sales	Total Cost ($)
Direct labor cost ($)	12,000	9600				21,600
Indirect labor cost ($)			4000			4000

(*Continued*)

Cost	Product X	Product Y	Overhead Expenses ($) Factory	Administrative	Sales	Total Cost ($)
Direct material cost ($)	8000	40,000				48,000
Indirect Material cost ($)			5500			5500
Direct Engineering cost ($)	1500	2500				4000
Indirect Engineering cost ($)			2500			2500
Direct expense ($)	1500	2000				3500
Other factory costs ($)			2500			2500
Administrative expense ($)				9000		9000
Sales and distribution costs						
Direct costs ($)	1500	1200				2700
Indirect costs ($)					4000	4000
Total cost ($)	24,500	55,300	17,500	9000	4000	110,300

2. For Problem 1, assume that the allocation base was machine hours instead of direct labor hours. Compute the overhead rate per machine hour. Using this rate, compute the unit cost for product X and product Y. Compare the costs thus calculated to those determined using machine hour proportioning.
3. An assembly area has floor space allocation as shown below:
   a. Drop area: 780
   b. Conveyor 1: 300
   c. Conveyor 2: 350
   d. Bench Area: 1100
   e. Total: 2530 square feet

   The total overhead costs are $12,500 for the area in question. Determine the overhead costs for each assembly production center.
4. Using the same scenario in Problem 3, determine the overhead costs for each assembly production center with the following constraints:
   a. Conveyor 1: 450 square feet
   b. Conveyor 2: 390 square feet
   c. Total Square Footage: 2720 square feet
   d. Total Overhead Costs: $16,800
5. Management at the Monroe Company plant is attempting to maintain an overhead rate of 183% for their product. The area is in the machining process, where the overhead rate is 237% based on

$35,000 direct labor. The total direct amount, excluding machining, is $93,000. What overhead rate will the company reach based on this information?

The following is cost data for an automobile part produced at Zoom Automotive, Inc.:

Element	Amount	Cost Factor	Variable Overhead Rate (%)
Direct material, steel	7 lb	$9.83/hwt	–
Plating, chrome	.9 lb	$23.71/hwt	105
Tooling	$1.05/unit	–	205
Direct labor	.30 hr	9.5/hr	295

6. Determine the total direct costs. Assume that tooling is a direct cost. If the overhead rate of the total direct cost is 58.12% and the variable overhead rates are found within the table above, determine the total cost for the part in question or the sum of direct cost and total overhead.
7. Using the table from Problem 6, if the overhead rate of the total direct cost is 74.30% and the variable overhead rates are found within the table above, determine the total cost for the part in question or the sum of direct cost and total overhead.

   Use the following information to answer the questions below.

   A job is routed through machining and finishing. Machining is mechanized with automated equipment and numerical control, leading to a high burden cost. Finishing has a variety of tools, thus having an incredibly low burden cost. Job 1 takes 2 hours of labor in machining and 13 hours in finishing. Job 2 takes 14 hours in machining and 6 in finishing. If a single rate based on labor hours is applied to both jobs, the burden allocation would be the same in both cases (15 hours for each job). Use the following table.

	Blanket Rate		Department Rate	
	Machining	Finishing	Machining	Finishing
Budgeted annual burden	$370,000	$26,000	$370,000	$26,000
Direct labor hours	17,000	17,000	17,000	17,000
Blanket rate/DLH	$11.78			
Department rate/DLH			$30	$1.90

8. What are the overhead costs for Job 2 using the blanket rate?
9. What are the overhead costs for Job 1 using the department rate? How does it compare with the overhead costs for Job 2?
10. Discuss the importance behind selecting the correct base.

6

Estimation of Capital Cost and Operating Cost

6.1 The Importance of Estimating Capital Costs and Operating Costs

In a production enterprise, certain costs need to be incurred in order to facilitate the production operation. Some of these costs need to be incurred to obtain the production equipment necessary to run the operation. Certain other costs need to be incurred to make sure that the production process is running as planned. Costs associated with procurement of equipment necessary to run a production process are referred to as capital costs.

For instance, in the case of a milling operation, a certain amount of money would have to be spent in order to procure the equipment necessary to set up the milling operation. Thus, the cost incurred in order to obtain the milling machine and auxiliary equipment can be classified as capital cost.

Similarly, the cost incurred to actually run the milling operation, such as cost of utilities, direct material cost, direct labor cost, etc., can be classified as operating cost. This chapter will illustrate through the use of examples how capital costs and operation costs are estimated.

6.2 Estimation of Capital Costs

A considerable amount of work has been done on capital cost estimation in the construction industry. Little of this work, however, may be used in the manufacturing sector. We will address capital cost estimation of a project in manufacturing and production industries using the five different techniques listed below.

1. Detailed design estimation: This method of cost estimation has the highest accuracy and can be used for detailed estimates.

2. Turnover ratio method: This method of cost estimation is used for obtaining conceptual estimates, as well as estimates of magnitude.
3. Single factor method: This method of cost estimation is also used for obtaining conceptual estimates as well as estimates of magnitude.
4. Knowledge and Experience (K&E) method: This method of cost estimation is used for individual equipment and processes.
5. AACE total capital requirements method: This method of cost estimation is used for obtaining budgetary estimates of plants and processes.

6.2.1 Detailed Design Estimation Method

The detailed design estimate is also often referred to as the definitive estimate. It is defined by the American Association of Cost Engineers (AACE) as "an estimate prepared from much defined engineering data." The "engineering data" referred to in the above definition can assume many forms. For example, it can include items such as completed plot plans, elevations, instrument diagrams, piping diagrams, equipment data, electrical diagrams, and equipment quotations. It can also include other items such as soil data and structural sketches. When used with respect to civil engineering applications, it can include sketches of major foundations and sketches of buildings as well as a comprehensive set of specifications. All estimates used to ascertain cost using detailed design estimation are expected to be considerably accurate. They are expected to range in accuracy from 15% to −5%.

Based on the examples mentioned above, it is obvious that this method of cost estimation is most commonly used in the field of building construction. However, it can also be useful in other industries. For instance, the same level of accuracy (+15% to −5%) is required if the method is used to estimate cost in the manufacturing and production industry. It should be kept in mind that most manufacturing operations are indeed housed within building structures. This method of cost estimation is indirectly applicable to the manufacturing industry as well. Examples of applications in the manufacturing industry could include changes in plant layout, designing a new work cell, a new foundry, new smelting works, and so on. Notice that each of the aforementioned structures is housed within some type of facility for which final approval is required.

6.2.2 Turnover Ratio Method

Turnover is a term often used by company management to evaluate the performance of the organization. It is an accounting term that may also be used to estimate the capital cost necessary to set up a business enterprise. By definition, the turnover ratio computes the maximum investment from expected revenue. The investment requirement is not computed as a function of equipment that is used to manufacture the product. The mathematical expression of the turnover ratio is expressed in Equation 1.

$$\text{Turnover ratio} = \frac{\text{Annual revenue}}{\text{Total investment required}} \quad \text{(Equation 1)}$$

The above equation can easily be solved to compute the total investment required. The modified form of the equation is expressed as follows:

$$\text{Total investment required} = \frac{\text{Annual revenue}}{\text{Turnover ratio}} \quad \text{(Equation 2)}$$

Each term appearing in the above equation is defined as follows:

Total investment required: Total investment required in the facility.

Annual revenue: Total annual revenue accruing as a result of total annual sales.

Turnover ratio: A ratio of two independent items. No unit is assigned to this ratio.

The turnover ratio is used with advantage in several industries ranging from construction to chemical to manufacturing. In the chemical industry, for instance, the commonly accepted value of the turnover ratio is considered to be 0.9. In the steel manufacturing industry, on the other hand, this value is taken to be 1.1. The turnover ratio method works best when the cost capacity index has a value of 1. If the capacity is increased, the cost increases as well. In order to obtain more accurate and relevant estimates, it is essential to designate the turnover ratio for a particular capacity level.

Consider Example 6.1 in order to gain a better understanding of the application of the turnover ratio in capital cost estimation.

EXAMPLE 6.1

XYZ Metals, Inc. is planning to setup a new foundry. The foundry is expected to produce 750,000 kilograms of castings annually. Each casting can be sold at the rate of $10 per kg. Estimate the expected investment necessary to build the foundry.

Solution

From the problem statement, we know the following data:

Annual sales volume = 750,000 kilograms
Selling price per kilogram = $10
Thus, expected annual revenue = Annual sales volume × Selling price
Using the above relationship, we have the expected annual revenue = 750,000 kilograms × $10/kg
Thus, annual revenue = $7,500,000

Using Equation 2, we can compute the total investment required as follows:

$$\text{Total investment required} = \frac{\text{Annual revenue}}{\text{Turnover ratio}} \quad \text{(Equation 2)}$$

Substituting the pertinent values into the above equation, we have the following:

$$\text{Total investment required} = \frac{\$7,500,000}{\text{Turnover ratio}}$$

The value of the turnover ratio most commonly used for a foundry operation is 0.7.

Substituting this value into the above equation, we obtain the following:

$$\text{Total investment required} = \frac{\$7,500,000}{0.7} = \$10,714,285.71.$$

A total investment of $10,714,285.71 is required in the foundry operation.

Solution

6.2.3 Single Factor Method

As its name implies, the single factor method uses one factor, namely, the delivered equipment cost, to estimate capital costs involved in setting up the production facility. The chemical industry, in particular, has used this factor with great success for several years. The single factor method offers an estimate that is within +50% to −30%. The chemical industry uses three sets factors for capital cost estimation. They are as follows:

- One set of factors used for solid processing plant
- One set of factors used for solid fluid processing plant
- One set of factors used for fluid processing plant

The aforementioned factors enable cost estimation of plant cost and fixed capital investment, as well as total capital investment. The factors are used for major additions to process plants as well as for smaller equipment modifications. It should be noted that delivered equipment costs represent as much as 18%–35% of total plant costs. As seen from the manufacturing and production industry, the factors representing solid processing plant are the most relevant. They are depicted in Table 6.1.

In the following table, direct plant costs include the following items:

- Cost of purchased equipment
- Installation of purchased equipment
- Procurement and installation of controls necessary to operate the facility
- Procurement and installation of piping

TABLE 6.1

Values of Factors for Capital Investment Based upon Delivered Equipment Cost

No	Description	Factor
1.	Total direct plant cost	2.64
2.	Indirect costs	0.72
	Total plant cost	3.36
3.	Fees and contingency	0.51
	Fixed capital investment	3.87
4.	Working capital	0.68
	Total capital investment	4.55

Source: Adapted from Peters, MS and Timmerhaus, KD. 1980. *Plant Design and Economics for Chemical Engineers,* Third Edition, McGraw Hill [13].

- Procurement and installation of electrical supplies
- Erection of buildings
- Installed service facilities
- Procurement of land

Remember that the total fixed capital investment may increase by a factor of as much as 100% if the plant is located at an entirely undeveloped site.

As with direct plant costs, the indirect plant costs include a variety of items. They are listed as follows:

- Costs associated with engineering
- Supervision and related costs
- Costs associated with construction

As previously mentioned, Table 6.1 depicts the various factors associated with estimating different types of costs as described above.

In order to better understand the application of the single factor method, consider Example 6.2.

EXAMPLE 6.2

Compute the expected cost of the plant and total fixed capital investment for a new plant that will manufacture die cast parts. The individual equipment costs are presented in the following table.

Item Description	Cost ($)
Melting unit	450,000
Die casting machine	850,000
Trimming machine	170,000
Total cost	1,470,000

Solution

From Table 6.1, the factor for total plant cost = 3.36. This factor is used to obtain the total plant cost.

The total cost of equipment necessary to undertake the die casting operation = $1,470,000.
Thus, the total plant cost estimate = $1,470,000 × 3.36 = $4,939,200.
Solution 1

From Table 6.1, the factor for fixed capital investment = 3.87. This factor is used to obtain the estimate for fixed capital investment.

Estimate of fixed capital investment = $1,470,000 × 3.87 = $5,688,900. **Solution 2**

For an estimate of total capital investment, we once again refer to Table 6.1. The factor for total capital investment in Table 6.1 is designated as 4.55. The estimate of total capital investment is obtained as follows:

Estimate of total capital investment = $1,470,000 × 4.55 = $6,688,500. **Solution**

6.2.4 The Knowledge and Experience Method (K&E Method)

The knowledge and experience method of estimating capital costs was first developed at Proctor and Gamble for pricing developmental equipment. In its early stages, it was entitled DE × CID. This was an acronym for the underlying formulas used for capital cost estimation:

Developmental equipment × Cost improvement database

The developmental equipment factor in this method is determined from knowledge as well as experience. This has led to coining of the term: "the knowledge and experience method."

In order to apply the K&E method, a five step procedure is followed in order to calculate the total delivered cost of a piece of equipment. These steps are enumerated below:

1. Ascertain the level of knowledge and experience
2. Ascertain the developmental equipment factor
3. Ascertain the number of units to be developed
4. Compute the replacement value of the equipment post development and manufacture
5. Compute the total delivered equipment cost using the aforementioned variables as input

Each step is described in detail within the following section.

1. *Ascertain the level of knowledge and experience*: Insofar as the determination of knowledge and experience is concerned, several crucial factors are considered in this estimation. These factors are listed below:
   a. Equipment base: This includes the types of equipment currently installed at the company.
   b. Vendor experience: This includes the level of experience possessed by the vendor in fulfilling the need for various types of equipment.
   c. Similarity of current products: This includes a direct comparison of the potential product line with the existing product line in order to look for similarities between the two.
   d. Design base.
   e. Type of concept: This includes evaluation of the type of concept being developed with a view to ascertaining the company's capability in terms of seeing such a project through to completion.

   The considerations mentioned above are presented in Table 6.1.
2. *Ascertain the developmental equipment factor*: The developmental equipment factor is computed using the knowledge and experience (K&E) curve depicted in Figure 6.1. It should be noted that the curve is unique to every company based on their own strengths and weaknesses. As such, it should be developed by each company using its own unique set of proprietary data. This data would be based on the level of knowledge and experience available at each company.
3. *Ascertain the number of units to be developed*: The number of units to be developed is quite different from production volume. At the outset, the number of units to be developed is generally equal to one for each individual product line. Once the design is finalized, it is then handed over to manufacturing for mass production. All mass produced items are based on this version of the fully developed design. In some cases, when there is a need to develop products across several product lines, the number of units to be developed could have a value greater than one.
4. *Computation of replacement value*: The replacement value of any equipment can be defined as "the value of an item computed by repricing the item on the basis of replacing it in new condition with another item having a similar capability or the same productive capacity." The new item should, however, apply contemporary economic design that has been adjusted for the physical deterioration of the existing equipment. We will deal with the concept of replacement analysis is considerable detail in Chapter 14 of this text.
5. *Determination of total equipment delivered cost (TDC)*: The total delivered cost is a function of the following variables:

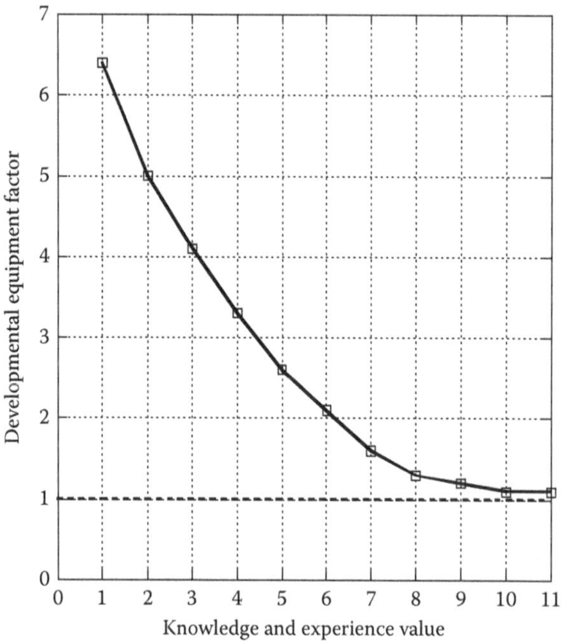

FIGURE 6.1
Relationship between developmental equipment factor and the value of knowledge and experience. (Adapted from Stamp GL. 1991).

a. Developmental equipment factor
b. Number of units under development
c. Replacement value

The total equipment delivered cost is expressed by the following equation:

Total equipment delivered cost (TDC) = DEF × NU × RV. (Equation 3)

wherein:
TDC: Total delivered cost of the equipment
DEF: Development equipment factor
NU: Number of units under development
RV: Replacement value of the equipment

In order to better comprehend the application of the Knowledge and Experience (K&E) method, consider Example 6.3.

EXAMPLE 6.3

A new production device is being developed for robotic material handling. This is a unique device which needs to be developed from scratch. The vendor has a slight degree of experience in developing such a product.

The replacement value for this device is estimated to be about \$800,000. The objective is to first develop one unit and then expand production. Compute the development cost of the equipment.

Solution

In light of the fact that the concept under development is a brand new concept that has to be developed from scratch and the fact that the vendor has only a slight experience in developing a product of this nature, we use material presented in Table 6.2.

It is evident that the level of knowledge and experience corresponding to the above qualitative information is 2.

Referring to Figure 6.2, the corresponding developmental equipment factor is 5.

Additionally, there is only one piece of equipment that needs to be developed.

Thus, all variables to estimate the total delivered cost of the product are in place. Substituting these values into Equation 3, we obtain the following:

$$\text{Total equipment delivered cost (TDC)} = \text{DEF} \times \text{NU} \times \text{RV}. \quad \text{(Equation 3)}$$

In this case, development equipment factor = 5
Number of units to be developed = 1
Replacement value = \$800,000

Substituting these values into Equation 3, the value of total delivered cost is as follows:

$$\text{Total delivered cost (TDC)} = 5 \times 1 \times \$800{,}000 = \$4{,}000{,}000.$$

Solution

6.2.5 The AACE Total Capital Requirements Method

The American Association of Cost Engineers (AACE) recommended a procedure to estimate capital costs. This method of cost estimation was prepared with a budget type estimate in mind. Such an estimate provides an accuracy of 30% to –15%. The different components of cost of using this method are depicted in Table 6.3.

It is evident from Table 6.3 that the total plant costs are subdivided into four main components:

- Process capital costs: These costs are further subdivided into direct costs and indirect costs (overhead costs)
- General facilities costs
- Home office overhead costs (and relevant fees)
- Contingency costs

TABLE 6.2

Knowledge and Experience (K&E) Levels as a Function of Different Factors

K&E Level	Type of Concept	Existing Equipment Similarity	Design Base	Equipment Base	Vendor Experience	Company Product Similarity	Company Experience
1	New	None	Scratch	Scratch	No experience	No similarity	No experience
2	New	None	Scratch	Scratch	Slight experience	No similarity	No experience
3	New	None	Scratch	Scratch	Slight experience	Existing	Little experience
4	Existing	Existing	Scratch	Scratch	Slight experience	Existing	Little experience
5	Existing	Existing	Some standard components	Scratch	Moderate experience	Existing	Little experience
6	Existing	Existing	Many standard components	Some standard components	Moderate experience	Existing	Little experience
7	Existing	Existing	Many standard components	Many standard components	Moderate experience	Existing	Moderate experience
8	Vendor	Existing	Major redesign	Major modification	Modified vendor design	Existing	Moderate experience
9	Vendor	Existing	Moderate redesign	Moderate modification	Standard vendor design	Existing	Moderate experience
10	Vendor	Existing	No redesign required	No modification	Standard vendor design	Existing	Considerable experience

Source: Adapted from Creese, R, Adithan, M and Pabla, B. 1992. *Estimating and Costing for the Metal Manufacturing Industries*, Marcel Dekker Inc., New York [5].

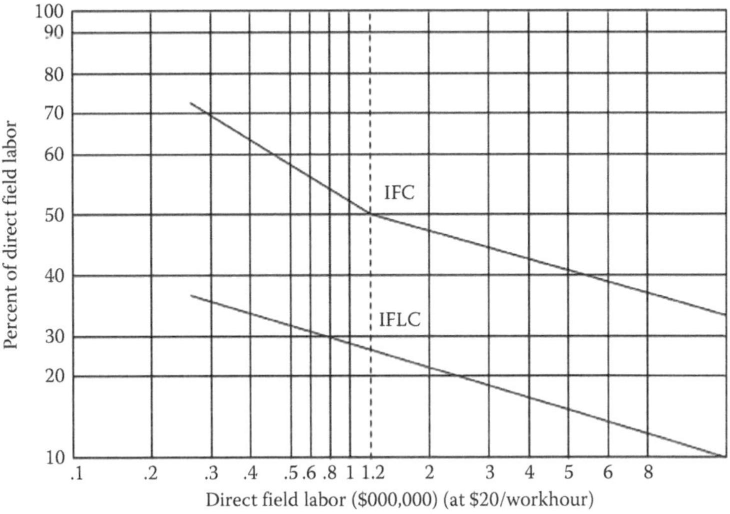

FIGURE 6.2
Indirect field cost and indirect field labor cost as a function of direct labor cost. (Adapted from AACE Recommended Practice, 1990.)

Each of the aforementioned elements of the AACE Method are elaborated upon in greater detail in the following paragraphs.

1. *Process capital costs*: The process capital costs are comprised of different types of direct costs such as direct material cost and direct labor cost. Direct material cost is further classified as purchased equipment costs and installation material costs.
   a. *Purchased equipment costs*: Purchased equipment costs are the foundation for most of the other costs. They serve as the foundation for the AACE method. Purchased equipment costs can be estimated by resorting to any of the following methods:
      i. Obtaining preliminary quotations from vendors
      ii. Comparing figures with cost of equipment used in previous projects
      iii. Obtaining cost data that has already been published
      iv. Interpolating and scaling up data for similar equipment
      v. Vendor data can be used to develop cost capacity equations
      vi. Obtaining firm bids or quotations from vendors
2. *Labor cost to handle and place bare equipment*: Costs associated with handling and placing bare equipment can be categorized as costs associated with activities such as unloading, removing equipment from the crate (similar to unpacking), and placing the equipment in

TABLE 6.3

Elements of the AACE Cost Estimation Method

I Total Cost of the Plant

Process capital

1. *Direct costs*
   a. Direct material costs
      1. Cost of purchased equipment
      2. Installation material cost
      Total direct material cost = a(1) + a(2)
   b. Direct labor costs
      1. Labor cost to handle and place bare equipment
      2. Labor required for installation of equipment
      Total direct labor cost = b(1) + b(2)
   Total direct cost = a(1) + a(2) + b(1) + b(2)
2. *Indirect costs*
   x. Indirect field costs for construction equipment, support, and tools
   y. Labor benefits
   z. Small tools <$500
   Total indirect field cost = 2x + 2y + 2z

Total process capital = Direct cost + Indirect cost = a(1) + a(2) + 2x + 2y + 2z

General facilities cost

Home office overhead and fees

Contingencies

1. Project contingencies
2. Process contingencies

Contingency cost = Project contingency cost + Process contingency cost

II Prepaid royalties

III Startup costs including startup materials

IV Spare parts related costs

V Cost of land acquisition

VI Working capital

Source: Adapted from AACE website: web.aacei.org

its intended place of location. These costs are expressed as a function of purchased equipment cost and are further elaborated upon in Table 6.4 for three classes of equipment.

3. *Installation material cost and installation labor required for equipment installation*: The AACE recommends several factors to deal with installation material and labor costs. They deal with installation activities such as foundations, buildings, electrical work, piping, painting, etc. A detailed list of factors is presented in Table 6.5. It should be kept in mind that the installation material cost should be treated separately from installation labor cost. This is because installation labor cost is a part of total labor cost, which also includes

TABLE 6.4

Factors Used to Estimate Labor Cost to Handle and Place Bare Equipment

Equipment Class	Equipment Description	Labor Factor Range	Labor Factor Average
1	Stationary equipment, for example: Chutes, hoppers, etc., containing no moving parts	0.05–0.15	0.10
2	Machinery involving linear motion, for example: Conveyers, feeders, etc.	0.10–0.20	0.15
3	Rotary equipment., for example: Compressors, pumps, fans, etc.	0.15–0.35	0.25

Source: Adapted from AACE website: web.aacei.org

TABLE 6.5

Material and Labor Installation Factors Represented as a Function of Purchased Equipment Cost

	Factor	
Item Being Installed	**Material**	**Labor**
Foundation	0.04	0.06
Buildings	0.02	0.02
Structural steel	0.04	0.02
Insulation	0.01	0.015
Instrumentation	0.04	0.016
Electrical	0.08	0.06
Piping	0.05	0.025
Painting	0.005	0.015
Miscellaneous	0.03	0.024
Total	0.315	0.255

Source: Adapted from AACE website: web.aacei.org

labor cost to place bare equipment. Also, total direct material cost is expressed as the sum of purchased equipment cost and installation material cost.

4. *Indirect costs*: According to the AACE system, indirect costs can be classified as follows:
   a. Indirect field costs
   b. Labor benefits
   c. Cost of small tools

The above components of indirect costs are also evident from Table 6.3.

Indirect field costs are comprised of the following elements:

a. Wages paid to indirect field labor, such as supervision, field engineering, accounting, warehousing, and staff engineering.
b. Construction support in the form of temporary buildings and roads, field communications, and field utilities.
c. Construction supplies, such as welding equipment, safety equipment, and office equipment among others.
d. Cleanup supplies and associated costs.
e. Benefits paid to labor such as fringe benefits, travel benefits, subsistence funds, etc. These benefits include benefits for indirect field labor and total direct labor costs.
f. Equipment and tool costs such as cost of construction equipment and cost of equipment servicing.
g. Payroll taxes and insurance.
h. Construction camp expenses.

Small tools cost are a separate item and are handled as such. They address costs wherein cost per tool is less than \$500. These costs are expressed as a percentage of total direct labor cost for various levels of direct labor cost. This data is presented in Table 6.6.

Similarly, relationships were developed for indirect field cost as well as the labor component of indirect field cost. The relationships were developed using "curves" depicted in Figure 6.2.

It is obvious from Figure 6.2 that the first range varies from 49% to 75% whereas the second range varies from 38% to 49%. The respective equations are depicted below.

The indirect field cost percentage for the first range (49%–75%) is expressed through Equation 4.

$$\text{IFC (\%)} = 51 \times (\text{T.D.L.C} \times 20/\text{LR})^{(-0.25)}. \qquad \text{(Equation 4)}$$

TABLE 6.6

Small Tool Cost Expressed as a Percentage of Total Direct Labor Cost

Total Direct Labor Cost of Project (TDLC)	Small Tool Cost (Percentage of TDLC)
<\$300,000	5%
\$300,000–\$3,000,000	3.5%
>\$3,000,000	2%

Source: Adapted from AACE website: web.aacei.org

wherein:

IFC: Indirect field cost expressed as a percentage of the total direct field labor (this value fluctuates between 49% and 75%)

T.D.L.C: Total direct field labor expressed in terms of million dollar units (the value usually ranges from 0.25 to 1.2 units, both extremes inclusive)

LR: Average direct field labor rate expressed in terms of \$/hr not including benefits

The indirect field cost percentage for the first range (38%–49%) is expressed through Equation 5.

$$\text{IFC (\%)} = 29 + 21 \times (\text{T.D.L.C} \times 20/\text{LR})^{(-0.34)} \qquad \text{(Equation 5)}$$

wherein:

IFC: Indirect field cost expressed as a percentage of the total direct field labor (this value fluctuates between 38% and 49%)

T.D.L.C: Total direct field labor expressed in terms of million dollar units (the value usually ranges from 1.2 to 15 units, both extremes inclusive)

LR: Average direct field labor rate expressed in terms of \$/hr not including benefits

The indirect field labor cost percentage can be computed using Equation 6 as follows:

$$\text{IFLC (\%)} = 21 \times (\text{T.D.L.C} \times 20/\text{LR})^{(-0.34)} \qquad \text{(Equation 6)}$$

wherein:

IFLC: Indirect field labor cost expressed as a percentage of the total direct field labor (this value fluctuates between 8% and 33%)

T.D.L.C: Total direct field labor expressed in terms of million dollar units (the value usually ranges from 1.2 to 15 units, both extremes inclusive)

LR: Average direct field labor rate expressed in terms of \$/hr not including benefits

From Table 6.3, it will be observed that the next term in the sequence is the labor benefit cost. This includes benefits on direct labor cost, as well as indirect field labor cost. As a general rule, it has been recommended that the labor benefit cost be assigned a value of 35% of the total direct and indirect labor cost. Since this is just a recommendation, the actual value could fluctuate.

From Table 6.3, we can deduce the expression for total indirect field cost as the sum of indirect field cost, labor benefits, and small tool costs. Similarly, total process capital is designated as the sum of total indirect field costs and total direct cost. This is expressed through Equation 7.

$$TPC = TDC + TIFC. \quad \text{(Equation 7)}$$

wherein:

TPC: Total process capital
TDC: Total direct cost
TIFC: Total indirect field cost

Referring to Table 6.3, it will be noticed that the next cost items in sequence are:

- General facilities cost
- Home office overhead and fees

A brief description of these items is presented in the following paragraph.

5. *General facilities, home office overhead, and fees*: General facilities cost and home office overhead and fees are expressed as functions of the total process capital as computed in Equation 7. The general facilities cost accounts for expenditure undertaken on roads, office buildings, and other facilities.

   The general facilities cost is usually approximately 5%–20% of total process capital (TPC).

   The home office overhead and fee, also known by its acronym HOOF, is a quantification of the amount of profit that must be earned by the workforce. It is expected that the home office will have people continually check the progress of the project.

   HOOF costs are usually about 7%–15% of the total process capital (TPC).

6. *Contingency cost*: From Table 6.3, it is obvious that there are two types of contingencies: project contingency and process contingency. The corresponding costs: project contingency cost and process contingency represent the additional cost that would have to be accrued due to a variety of factors such as lack of knowledge of the underlying technology, inconsistencies in the budget estimate, etc. The process contingency costs are expressed as a percentage of total process cost. These percentages are depicted in Table 6.7.

TABLE 6.7

Process Contingency Ratios for Various Levels of Technology Development

State of Technology Development	Data Availability	% of TPC
New concept	Limited data	>40%
Concept	Bench scale data	30%–70%
Pilot plant	Pilot plant data	20%–35%
Modules	Full size modules	5%–20%
Commercial	Process in use commercially	0%–10%

Source: Adapted from AACE website: web.aacei.org

The project contingency cost, on the other hand, covers costs that would have to be incurred in the event that a detailed costing structure was adhered to. Project contingency costs are usually expressed as 15%–35% of the sum of total plant cost, home office overhead, and fees and any process contingencies.

From Table 6.3, we can obtain the equation for total plant cost as follows:

$$TPCT = TPC + GF + HOOF + CPP \qquad \text{(Equation 8)}$$

wherein:

TPCT: Total plant cost
TPC: Total process cost
GF: General facilities cost
CPP: Project and process contingency cost

7. *Prepaid royalties, startup costs, and spare parts costs*: Royalty charges are generally calculated for proprietary information and processes. Prepaid royalties are calculated as 0.5%–1% of total process capital. Any additional royalties should be considered as part of the operating costs.

   Startup costs are defined as the expenses incurred to support activities that involve startup operations. For instance, activities such as plant modifications, operator training, additional maintenance, etc., can be classified as startup operations. According to the AACE, the following guidelines apply to startup cost computation.

   a. If the plant is operating at full capacity, startup costs should be assigned as one month's worth of total annual operating costs.
   b. An additional cost equal to 25% of fuel costs at full capacity for one month of operation.
   c. Any changes necessary to reach full capacity as well as equipment modifications to help achieve this goal can be covered by assigning 2% of total plant costs.

It will be obvious from the above discussion that startup costs cannot be accurately determined until operating costs have been thoroughly evaluated.

Spare parts are necessary to ensure smooth functioning of equipment and plant machinery. Thus, spare part costs are an indispensable component of costing. Spare part costs are calculated as 0.5% of total plant costs.

Depreciable capital expenses are defined as the sum of total plant cost and prepaid royalties, startup costs, and spare part costs. This is depicted in Equation 9.

$$\text{DCE} = \text{TPCT} + \text{RSS} \qquad \text{(Equation 9)}$$

wherein:

DCE: Depreciable capital expenses
TPCT: Total plant cost
RSS: Prepaid royalties, startup costs, and spare part costs

8. *Land and working capital*: Land is a prerequisite to setting up a new manufacturing or production plant. The cost of land depends on the specific location of the site as well as land requirements for the facility (such as square footage, etc). Fixed capital investment is defined as the sum of depreciable capital expenses (as depicted in Equation 9) and the cost of land. This relationship is expressed in Equation 10.

$$\text{FCI} = \text{DCE} + \text{LC} \qquad \text{(Equation 10)}$$

wherein:

FCI: Fixed capital investment
DCE: Depreciable capital expenses
LC: Cost of land

9. *Working capital*: Every company needs a certain amount of funds to start the operation and keep it running. This stash of funds is referred to as working capital. Instances of working capital include items like cash and inventories as well as accounts receivable minus accounts payable. These funds are referred to as liquid assets because they can be liquidated and readily converted to cash. According to AACE requirements, two months of operating cost at full operation should be set aside as working capital. This brings us to total capital investment.

$$\text{TCI} = \text{FCI} + \text{WC} \qquad \text{(Equation 11)}$$

wherein:

TCI: Total capital investment
FCI: Fixed capital investment
WC: Working capital

Let us consider an example in order to gain a better understanding of the AACE method of capital cost estimation.

EXAMPLE 6.4

A new production device is being developed for a die casting operation. This is a unique device which needs to be developed from scratch. The vendor has a slight degree of experience in developing such a product. The replacement value for this device is estimated to be about \$800,000. The objective is to first develop one unit and then expand production. Compute the development cost of the equipment using the AACE method (total plant cost). In order to estimate the labor factor to handle bare equipment costs, all equipment should be considered as class three equipment. The average field labor cost is \$20/hr.

Solution

From the data given in the problem, the total purchased equipment cost = \$800,000.

Referring to Table 6.3, this corresponds to item a (1).

Computation of Installation Material Cost: a (2)

Installation material cost can be computed by using the factors presented in Table 6.5. The installation of the equipment involves undertaking all activities outlined in Table 6.5. Thus, the pertinent cost is computed as follows:

Installation material cost = $0.315 \times$ a (1)
Installation material cost corresponds to item a (2) in Table 6.3
Thus, a (2) = $0.315 \times \$800{,}000 = \$252{,}000$

From Table 6.3, we know that total direct material cost is the sum of cost of purchased equipment and installation material cost.

Thus, total direct material cost = $\$800{,}000 + \$252{,}000 = \$1{,}052{,}000$

Computation of Labor Cost to Handle and Place Bare Equipment: b (1)

In order to compute the labor cost to handle and place bare equipment, we use the labor factors presented in Table 6.4. From the information presented in the problem, all equipment is to be considered as class 3 equipment. The labor factor for all class three equipment is taken to be 0.25 on average.

Thus, labor cost to handle and place bare equipment (corresponding to item b (1) in Table 6.3) = $0.25 \times \$800{,}000 = \$200{,}000$.

Computation of Labor Required for Installation of Equipment: b **(2)**

From Table 6.5, we notice that the factor corresponding to labor requirement for installation of equipment is 0.255. We use this factor to compute corresponding cost.

Labor required for installation of equipment (b (2)) = 0.255 × cost of purchased equipment.

Thus, b (2) = 0.255 × $800,000 = $204,000
From Table 6.3, we know that the total direct labor cost = b (1) + b (2)
Thus, total direct labor cost = $200,000 + $204,000 = $404,000

Computation of Total Direct Cost (TDC)

From Table 6.3, we know that total direct cost (TDC) = a (1) + a (2) + b (1) + b (2).

Substituting the respective values into the above equation, we obtain the value for total direct cost as follows:

Total direct cost (TDC) = $800,000 + $252,000 + $200,000 + $204,000
Thus, total direct cost (TDC) = $1,456,000

Computation of Indirect Field Costs: **2**(*x*)

From the preceding discussion, we know that direct field labor cost amounts to $404,000.

This value falls in the range: 0.25–1.2 in Figure 6.2.

Thus, we use Equation 4 to compute the percentage of indirect field labor costs.

$$\text{IFC (\%)} = 51 \times (\text{T.D.L.C} \times 20/\text{LR})^{(-0.25)}. \qquad \text{(Equation 4)}$$

Substituting the value of T.D.L.C = 0.404 in Equation 4, and using the value of labor rate as $20/hr. as given in the problem statement, we obtain the value of IFC as follows:

$$\text{IFC (\%)} = 51 \times (0.404 \times 20/20)^{-0.25}.$$

Thus, IFC = 63.96% = 0.6396.

Using this percentage, we compute the value of indirect field cost as follows:

Indirect field cost (IFC) = 0.6396 × ($404,000) = $258,437.

The indirect field cost percentage is computed from Equation 6 as follows:

$$\text{IFLC (\%)} = 21 \times (\text{T.D.L.C} \times 20/\text{LR})^{(-0.34)}. \qquad \text{(Equation 6)}$$

Substituting the values into the above equation, we obtain the following:

$$\text{IFLC } (\%) = 21 \times (0.404 \times 20/20)^{(-0.34)}. \qquad \text{(Equation 12)}$$

Thus, IFLC = 28.579%.
The indirect field labor cost (IFLC) is computed as follows:

Indirect field labor cost (IFLC) = 0.28579 × 404,000 = $115,460.

Computation of Labor Benefits: 2(y)

In order to compute the labor benefits we use the recommended value of 35%.

Labor benefits = 0.35 × ($115,460 + $404,000).

Thus, labor benefits = 0.35 × ($519,460) = $181,811.

Computation of Small Tool Cost: 2(z)

We have already computed the total direct labor cost and know it to be $404,000.

In order to compute the small toll cost, using the information from Table 6.6, we know that the small tool cost amounts to 3.5% of total direct field labor cost.

Thus, small tool cost = 0.035 × $404,000 = $14,140.

From Table 6.3, we know that total indirect field cost = 2x + 2y + 2z.

Substituting the respective values into the above equation, we obtain the value for total indirect field cost as follows:

Total indirect field cost = $258,437 + $181,811 + $14,140 = $454,388.

Computation of Total Process Capital

The total process capital can be computed as the sum of direct cost and indirect cost.

Thus, total process capital = $1,456,000 + $ 454,388 = $1,910,388.

Computation of General Facilities Cost

We know from previous discussions that general facilities cost is estimated to be around 15% of total process capital. We also know that total process capital amounts to $1,910,388.

Thus, general facilities cost = 0.15 × $1,910,388 = $286,558.

Computation of Home Office Overhead Fees (HOOF)

Similar to general facilities cost, the home office overhead fees (HOOF) are also estimated to be 15% of total process capital.

Thus, the value of HOOF = \$286,558.

Computation of Process Contingency Fees

The process contingency fee ranges from 0% to 10%. For this example, we will use a fee of 5% of total process capital.

$$\text{Process contingency fee} = 0.05 \times \$1,910,388 = \$95,520.$$

Computation of Project Contingency Fees

The Project contingency fee = 25% of (total process capital + HOOF + process contingency fee).

Substituting the respective values into the above equation, we obtain the following:

$$\text{Project contingency fee} = 0.25 \times (\$1,910,388 + \$286,558 + \$95,520)$$

Thus, project contingency fee = 0.25 × (\$2,292,465) = \$573,116.

Computation of Total Contingency Cost

Total contingency cost = Process contingency cost + Project contingency cost.

Substituting the respective values into the above equation, we obtain the following:

$$\text{Total contingency cost} = \$95,520 + \$573,116 = \$668,636.$$

Computation of Total Plant Cost

We compute total plant cost using Equation 8 as follows.

$$\text{TPCT} = \text{TPC} + \text{GF} + \text{HOOF} + \text{CPP}. \qquad \text{(Equation 8)}$$

Substituting the respective values into the above equation, we obtain the following.

$$\text{Total plant cost} = \$1,910,388 + \$286,558 + \$286,558 + \$668,636.$$

Thus, total plant cost = \$3,152,140. **Solution**

6.3 Estimation of Operating Cost

We dealt with operating cost estimation in detail in Chapter 2. It should be clear at this point that estimation of operating costs is essential in determining items such as startup costs as well as working capital. In general, annual operating costs are classified into four categories as follows:

1. *Cost of raw materials*: We have already seen that raw material costs comprise a significant chunk of total cost. Raw material costs are also a function of process yield, quality requirements, product design parameters, etc.
2. *Cost of utilities and fuel:* It is obvious that utilities and fuel are necessary to provide energy for the manufacturing and production operation. They also provide energy for other purposes such as lighting and heating the facility. Furthermore, fuel needs to be expended in order to provide for transportation of raw materials, work in process, and the finished goods.
3. *Labor cost*: With the advent of automation in modern industry, labor costs are largely confined to maintenance and operations that involve indirect labor. This has come mostly at the expense of direct labor, which is being increasingly displaced by automated machines. The following are some recommendations made by AACE when estimating indirect labor costs.
   a. Direct supervision labor cost is estimated to be about 15% of direct labor cost.
   b. Maintenance labor cost is estimated to be about 3% of total plant cost.
   c. Indirect labor cost is estimated to be about 75% of the sum of direct labor cost, direct labor supervision cost, and maintenance labor cost.
4. *Miscellaneous costs*: Miscellaneous costs are a mélange of the following cost items:
   a. Labor benefit cost: This item is taken to be about 35% of total labor cost.
   b. Property taxes and insurance: This item is taken to be about 2% of total plant cost.
   c. Selling expenses: This item is taken to be about 10% of total sales.
   d. Administration and corporate expenses: This item is taken to be about 60% of total labor cost.
   e. Expenses incurred for maintenance material and equipment: This item is taken to be about 3% of total plant cost.

6.4 Conclusion

Various methods of capital cost estimation were discussed at length in this chapter. It should be noted that very little work has been done in this area, especially with respect to capital cost estimation for manufacturing and production industries. Capital expenditures form the backbone of new operations as well as capacity expansions for existing industries. Therefore, the importance of this chapter in setting up a manufacturing operation cannot possibly be overstated. This chapter also addressed the basic concepts of operation costs. This topic has been dealt with in more detail elsewhere in this book. In view of the practical significance of this chapter, its usefulness to the academic as well as the practitioner is obvious.

PROBLEMS

1. What is the Turnover Ratio Method and when is it most advantageous to use it?
2. What steps need to be completed in order to apply the K&E Method correctly?
3. ABC Castings, Inc. is planning to setup an assembly plant. The plant is expected to produce 1,550,000 kilograms of castings annually. Each casting can be sold at the rate of $25 per kg. Estimate the expected investment necessary to build the foundry.
4. A new investment casting plant is expected to cost $30,000,000 to set up. Each casting can be sold at the rate of $40 per kg. Estimate the expected output per year.
5. UVW, Inc. manufactures castings. They set up a new plant earlier this year. The plant cost $25,000,000 to set up. The plant is expected to produce 1,750,000 kilograms of castings annually. Estimate the selling price per casting.
6. Compute the expected cost of the plant and total fixed capital investment for a new plant that will manufacture die cast parts. The individual equipment costs are presented in the following table. Use the single factor method.

Item Description	Cost ($)
Melting unit	750,000
Die casting machine	1,250,000
Trimming machine	500,000
Total cost	2,500,000

7. A new production device is being developed for a specialized application. This is a unique device which needs to be developed from scratch. The vendor has a little amount of experience in developing such a product. The replacement value for this device is estimated to be about $1,700,000. The objective is to first develop three units and then expand production. Compute the development cost of the equipment using the K&E method.

8. A new production device is being developed for a casting operation. The device needs to be developed from scratch. The vendor has a little amount of experience in developing such a product. The replacement value for this device is estimated to be about $975,000. The objective is to first develop one unit and then expand production. Compute the development cost of the equipment using the AACE method (Total Plant Cost). In order to estimate the labor factor to handle bare equipment costs, all equipment should be considered as class three equipment. The average field labor cost is $40/hr.
9. CAST, Inc. is planning to set up an assembly plant. The plant is expected to produce 2,50,000 kilograms of castings annually. Each casting will be sold at the rate of $41 per kg. Estimate the expected investment necessary to build the foundry.
10. Calculate the expected cost of the plant and total fixed capital investment for a new plant that will manufacture castings using the single factor method. The individual equipment costs are shown in the table below.

Item Description	Cost ($)
Melting unit	250,000
Die casting machine	730,000
Trimming machine	180,000
Total cost	1,140,000

11. Enthusiast Metals is planning to set up a new foundry. The foundry is expected to produce 400,000 kilograms of castings annually. Each casting will be sold at the rate of $109 per kg. Estimate the expected investment necessary to build the foundry.
12. A new device is being developed for material handling. Even though the device would need to be built from scratch, the vendor has some experience in developing such a product. The replacement value for this device is estimated to be $400,750. The objective is to develop one unit then expand production. All equipment is class three equipment. The average field labor cost is $27.25 hourly.
13. Calculate the expected cost of the plant and total fixed capital investment for a new plant that will manufacture castings using the single factor method. The individual equipment costs are shown in the table below.

Item Description	Cost ($)
Melting unit	400,720
Die casting machine	690,000
Trimming machine	214,000
Total cost	1,700,000

7

Costing of Material Handling Operations

7.1 What Is Material Handling?

The material handling industry of America defines material handling as "the movement, storage, protection, and control of materials throughout the manufacturing and distribution process including their consumption and disposal." Materials must be handled safely, efficiently, and inexpensively. Moreover, it is important to deliver the right materials in the right quantities to the right locations on time. Also, the materials should not be damaged during the handling process. Material handling is an all too important component in the field of manufacturing and production. However, it is often overlooked. This chapter seeks to address the issue of costing as it relates to material handling.

It is important to assign and monitor the costs related to material handling. This is because it is estimated that the cost of material handling constitutes a significant portion of total production costs. Material handling costs can amount to as much as 20%–25% of total manufacturing labor cost. This proportion can vary however. It is a function of the type of production and the degree of automation inherent in the material handling system.

Logistics is often used to describe material handling activities. In general, material handling is concerned with the acquisition, movement, storage, and distribution of in process materials as well as finished products. It also includes planning and control of the aforementioned processes in order to satisfy customer demand.

There are two types of logistics operations:

- *External Logistics*: External logistics deals with all transportation and related activities that occur outside a facility. For instance, the modes of transportation such as rail, ship, truck, air, and pipeline are examples of external logistics. External logistics involve the movement of materials between sender and receiver located in different geographical locations.

- *Internal Logistics*: Internal logistics involves the movement as well as storage of materials inside a facility and is referred to as material handling. This chapter will present the different aspects of internal logistics or material handling.

A typical material handling system is comprised of several components. They are listed as follows:

1. *Equipment used to transport material*: Material transport equipment is used to move work in process materials as well as finished products inside the confines of a facility such as a factory, warehouse, etc.
2. *Storage Systems*: In most manufacturing operations, material storage during process and post process is unavoidable. Material needs to be stored albeit temporarily. Equipment used for storage can be classified into two categories as follows:
    - *Conventional Storage Methods*: As a general rule of thumb, conventional storage methods tend to be labor intensive. Examples of this method include rack systems, bulk storage, and shelving and drawer storage. In this method, workers put material into storage and retrieve it from storage.
    - *Automated Storage Systems*: Automated storage systems reduce or eliminate manual labor from the material handling operation. Automated storage systems can be further classified as: 1. Automated storage and retrieval systems and 2. Carousel systems.
3. *Unitizing Equipment*: Unitizing equipment consists of the following:
    - Containers used to hold each item during the handling process: Examples of containers include pallets, boxes, barrels, drums, etc. Containers enable material movement in the form of a unit load. This can be contrasted with moving the same material as individual items that can be quite inefficient. Most types of containers can be handled using forklifts, which are commonly used in most factories, warehouses, and distribution centers. If automatic material handling and storage equipment is used to handle loads, it is important that a facility standardize on a specific size and type of container.
    - The unit load principle is universally used in material handling. A unit load can be defined as a mass that needs to be moved or handled. It could consist of one part only or a container filled with parts or a pallet containing several containers. The unit load should be as large as the particular material handling system is capable of moving and storing.

TABLE 7.1
Standard Pallet Sizes for Use in Material Handling

Depth ("X" Dimension)	Width ("Y" Dimension)
32 inches	40 inches
36 inches	48 inches
40 inches	48 inches
42 inches	42 inches
48 inches	48 inches

Source: Tompkins et al. (2003), Facilities Planning. 3rd Edition.

- Equipment used to load and pack the containers in order to facilitate material handling: This type of equipment, as the name suggests, is designed to automatically load cartons onto pallets. It can also shrink wrap plastic film around the pallets as they are being prepared for shipping. A second type of equipment, namely depalletizers, are designed to facilitate unloading cartons from pallets. Table 7.1 depicts the standard pallet sizes often used in factories and warehouses.

Table 7.2 identifies the different physical characteristics of materials in materials handling.

4. *Identification and Tracking Systems*: In order to build an efficient material handling system, it is necessary to keep track of the different materials being moved and stored. Thus, it is necessary to use some type of labeling system that is able to uniquely identify each container and/or item. Bar code identification is most commonly used for this purpose. Radio frequency identification (RFID) is rapidly growing in importance as well. The following section will describe material transport equipment in detail.

TABLE 7.2
Physical Characteristic of Materials in Materials Handling

Attribute	Description
Physical condition	Solid, liquid, gas
Physical size	Length, width, height, volume
Weight	Weight per unit volume, weight per unit
Shape characteristic	Circular, square, rectangular, long, flat, etc.
Special condition	Brittle, strong, fragile, etc.
Safe handling	Flammable, explosive, toxic, corrosive, etc.
Temperature	Hot, cold
Surface texture	Sticky, rough, dirty, etc.

7.2 Equipment Used to Transport Material

Material transport equipment can be generally classified into five different categories as follows:

- Manual and powered industrial trucks
- Automated guided vehicles
- Monorail and rail guided vehicles
- Conveyors
- Cranes and hoists

The principal features and specific applications of each type of vehicle are presented in Table 7.3.

TABLE 7.3

Different Types of Material Handling Equipment, Their Features, and Areas of Application

Material Handling Equipment	Equipment Features	Areas of Application
Manual industrial trucks	• Low cost equipment • Low rate and number of deliveries per hour	• To move light loads in a factory setting
Powered industrial trucks	• Medium cost	• To move pallet loads and palletized containers in a factory or a warehouse
Automated Guided Vehicle Systems (AGVs)	• High cost equipment • Vehicles use battery power • Allows for flexible routing • Use nonobstructive pathways	• To move pallet loads in a warehouse or factory • To move work in process (WIP) between machines in a low and medium production environment
Monorails and rail guided vehicles	• High cost equipment • Allows for flexible routing • Can be mounted on the floor or overhead to maximize space utilization	• To move products and pallet loads long variable routes inside a facility • To move large quantities of items along a fixed route
Powered conveyors	• Large variety of equipment • Versatile: Can be mounted in floor, on the floor or overhead • Mechanically powered to move loads	• To move products along an assembly line • To move items in a distribution center for sorting
Cranes and hoists	• Can lift loads in excess of 100 tons	• Moving heavy and large items in a factory and warehouse

- *Industrial Trucks*: Industrial trucks can be powered and nonpowered. Nonpowered trucks are also referred to as hand trucks and are pushed or pulled by workers. Given the obvious limit to which they can be pushed or pulled, the amount of material that can be moved as well as the distance over which it can be moved is quite limited. Hand trucks can be either two wheeled or can have multiple wheels. Two wheeled trucks are easier to maintain but can only be used to move lights loads. Multiple wheel trucks are classified as dollies and pallet trucks. Pallet trucks have two forks that are inserted through the openings in a pallet in order to lift or lower the pallet. Lifting and lowering is accomplished by engaging the lift mechanism.

Powered trucks, on the other hand, are self-propelled. There are three types of powered trucks as follows:

1. *Walkie Trucks*: Walkie trucks are operated using a battery and utilize wheeled forks that can be inserted into pallet openings. There is no provision for the worker to ride the truck. Steering mechanism consists of a control handle at the front of the truck.
2. *Forklift Rider Trucks*: Forklift rider trucks are equipped with a small cab to enable the worker to drive the truck. The load carrying capacity of forklift trucks ranges from 1000 lb. to 10,000 lbs. They can operate on internal combustion engines or electric motors.
3. *Towing Tractors*: Towing tractors for industrial use pull one or multiple trailing carts over somewhat smooth surfaces within the confines of a factory or warehouse. Towing tractors move a large amount of materials between collection and distribution centers. Similar to forklift rider trucks, towing tractors can derive their power from internal combustion engines or electric motors.
   - *Automated Guided Vehicles*: An automated guided vehicle (also designated by the acronym AGV) uses independently operated, self-propelled vehicles that move along well defined pathways. They derive their power from onboard batteries that can run for as long as 8–16 hours before being recharged. AGVs are much like rail guided vehicles and conveyor systems. The only difference is that in the case of AGVs, their pathways are not obtrusive. AGVs are typically used when materials have to be moved from different load points to a variety of upload points. Thus, AGVs are most suitable for use in batch manufacturing as well as mixed model production. AGVs can be further classified into three types as follows:
     1. *Driverless Trains*: This is a type of AGV in which a towing vehicle pulls one or more trailers just like a train. It is used to move heavy payloads over long distances within the confines

of a facility such as a factory or a warehouse. There may or may not be multiple pickup and drop off points en route.

2. *Pallet Trucks*: Within the context of AGVs, pallet trucks move palletized loads along fixed routes. A worker typically backs up the vehicle into a loaded pallet and uses the forks to slightly elevate the load. The vehicle is then driven to its predetermined guide path. The truck drives itself to its destination after the worker has programmed it in.
3. *Unit Load Carriers*: A unit load carrier moves unit loads from one station to another. Pallets can be automatically loaded and unloaded by means of different devices such as moving belts, powered rollers, mechanized lift platforms, etc. Unit load carriers can be further classified as light load AGVs and assembly line AGVs.

- *Monorails and Rail Guided Vehicles*: In the context of material handling systems, monorails and rail-guided vehicles consist of motorized vehicles that use a fixed rail as a guide. If there is one rail, the system is referred to as a monorail. In a factory or a warehouse, monorails are usually suspended from the ceilings in order to maximize space utilization. Rail systems using parallel fixed rails are usually mounted on the ground. The vehicles are driven by an on board electric motor. They pick up electric power from an electrified rail and do not carry their own power supply.
- *Conveyor Systems*: Conveyors are used to move material in relatively large quantities between prespecified locations in a facility. Conveyors can be mounted in the floor, above the floor or overhead. There are two types of conveyors: powered and nonpowered. In the case of the former, the power mechanism is contained in the fixed path. Such systems are generally used in automated material transport systems. In the case of nonpowered conveyors, on the other hand, material movement is facilitated by workers pushing the load manually along the fixed path or by gravity from a higher elevation to a lower elevation. The different types of powered conveyors are described below:

1. *Roller Conveyors*: In the case of roller conveyors, the pathway of the conveyor is a series of rollers that are arranged perpendicular to the direction of travel. The load must be large enough that the surface area spans several adjacent rollers. This type of conveyor is used to move loads such as pallets and cartons. They are used in applications related to manufacturing, assembly, packaging, sorting, and distribution.
2. *Skate Wheel Conveyors*: Skate wheel conveyors use skate wheels rotating on shafts that are in turn connected to a frame. In

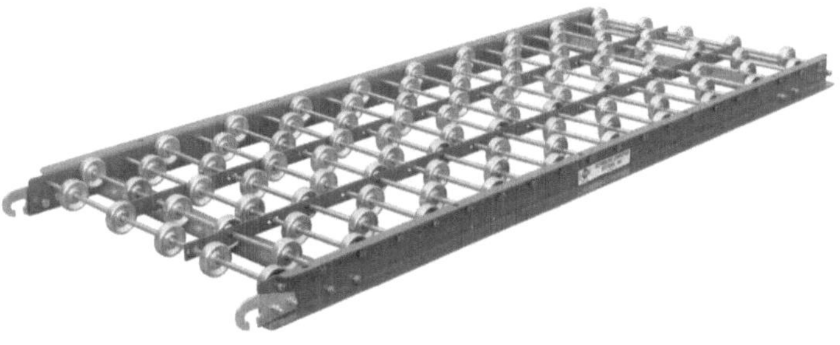

FIGURE 7.1
Skate wheel conveyor. (Automated Conveyor Systems, Inc.)

this case, the load must be lighter due to concentrated points of contact between the load and the conveyor. An example of a skate wheel conveyor is presented in Figure 7.1.

3. *Belt Conveyors*: Belt conveyors are a form of continuous loop conveyor. Half the length of the belt is used to deliver material and the other half is the return run. The belt is made of a highly elastic material that possesses a high degree of flexibility but limited extensibility. The conveyor belt is powered at one end by a drive roll. The belt is supported by a frame containing rollers along the forward loop. Belt conveyors are further classified as flat belt conveyors (for pallets) and trough belt conveyors (for bulk materials). An example of a belt conveyor is presented in Figure 7.2.

FIGURE 7.2
Belt conveyor. (Automated Conveyor Systems, Inc.)

4. *Overhead Trolley Conveyors*: In the context of material handling, a trolley can be defined as a wheeled carriage that runs on an overhead rail. Loads can be suspended from a trolley. An overhead trolley conveyor comprises of several equally spaced trolleys along a fixed track. The trolleys are connected together and moved along the track by means of a chain or a cable that runs in a continuous loop. A drive pulley pulls the chain at a constant velocity. This system is used in factories to move assemblies and subassemblies between major production departments.
5. *Cart on Track Conveyors*: A cart on track conveyor system is usually mounted a few feet above floor level. It consists of individual carts riding on a track. The cart speed is controlled by regulating the angle of contact made by the drive wheel and a spinning tube entrusted with driving the cart. When the angle made by the drive wheel is 45 degrees, the cart moves forward. On the other hand, when the axis of the drive wheel is parallel to the tube, the cart is stationary. One of the major advantages of this system is the high level of accuracy with which carts can be positioned. These systems are used for robotic spot welding lines in several manufacturing applications.

- *Cranes and Hoists*: In an industrial setting, a crane is used to enable horizontal movement of material in a facility. A hoist, on the other hand, is used to lift material in the vertical direction. Thus, a crane usually includes a hoist, which lifts the load, and the crane moves it horizontally. Cranes are capable of lifting very heavy material, up to 100 tons. Cranes can be further classified into three types as follows:

1. *Bridge Cranes*: A bridge crane assembly consists of one or two horizontal girders suspended between fixed rails on either end. Each end is connected to the building structure. The hoist trolley is moved along the length of the bridge. The bridge can be moved along the length of the rails. The girders of large bridge cranes span as much as 120 ft and can carry loads up to 10 tons. They are used for fabricating heavy machinery, steel manufacture, and in power generating stations. An example of a bridge crane is presented in Figure 7.3.
2. *Gantry Cranes*: One or two vertical legs act as a support for the horizontal bridge in the case of a gantry crane. Typically, the span of a gantry crane can extend to about 150 ft and it is capable of lifting loads up to 150 tons. By definition, a double gantry crane has two legs. In the case of a cantilever gantry crane, a bridge extends beyond the span created by the support legs, thus forming a cantilever. An example of a gantry crane is presented in Figure 7.4.

FIGURE 7.3
Bridge crane. (Beaton Industrial.)

3. *Jib Cranes*: In the case of a jib crane, a hoist is supported by a horizontal beam that is cantilevered from a wall support. The horizontal beam can rotate around the vertical support in order to provide a horizontal sweep for the crane. This makes for a circular or semicircular horizontal area of application. The hoist can raise and lower a load up to 5000 kgs. Due to limited freedom of movement, wall mounted jib cranes can achieve swings to 180 degrees whereas floor mounted jib cranes can

FIGURE 7.4
Gantry crane. (Beaton Industrial.)

FIGURE 7.5
Jib crane. (trademark-hoist.com.)

achieve a full sweep of 360 degrees. This is because the floor mounted variety is not impeded by any obstructions to its movement. An example of a jib crane is presented in Figure 7.5.

7.3 Analyzing Material Handling Systems

The preceding sections have clearly shown the important role played by material handling devices in the factory, warehouse, and other production facilities. It is, therefore, essential to analyze such systems from the cost perspective. Large material handling systems such as AGVs, material handling trucks, and cranes and hoists involve a large amount of capital expenditure upfront. We have already dealt with these topics elsewhere in this book. Capital cost estimation was dealt with in detail in Chapter 6. Breakeven and benefit cost analysis will be dealt with in Chapter 12. Other investment topics such as present value and equivalence were addressed in Chapter 1. Also, the issue of equipment replacement will be dealt with in Chapter 14. As far as the remainder of this chapter is concerned, we will analyze material handling systems in terms of the metrics such as amount of time required to transport material between two points, cycle time per delivery per vehicle, number of carriers required in order to efficiently transport a given amount of production volume, work in process, etc. It is important to determine the aforementioned metrics in light of the fact that material handling, inherently, is a non-value adding activity. Thus, any time spent in excessive or inefficient material handling often comes at the expense of value adding activities such as material processing and finishing as well as assembly. This will in turn adversely affect production volume. An accurate analysis of the capability of material handling systems is essential to maintain and enhance a company's profitability. In this section, we will analyze two types of material handling systems: Vehicle based systems and conveyor systems.

7.3.1 Vehicle Based Systems Analysis

Vehicle based material handling systems are comprised of the following types of equipment:

- Manual and powered industrial trucks
- Automated Guided Vehicles (AGVs)
- Monorails and rail guided vehicles

The "From-To chart" is commonly used to display and analyze deliveries between two points (also referred to as origination and destination points). Table 7.4 depicts an example of a "from-to chart." It will be clear from the composition of the chart that the vertical column on the left hand side of the chart represents origination points while the horizontal row at the top of the chart represents destination points. Material is moved from the origination point to the destination point.

Each material handling vehicle is assumed to operate at a constant velocity throughout the operation. Any acceleration, deceleration, and speed alterations along the way are ignored. The time to complete a delivery cycle consists of the following elements:

1. Picking up the load and loading at the pickup station
2. Amount of time required to travel to reach the destination (drop off station)

TABLE 7.4

A "From–To Chart" Depicting Flow Rates, Loads per Hour, and Distance Traveled

	To	1	2	3	4	5
	1	0	10/45	8/120	10/200	0
	2	0	0	0	0	12/75
From	3	0	0	0	5/75	8/130
	4	0	0	0	0	7/80
	5	0	0	0	0	0

Loads per hour: Value before the slash.
Distance Traveled: Value after the slash.

3. Amount of time required to unload the vehicle at the drop off station
4. Amount of time the vehicle travels while empty between deliveries

The total cycle time per delivery per vehicle is expressed as follows:

$$T_c = T_L + L_d/V_c + T_U + L_e/V_o. \qquad \text{(Equation 1)}$$

wherein:

T_c: Delivery cycle time expressed in terms of minutes per delivery

T_L: Amount of time taken to load the vehicle at load station, expressed in minutes

L_d: Average value of distance traveled by the vehicle between load and unload stations, expressed in ft

V_c: Velocity of the vehicle, expressed in ft/min

T_U: Amount of time taken to unload the vehicle at unload station, expressed in minutes

L_e: Average value of distance traveled by the vehicle when empty until the next delivery cycle starts, expressed in ft

It should be borne in mind that all delivery cycles are not identical. The origination and destination points for each trip could be different, thus affecting the amount of distance traveled by the vehicle. The delivery cycle time can be used to ascertain the following metrics of interest in material handling:

- Rate of deliveries per vehicle
- Number of vehicles required to satisfy a particular delivery assignment

In order to compute the above metrics, it is essential to take certain time losses into consideration. Thus, times losses for availability, traffic congestion, and efficiency of drivers (in case the truck is being driven manually) need to be considered. This is accomplished as follows:

- F_t: Traffic Factor. This factor deals with delays and time losses due to congestion. If there are no time losses due to congestion (such as blocking of vehicles, waiting at intersections, etc.), the value of this factor is equal to 1. The value of this factor decreases with increasing blocking. Blocking is mostly affected by the number of vehicles in the system. Thus, only one vehicle in the system would imply no blocking. This would result in a traffic factor value of 1. A higher number of vehicles in the system is likely to lead to a greater amount of blocking. For most AGV systems, the value of this factor ranges from 0.85 to 1.

In the case of industrial trucks that are operated by human beings, traffic congestion is mostly a result of operator efficiency. In this context, worker efficiency is defined as the actual work rate of an operator as compared to the work rate expected from that operator under normal working conditions. Worker efficiency is designated by the notation: E_w.

Available time is expressed by Equation 2 as follows:

$$AT = 60\ A\ F_t\ E_w. \quad \text{(Equation 2)}$$

wherein:

AT: Available time, expressed as minutes/hour per vehicle
A: Availability
F_t: Traffic factor
E_w: Worker efficiency

The rate of deliveries per vehicle is expressed using Equation 3 as follows:

$$R_{dv} = AT/T_c. \quad \text{(Equation 3)}$$

wherein:

R_{dv}: Hourly delivery rate per vehicle, expressed as deliveries per hour per vehicle
T_c: Delivery cycle time
AT: Available time in one hour, adjusted for time losses

The total number of vehicles required to satisfy a particular flow rate of total deliveries per hour is expressed by Equation 4.

$$n = WL/AT \quad \text{(Equation 4)}$$

wherein:

n: Number of vehicles required
WL: Workload, expressed in minutes per hour
AT: Available time in one hour, adjusted for time losses

The workload can be computed using Equation 5 as follows:

$$WL = R_f \times T_c \quad \text{(Equation 5)}$$

wherein:

WL: Workload, expressed in minutes per hour
R_f: Specified flow rate of deliveries per hour, expressed as deliveries/hour
T_c: Delivery cycle time, expressed as minutes/delivery

Substituting the expressions for the workload and available time in Equation 4 can further modify the equation for the number of vehicles required. The modified equation is expressed as follows:

$$n = R_f/R_{dv}. \qquad \text{(Equation 6)}$$

Remember that real world applications and the nature of load and unload demands are random. This can lead to workstations having to wait while delivery vehicles are busy making other deliveries. This can have an impact on operating cost. The equations presented here do not take this into consideration. In order to minimize station idle time, it is recommended that more vehicles be used than those obtained by using Equations 4 and 6.

In order to better understand the concepts presented in the preceding section, consider Example 7.1.

EXAMPLE 7.1

An AGVs layout is presented in Figure 7.6.

In the layout, vehicles travel counterclockwise delivering loads. Loading time at the load station is about 0.80 minutes. It takes about 0.60 minutes at the unload station to unload the vehicles. A total of 45 deliveries must be

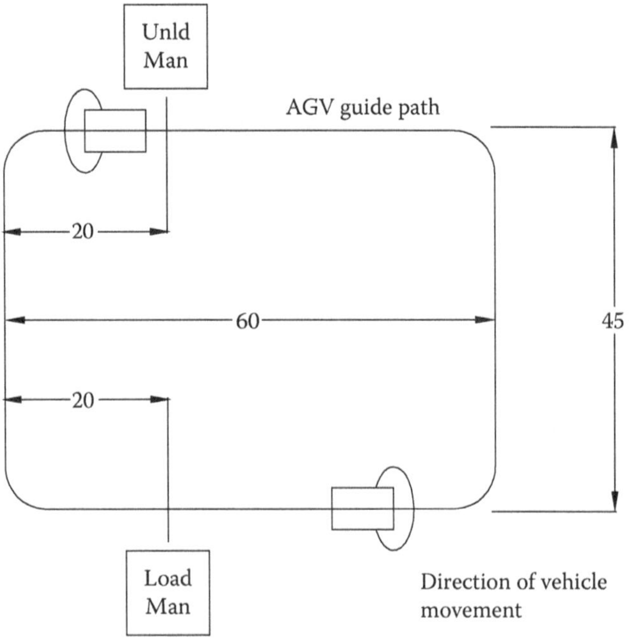

FIGURE 7.6
AGVS loop layout for Example 7.1. AGVS loop layout. Key: Unld: Unload, Man: Manual operation. All dimensions in meters.

made per hour. Vehicles travel at the speed of 60 m/min and are available 92% of the time. Traffic factor can be taken as 0.92. Since this is a fully automatic layout, the value of E_w (operator efficiency) can be taken as 1. Based on this information, answer the following questions:

1. What is the travel distance per vehicle loaded and empty?
2. What is the ideal delivery cycle time?
3. How many vehicles would be required to satisfy the demand?

Solution

From the data given in the problem, we know the values of the following parameters:

T_L: Loading time at the station = 0.80 minutes
T_U: Unloading time at the unloading station = 0.60 minutes
V_c: Vehicle velocity = 60 m/min
R_f: Specified flow rate = 45 del/hour
A: Availability = 0.92
F_t: Traffic factor = 0.92

1. From Figure 7.6, the travel distances loaded and empty can be easily read as follows:
   L_e: Distance traveled empty by the vehicle = 85 m ignoring the distance on the curves.
   L_d: Distance traveled by the vehicle between load and unload station = 125 m ignoring the distance on the curves.
2. Ideal delivery cycle time can be computed using Equation 1 as follows:

$$T_c = T_L + L_d/V_c + T_U + L_e/V_o. \qquad \text{(Equation 1)}$$

   Substituting the values of each variable into Equation 1, we obtain the following:

$$T_c = 0.80 + 125/60 + 0.60 + 85/60. \qquad \text{(Equation 7)}$$

   Thus, ideal delivery cycle time = 4.899 minutes.
3. Number of vehicles required to satisfy the demand of 45 deliveries/hour.

   In order to compute the number of vehicles required to satisfy the demand, we use Equation 4 as follows:

$$n = WL/AT. \qquad \text{(Equation 4)}$$

$$WL = R_f \times T_c. \qquad \text{(Equation 5)}$$

   Substituting the respective values into Equation 5, we obtain the value of WL as follows:

$$WL = 45 \times 5.899 = 220.455 \text{ min/hour}.$$

Also,

$$AT = 60\ A\ F_t\ E_w. \qquad \text{(Equation 2)}$$

Thus,

$$AT = 60 \times 0.92 \times 0.92 \times 1 = 50.784 \text{ min/hour per vehicle}$$

Substituting the respective values into Equation 4, we obtain the number of vehicles requires as follows:

$$n = 220.455/50.784 = 4.34.$$

Rounding up to the nearest integer, a total of five vehicles are required.

Consider Example 7.2 in order to solve a more complex problem.

EXAMPLE 7.2

Consider the AGVS layout depicted in Figure 7.7. Also consider the "from-to chart" presented in the Table 7.5. Parts are loaded onto the AGVS at station 1. They are processed at any of three stations: station 2, 3, and 4. Parts are unloaded at station 5. It takes 0.6 minutes each to load parts

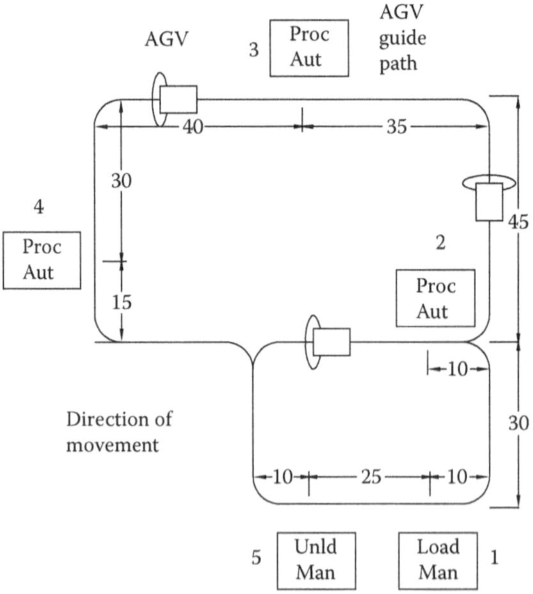

FIGURE 7.7
AGVS loop layout for Example 7.2. AGVS layout for production system. Key: Proc: Processing operation, Aut: Automated, Unld: Unload, Man: Manual operation. All dimensions in meters.

TABLE 7.5

"From–To Chart" for Example 7.2

	To	1	2	3	4	5
	1	0	10/50	8/120	10/190	0
	2	0	0	0	0	12/75
From	3	0	0	0	5/70	8/155
	4	0	0	0	0	7/85
	5	0	0	0	0	0

(at station 1) and unload the parts (at station 5). Production rates at each workstation are depicted by the delivery requirements as shown in Table 7.5. Based on these constraints, compute the average delivery distance, L_d.

Solution

In order to compute the average delivery distance, we compute a weighted average based on the number of trips as well as corresponding distances as read from the chart depicted in Table 7.5.

Thus,

$$L_d = \frac{10(50)+8(120)+10(190)+5(70)+12(75)+8(155)+7(85)}{(10+8+10+5+12+8+7)}$$

Thus,

$$L_d = 6445/60 = 107.417 \text{ m.}$$

7.3.2 Analysis of Conveyor Systems

Different types of conveyor systems used for material handling have already been discussed in a previous section of this chapter. In this section, we will analyze three types of conveyors as follows:

- Single Direction Conveyors
- Continuous Loop Conveyors
- Recirculating Conveyors

Single Direction Conveyors: In the case of single direction conveyors, the conveyor moves in one direction with a load station at one end and an unload station at the other end. Materials are loaded on to the conveyor at the load station and unloaded at the unload station. If the conveyor travels at a constant speed, the amount of time required to move materials between the two aforementioned stations can be expressed using Equation 7.

$$T_d = L_d/V_c. \qquad \text{(Equation 8)}$$

wherein:

T_d: Delivery time, expressed in minutes

L_d: Length of conveyor between load and unload stations, expressed in feet or meters

V_c: Velocity of the conveyor, expressed in feet/min or meters/min

The rate of flow of materials on the conveyors can be computed using the rate of loading at the load station. The rate of loading, in turn, is the reciprocal of the time required to load all the materials on to the conveyor. If the conveyor speed is known, the loading rate determines the spacing of materials on the conveyor. Thus, we have Equation 8 as follows:

$$R_f = R_L = V_c/S_c \leq 1/T_L. \quad \text{(Equation 9)}$$

wherein:

R_f: Flow rate of materials, expressed as parts per minute

R_L: Loading rate of parts onto the conveyor, expressed as parts per minute

S_c: Spacing of parts on the conveyor, expressed as center to center distance, measured in feet per part or meters per part

V_c: Velocity of the conveyor, expressed in feet/min or meters/min

T_L: Amount of time taken to load the parts onto the conveyor, expressed as minutes per part

An important constraint in this respect is that the time required to unload the conveyor should be equal to or less than the reciprocal of the material flow rate. Thus, we have Equation 9 as follows:

$$T_U \leq 1/R_f. \quad \text{(Equation 10)}$$

wherein:

T_U: Amount of time required to unload material from the conveyor, expressed as minutes per part

R_f: Flow rate of materials, expressed as parts per minute

Using the unit load principle, the flow rate can be computed using Equation 10 as follows:

$$R_f = \frac{n_p v_c}{s_c} \leq T_L. \quad \text{(Equation 11)}$$

wherein:

R_f: Flow rate of materials, expressed as parts per minute

n_p: Number of parts per carrier

s_c: Spacing of parts on the conveyor, expressed as center to center distance, measured in feet per part or meters per part

T_L: Amount of time taken to load the parts on to the conveyor, expressed as minutes per part

Continuous Loop Conveyors: In a continuous loop conveyor, the pathway of the conveyor is formed by a continuous chain moving in tracks along predetermined pathway. The conveyor moves parts in a carrier between the load and unload stations. The entire loop consists of two parts: A delivery loop, also referred to as the forward loop, which carries parts from the load station to the unload station and a return loop. During the return loop, the carrier travels empty. Thus, we have Equation 11 as follows:

$$L = L_d + L_e. \quad \text{(Equation 12)}$$

wherein:
- L: Total length of the conveyor, expressed in feet or meters
- L_d: Length of the delivery loop, expressed in feet or meters
- L_e: Length of the return loop, expressed in feet or meters

The total time required to travel the entire loop is given by Equation 12 as follows:

$$T_c = L/V_c. \quad \text{(Equation 13)}$$

wherein:
- T_c: Total cycle time, expressed in minutes
- V_c: Velocity of the conveyor, expressed in feet/min or meters/min
- L: Total length of the conveyor, expressed in feet or meters

The amount of time spent by a load in the forward loop is computed using Equation 13 as follows:

$$T_d = L_d/V_c. \quad \text{(Equation 14)}$$

wherein:
- T_d: Total cycle time, expressed in minutes
- V_c: Velocity of the conveyor, expressed in feet/min or meters/min
- L_d: Total length of the conveyor, expressed in feet or meters

Assuming that we are discussing a continuous loop conveyor such as an overhead trolley in which carriers are equally spaced along the chain, the number of carriers in the loop is expressed by Equation 14.

$$n_c = L/s_c. \quad \text{(Equation 15)}$$

wherein:
- n_c: Number of carriers, expressed as an integer
- L: Total length of the conveyor, expressed in feet or meters
- s_c: Spacing of parts on the conveyor, expressed as center to center distance between carriers, measured in feet per part or meters per part

Consistent with the description of the continuous conveyor, on the return trip, the carrier is empty. It holds no parts during the return trip. Thus, the maximum number of parts in the system at any given moment in time can be computed using Equation 15 as follows:

$$\text{Total number of parts in the system} = \frac{n_p n_c L_d}{L} \quad \text{(Equation 16)}$$

The maximum flow rate between load and unload stations is expressed by Equation 16.

$$R_f = \frac{n_p v_c}{s_c}. \quad \text{(Equation 17)}$$

Recirculating Conveyors: Kwo (1958, 1960) analyzed the problem of the recirculating conveyor. Based on his research, when designing a recirculating conveyor system, three rules must be adhered to. They are listed as follows:

1. *Speed Rule*: The speed rule states that the operating speed of a conveyor should be within a range determined by specific parameters. Accordingly, the lower limit of this range is constrained by the specified loading and unloading rates at respective workstations. Thus, the conveyor speed must be at least equal to these constraints. We have Equation 17 as follows:

$$\frac{n_p v_c}{s_c} \geq \text{Max}[R_L, R_U]. \quad \text{(Equation 18)}$$

wherein:

R_L: Prescribed loading rate at the loading station, expressed as parts per minute

R_U: Prescribed unloading rate at the unloading station, expressed as parts per minute

The upper limit of the speed range is constrained by the physical capacity of material handlers to perform their respective tasks. Thus, we have Equation 18 as follows:

$$\frac{v_c}{s_c} \leq \text{Min}[1/T_L, 1/T_U]. \quad \text{(Equation 19)}$$

wherein:

T_L: Amount of time taken to load the parts on o the conveyor, expressed as minutes per part

T_U: Amount of time required to unload material from the conveyor, expressed as minutes per part

2. *Capacity Constraint*: This constraint states that the flow rate capacity of the conveyor should be at least equal to the flow rate requirement that is necessary to accommodate reserve stock as well as to allow for the time elapsed between loading and unloading. The time elapsed between loading and unloading is a function of the delivery distance. The capacity constraint can be expressed using Equation 19 as follows:

$$\frac{n_p v_c}{s_c} \geq R_f. \qquad \text{(Equation 20)}$$

3. *Principle of Uniformity*: According to the principle of uniformity, parts should be distributed uniformly throughout the length of the conveyor. Thus, all sections of the conveyor will be used optimally. This will avoid unusually long waiting times at both the load and unload stations.

In order to better understand the concepts presented in this section, consider Example 7.3.

EXAMPLE 7.3

A roller conveyor follows a path that is 45 meters long between the production and assembly department of a manufacturing facility. It travels at a speed of 50 meters per minute. Two operators work at the loading station to load parts onto the conveyor. It takes 30 seconds for the first worker to load parts into a pan. Each pan can hold no more than 22 parts. The pans are loaded onto the conveyor by the second worker at the rate of 10 seconds per pan. Based on this information, answer the following questions:

1. What should be the spacing between pans on the conveyor?
2. What is the maximum possible flow rate?
3. How much time would it take to unload the pan at the assembly department?

Solution

1. In order to compute the spacing between pans on the conveyor, we use the information provided in the problem pertaining to loading time. It takes 30 seconds to load parts into the pan and an additional 10 seconds to load the pan onto the conveyor. Thus, the loading cycle is constrained by the 30-second time limit. The conveyor is traveling at a speed of 50 meters per minute. The spacing can be computed as follows:

$$s_c = (30/60) \times (50\ \text{m/min}) = 25\ \text{meters}.$$

2. In order to compute the flow rate, we use Equation 16 as follows:

$$R_f = \frac{n_p v_c}{s_c}. \qquad \text{(Equation 17)}$$

Based on the problem statement, we know the following values:
$n_p = 22$ parts
$V_c = 50$ meters/min
$s_c = 25$ meters
Substituting these values into Equation 16, we obtain the value of the flow rate as follows:

$$R_f = \frac{22 \times 50}{25} = 44 \text{ parts/min.}$$

3. The minimum amount of time required to unload a pan at the assembly department depends on the flow rate of pans on the conveyor. We know this rate to be one pan every 30 seconds. Thus the time required to unload the pan at the assembly department should be at least equal to 30 seconds.

EXAMPLE 7.4

A recirculating conveyor is 450 meters long. It travels at a speed of 90 m/min. Carriers are spaced 18 meters apart along its path. Three parts can be contained in each carrier. 0.30 minutes are required to load the parts into each carrier. The loading time is equal to the unloading time. The flow rate is 6 parts/min. Evaluate the conveyor system using the three principles outlined in this chapter.

Solution

In order to evaluate the conveyor system based on the three principles, we check the data presented in the problem against the speed rule, capacity constraint, and the uniformity principle. This is achieved as follows:

Speed Rule: The lower speed limit is computed using Equation 17 as follows:

$$\frac{n_p v_c}{s_c} \geq \text{Max}(R_L, R_U). \qquad \text{(Equation 18)}$$

Thus,

$$\frac{3\dfrac{\text{parts}}{\text{carrier}} \times 90\,\text{m/min}}{18\,\text{m/carrier}} = 15 \text{ parts/min.}$$

This is greater than the flow rate of 6 parts per minute as specified in the problem statement.

Similarly, as far as the upper limit is concerned, we have the following:

$$\frac{90\,\text{m/min}}{18\dfrac{\text{m}}{\text{carrier}}} = 5\ \text{carriers/minute.}$$

The reciprocal of the loading (and unloading) time is 3.33 minutes.

Based on this, we can observe that the speed rule has been satisfied.

Capacity Constraint: The flow rate of the conveyor is 15 parts/min. This is obviously greater than the required delivery rate of 6 parts per minute, therefore, we observe that the capacity constraint has been satisfied as well.

Uniformity Principle: As longs as the conveyor is loaded uniformly throughout its length, the uniformity principle will be satisfied. We can reasonably conclude this to be the case in light of the fact that the loading and unloading rates are the same and that the flow capacity is greater the loading and unloading rates.

7.4 Conclusion

Material handling systems are an important component of any manufacturing and production operation. They serve to move material and work in process between different workstations and departments in order to facilitate the production process. The objective of any material handling system should be to achieve its objective efficiently and in as little time as possible. This time is dependent on other factors as we have already seen in this chapter. To that end, this chapter provided a brief overview of the types of material handling systems as well as their analysis. This is important from the point of view of achieving production goals and minimizes production cost.

PROBLEMS

1. The truck your company is using is 34.1 ft long, 96.5 in wide and 81.8 in high. Your packaged product is 32 in × 26.7 in and 11.5 in high. If you were to use a pallet, what size would you use? Explain your answer, including packages per truckload.
2. A $48,000 heavy duty tractor truck has an expected life of 10 years. It is operated 300 days of the year with a daily travel average of 25,000 feet. With fuel costs nearly $12 daily and maintenance costs $1.15 an hour, what is the cost per foot for operating the truck? The operator cost is $10.72 per hour.

3. A factory produces 600 lamps daily. The dimensions of the lamp are 17.7 × 11 × 12.5 inches and it is 25.8 inches high. Management wants to have a 7-day supply in stock. 50 lamps are packed per box that creates a unit to be stacked seven high. How much floor space is required to store the supply desired by management? Allow aisles of 8%.
4. Using the same scenario in problem 3, what if 20 lamps were packed per box and stacked five high to meet management's goal of a ten day supply?
5. A robot is used for material handing in a system of three machines: N, O, and P. Q and S are input and output stations. The loading and unloading times are 0.5 minutes each. Movement takes 0.4 minutes. Processing times for N, O, and P are 7, 8, and 9 minutes, respectively. Calculate the cycle time for each system.
6. Using the scenario in problem 5, what would the cycle time be for each system if the following changes were applied?
   - Processing times for N, O, and P are 11.3, 13, and 14 minutes, respectively
   - The loading and unloading times are 1.3 minutes each
   - Movement takes 0.7 minutes
7. Three material handling units are available for moving items. The costs of these are $6000, $5400, and $4300, respectively. These units make six moves. Data on the operating cost, X, and the operating time, Y, are shown below. Which type of equipment should be used based upon the data?

Move	X1 ($)	Y1	X2 ($)	Y2	X3 ($)	Y3
1	110	1.1	100	0.8	140	1.4
2	120	1.0	130	0.6	110	0.8
3	130	0.6	M	—	90	1.1
4	140	0.7	160	0.5	120	0.6
5	160	0.4	80	0.1	100	0.3
6	90	0.2	M	—	M	—

8. The storage area at the Cone Company is filled to capacity with 45,000 items. The plant has recently increased its manufacturing capability by 120% and the finished goods inventory is expected to increase by the same ratio. The storeroom is 190′ × 345′ with a ceiling height of 40′. Due to the product line, the company does not want to stack the 15″ × 25″ × 12″ cases more than seven high without racks, which the company does not use in the area. What methods would you suggest for increasing the storage area to accommodate the desired increase in finished goods inventory area?

8

Inventory Analysis and Costing

8.1 What Is Inventory?

Inventory can be defined as an itemized list of tangible goods that are held by a manufacturing organization for a variety of reasons. The following list depicts the main reasons for the existence of inventory:

- Inventory can exist in order to support production and manufacturing processes. Raw material, subassemblies, and work in process are examples of inventories that are held in stock in order to support production operations.
- Inventory can exist in order to support other activities such as repair and maintenance. Thus, all tangible goods that support repair and maintenance activities such as tools, spare parts, lubricants, related equipment, etc., can be regarded as inventory.
- Inventory can exist so it can be sold to the customer. Thus, finished goods constitute inventory since they can be sold to the customer.

Carrying inventory is an expensive proposition. That is why a significant number of companies seek to control inventory levels. Why? It helps them:

- *Minimize the Inventory Carrying Cost*: Inventory carrying costs can be minimized by reducing the inventory on hand as much as possible. In the most extreme case, this level can be reduced to zero and new inventory can be ordered as and how the need arises.
- *Serve Customers to the best of their ability, thus maximizing customer satisfaction*: Customer satisfaction can be maximized by holding a large amount of inventory on hand. This would facilitate customer orders to be filled in quickly and efficiently, thus leading to a positive customer experience.

Inventory carrying costs can be classified into three categories as follows:

- Investment costs
- Storage costs
- Costs incurred due to obsolescence or spoilage of existing inventory

Collectively, the aforementioned costs are also referred to as carrying costs or holding costs. Of the three costs, investment costs constitute the largest element of carrying costs. Investment cost consists of interest that must be paid on borrowed money. Money is borrowed in order to invest in materials and supplies. These are processed in the factory in order to fill a customer order. It should be borne in mind that some time always elapses between start of production and fulfillment of the order. The interest that must be paid on borrowed money during this time is referred to as investment cost. We have dealt with this concept in Chapter 1 and throughout this book.

Minimizing inventory constitutes one of the more obvious ways in which inventory carrying costs can be minimized. However, inventory minimization can often occur at the expense of customer satisfaction. Customer orders could take longer to be filled due to low inventory levels. This dissatisfaction could lead to loss of business which could prove detrimental to a company's long term prospects. Thus, the question is bound to arise: "How do I minimize inventory carrying costs and still satisfy the customer?" In order to solve this dilemma, the concept of order point inventory systems is used when dealing with independent demand items. This concept is introduced in the following section.

8.2 Order Point Inventory Systems

Order point inventory systems are applicable to managing inventories of independent demand items. They seek to answer the following questions:

Q1: How many units of the item should be ordered?

Q2: When should we place the order for the units so as to replenish our inventory?

The number of items to be ordered to replenish depleted inventory usually corresponds to the value of a metric known as economic order quantity (EOQ). The answer to the second question is arrived at by using reorder point methods. Both of these concepts will be addressed at length in this section.

Economic Order Quantity (EOQ): Economic order quantity corresponds to the most appropriate quantity that must be ordered or produced when an item is

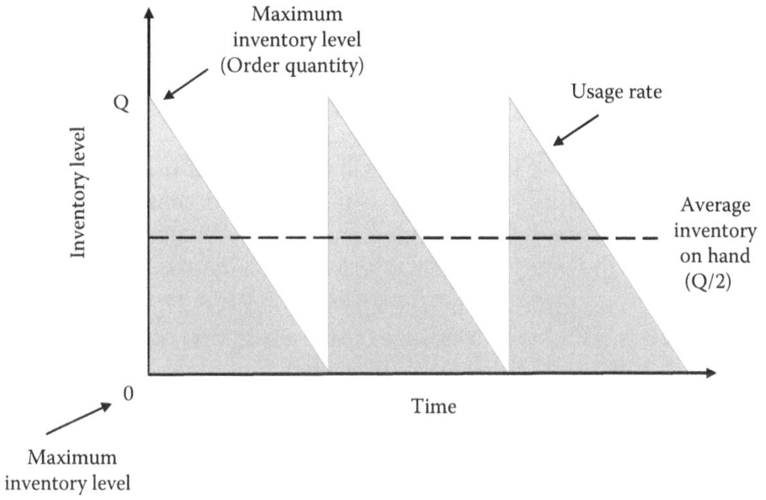

FIGURE 8.1
Typical sawtooth pattern used to depict inventory reduction over time and compute EOQ.

demanded as a fairly constant rate. Also, it is assumed that the production rate for said item is significantly greater than the rate at which it is demanded. Such a scenario is also referred to as the "make to stock" scenario. When an item is demanded at a steady rate, the inventory of goods on hand tends to gradually reduce over time. It is then replenished to a predetermined level by placing an order equal to the economic order quantity. This replenishment occurs almost instantaneously when the orders of items are received. The saw tooth pattern that results as a consequence of gradual inventory reduction over a period of time followed by a sudden replenishment is depicted in Figure 8.1.

The total inventory cost can be computed by adding the carrying cost and setup cost for the inventory model depicted in Figure 8.1. It will be observed from Figure 8.1 that the average level of inventory is actually equal to the half the maximum level (Q). The total annual inventory cost is expressed by Equation 1 as follows:

$$TIC = \frac{Q \cdot C_h}{2} + \frac{C_{su} D_a}{Q}. \qquad \text{(Equation 1)}$$

wherein:

TIC: Total annual inventory cost, expressed in \$/year

Q: Order quantity, expressed in number of units/order

C_h: Carrying cost of holding cost, expressed in \$

C_{su}: Setup cost or ordering cost for an order, expressed as \$/setup or \$/order

D_a: Annual demand for the item

The ratio: D_a/Q corresponds to the number of orders or batches that are produced annually. It therefore corresponds to the number of setups that must be accomplished every year.

The Holding cost (C_h) is comprised of two elements:

- *Investment Cost*: The investment cost is a result of funds invested in the inventory before it is sold to customers. The inventory investment cost is ascertained by multiplying the inventory cost by the interest rate paid by the company. The interest rate is expressed as "i."
- *Storage Cost*: Tangible goods take up storage space. This space must be paid for. It is obvious that storage cost is a function of the part size as well as the volume of space occupied by it. For all practical purposes, the storage cost is computed by multiplying the inventory cost by the storage rate, "s."

The storage rate and interest rate can be combined into one factor as follows:

$$h = i + s. \qquad \text{(Equation 2)}$$

wherein:

h: Holding cost rate, expressed as a fraction
i: Interest rate paid on borrowed money, expressed as a percentage
s: Storage rate, expressed as a fraction

Thus, holding cost is expressed by Equation 3 as follows:

$$C_h = h \cdot C_{pc}. \qquad \text{(Equation 3)}$$

wherein:

C_h: Carrying cost of holding cost, expressed in \$
C_{pc}: Cost of the item per unit, expressed as \$/unit
h: Holding cost rate, expressed as a fraction

Next, we compute the setup cost. This is calculated as the cost of idle production equipment during the changeover time that occurs between batches. This relationship is expressed by Equation 4 as follows.

$$C_{su} = T_{su} \cdot C_{dt}. \qquad \text{(Equation 4)}$$

wherein:

C_{su}: Setup cost or ordering cost for an order, expressed as \$/setup or \$/order
T_{su}: Setup time or changeover time between batches, expressed as hours/setup or hours/order
C_{dt}: Cost rate of downtime of the machine during the changeover process, expressed a \$/hour

If parts are ordered from an external vendor, the quoted price includes the setup cost either directly or in the form of quantity discounts. Next, we compute the annual cost of part production. This is expressed by Equation 5 as follows:

$$TC = D_a \cdot C_{pc} + \frac{Q \cdot C_h}{2} + \frac{C_{su} D_a}{Q}. \qquad \text{(Equation 5)}$$

wherein:

$D_a \cdot C_{pc}$: Product of annual demand and cost per item
TC: Total cost, expressed as $

In order to minimize the total cost, we take the first derivative of Equation 5 and set it equal to zero. The equation is solved for the unknown value of "Q." "Q" refers to the economic order quantity when the total cost has been minimized. The expression for EOQ is presented in Equation 6 as follows:

$$Q = EOQ = \sqrt{\frac{2 \cdot D_a C_{su}}{C_h}}. \qquad \text{(Equation 6)}$$

wherein:

EOQ: Economic Order Quantity, expressed as number of parts or items to be produced per batch

In order to better understand the concept of economic order quantity, consider Example 8.1.

EXAMPLE 8.1

The annual demand for a certain made-to-stock item is 25,000 pieces per year. Each unit of the item costs $35. The holding rate is taken to be 20% per year. It takes 7 hours to setup the equipment in order to produce a batch of pieces. The cost of equipment downtime plus labor is $200/hour. What is the economic order quantity and total inventory cost?

Solution

Based on the data given in the problem, we can compute the values of the following variables:

Set up cost, $C_{su} = 7 \text{ hours} \times \$200/\text{hour} = \$1400$

Holding cost per unit $= 0.20 \times \$35 = \$7$

We can compute the value of economic order quantity by using Equation 6 as follows:

$$Q = EOQ = \sqrt{\frac{2 \cdot D_a C_{su}}{C_h}}. \qquad \text{(Equation 6)}$$

Thus,

$$\text{EOQ} = \sqrt{\frac{2 \times 25{,}000 \times \$1400}{\$7}} = 3162.27\,\text{units} = 3163\,\text{units}.$$

Next, we compute the total inventory cost using Equation 1 as follows:

$$\text{TIC} = \frac{Q \cdot C_h}{2} + \frac{C_{su} D_a}{Q}. \qquad \text{(Equation 1)}$$

Thus,

$$\text{TIC} = (\$7).$$

$$(3163)/2 + \$1400\ (25{,}000/3163) = \$11{,}070.5 + \$11{,}065.44 = \$22{,}135.94.$$

If actual production costs were included in the annual cost computation, we would have the following:

$$\text{Total cost} = 25{,}000.\ (35) + \$22{,}135.94 = \$897{,}135.94.$$

In a production environment, the formula to calculate economic order quantity has been widely used in order to determine the optimal batch size. However, in practice, the EOQ value cannot always be determined easily. This is because it is sometimes difficult to obtain accurate values for the parameters that are used to compute the EOQ. For instance, it is difficult to obtain accurate values of metrics such as setup cost and inventory carrying cost. In a production environment, these metrics are comparatively difficult to evaluate. Their impact on the value of the economic order quantity is obvious from our preceding discussion.

The importance of the total inventory cost (TIC) should always be borne in mind when computing the value of EOQ. This is because EOQ is derived from TIC. By definition, the economic order quantity is the batch size that must be ordered so as to minimize total inventory cost. We have already witnessed this principle in the derivation of the EOQ formula. A careful examination of the equation for total inventory cost will reveal that reduction in cost and batch size can be achieved by adopting the following measures:

- Decreasing the value of holding cost
- Decreasing the setup cost

As we have already discussed, the holding cost is largely determined by the prevalent interest rate. As such, it is beyond the scope of influence of the production manager. However, reducing setup cost is definitely an achievable goal. Setup cost can be reduced by reducing the amount of time necessary to change over production equipment. Reducing change over time is one of the necessary ingredients for achieving just in time production. This method of production will be discussed in detail in the next section.

Reorder Point Systems: We have discussed the concept of inventories and related costs as length in this chapter. We have also been introduced to the

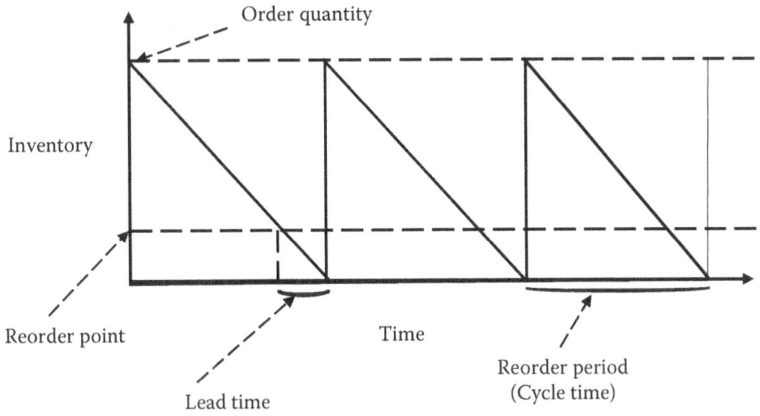

FIGURE 8.2
Reorder point and reorder lead time.

concept of economic order quantity. However, EOQ constitutes only one part of the puzzle. It informs the production manager how much inventory should be ordered so as to replenish the stock of items. The question still remains: "When should this order be placed?" The reorder point system is a method commonly used in order to answer this question. The exact point in time when a reorder to replenish inventory should be placed is difficult to ascertain in the case of variable demand rates. It can be determined with a fair degree of precision if the demand rate was a constant value though. In the case of reorder point systems, an order to restock the item is placed when the inventory held in stock for that item falls to a prespecified level. This reorder level is high enough to minimize the probability of a possible stock out during the time between when the reorder point is reached and the new stock is received. Computerized inventory control systems can be used to continuously monitor inventory levels and can be used to automatically trigger an order equal to EOQ the moment that inventory falls to or below the reorder point. The concept of reorder point and reorder lead time is depicted in Figure 8.2. In the next section, we will investigate the concept of just in time manufacturing.

8.3 Just In Time (JIT) Production Systems

The concept of just in time production was developed with the express purpose of minimizing inventories. The philosophy of JIT is widely used in the Toyota production system. Thus, it especially emphasizes minimizing work in process, also referred to as WIP. This is because excessive work in process in considered equivalent to waste in the Toyota production system. In its ideal state, the JIT system delivers the exact number of each component to a downstream manufacturing station at precisely the point in time when it is required. This adherence to discipline and commitment to minimizing waste also minimizes work in process and manufacturing lead time as well as floor space and funds invested in work in process. The JIT system can be extended to the company's suppliers in addition to being universally adopted within the company itself. This is precisely what happened at Toyota.

It is worth noting that not only Toyota but many manufacturers have embraced the JIT philosophy not only in the US, but globally as well. In the US, the term "continuous flow manufacturing" is often substituted for JIT. There are some subtle differences between JIT and continuous flow manufacturing however. In continuous flow manufacturing, parts are processed and transported to the next workstation in sequence a single unit at a time. In order to accomplish this, it is necessary that each preceding process concludes just before the beginning of the following process. In essence this is the JIT philosophy so that except in the case of continuous flow production, the batch size is equal to one unit.

The efficiency of the just in time manufacturing philosophy and the Toyota production system as a whole is quite legendary. It is highly effective in high volume repetitive operations. This is a very important point to bear in mind. WIP accumulation can happen very quickly in such operations due to the sheer volume of products manufactured as well as the large number of components present in every product assembly. The JIT philosophy uses the following principles in order to minimize inventories:

1. Use the pull system of production instead of the push system
2. Reduce the setup time for smaller batch sizes
3. Develop stable and reliable production operations

Each of these three principles will be addressed in detail in the following paragraphs.

8.3.1 JIT Uses Pull System of Production Control

Most part production consists of a sequence of operations. For instance, production of a shaft could involve different processes such as turning, milling, drilling, and thread cutting. Not all operations can be performed at a single workstation. Turning can only be performed on a lathe (turning center) and milling can be performed on a milling machine. In the case of a pull type production system, the order to produce and deliver parts at a particular workstation is issued by a downstream workstation that will use that part. The order to replenish the supply is issued by the workstation when its inventory of parts starts to run low. The upstream workstation can only start producing parts upon receipt of this order. This system rigorously implemented throughout the production line produces the effect of "pulling" parts towards the destination. Thus, parts are produced only when required, not otherwise. This results in waiting times that are short or nonexistent, thereby enhancing the efficiency of the system.

The pull type system can be contrasted with the push type system. In the case of the push type system, parts are produced at each workstation regardless of the specific need of downstream workstations. Thus, parts are effectively "pushed" through the system. Materials requirement planning (MRP) is an example of a push system of production. The main disadvantage of a push type production system is that the factory often gets overwhelmed by the volume of parts produced. This can result in long queues of parts at every machine and a situation in which machines are unable to cope with the demand.

In the Toyota production system, the "pull" is implemented by means of kanbans. The work "kanban" is a Japanese word for signal card. The kanban cards authorize the following actions:

1. Part production at the workstation
2. Parts delivery in the plant

Thus, there are two types of kanban cards: production kanbans (to authorize production) and transport kanbans (to authorize parts delivery). The production kanban is also referred to as the P-kanban. The transport kanban is referred to as the T-kanban. After the parts have been produced, they are placed in containers so as to just fill the container. Production in excess of this predetermined quantity is not allowed in the kanban system. As its name implies, the transport kanban authorizes delivery of parts to the downstream station. Accuracy of shop floor data can be enhanced by implementing modern bar code technology. This can also result in reduced transaction times.

In order to better understand the functioning of a kanban system, assume two workstations: "N" (upstream station) and "N + 1" (downstream station) in a sequence of workstations. Work flows from the upstream station to the downstream station. The following is a typical sequence of steps followed in a pull-kanban system:

1. The next p-kanban is removed from the dispatching rack by station "N + 1." This kanban, as we have already seen, authorizes station "N + 1" to process a container of part "X." Simultaneously, an incoming container of parts arrives at station "N + 1." A material handling worker removes a T-kanban from this container and takes it back to station "N."

2. At station "N," the material handling worker locates the container of part "X," removes the P-kanban, and replaces it with the T-kanban. The P-kanban is put in the dispatching rack at station "N."
3. The T-kanban put on the container of parts "X" authorizes the delivery of this container to station "N + 1." This is accomplished and the container is moved accordingly. Recall the P-kanban that was put in the dispatching rack at station "N." It authorizes this station to process a new container of part "X." However, it has to wait in line in the rack for the other P-kanbans ahead of it to be processed. Work is scheduled at each workstation based on the order of the P-kanbans. During this time, processing of parts at the downstream station "N + 1" has been accomplished. It removes the next P-kanban from the dispatching rack and begins to process that container of parts.

Notice that in the above production sequence, paperwork is reduced to a minimum. The kanban cards are used recursively. Some amount of manual labor is involved in transporting the cards and containers, but it is supposed to be good for employee morale and builds camaraderie. In the next section, we will examine setup time reduction for smaller batch sizes in a JIT system.

A comparison between the pull system and push system of inventory control is presented in Table 8.1.

8.3.2 Reduce Setup Time for Smaller Batch Sizes

It is essential to minimize batch sizes and setup times in order to minimize work in process inventories in a manufacturing environment. Notice from the formula for total inventory cost (Equation 1), average inventory level is equal to half the batch size. It is clear from this relationship that in order to

TABLE 8.1

A Comparison between the Pull System and Push System of Inventory Control

Push System of Inventory Control	Pull System of Inventory Control
• Tries to produce as much as possible • Approximates production numbers to meet its goals • Incorporates large lot sizes • Involves high levels of inventory • Results in high amount of waste • Involves poor communication between internal customers and vendors at best and no communication at worst • Manages inventory by resorting to firefighting	• Tries to produce only what is needed when it is needed • Requires precise production numbers and precision in production • Incorporates small lot sizes • Involves low levels of inventory • Accomplishes waste reduction by producing only what is needed when it is needed • The entire philosophy of the pull system of inventory control is based on a high degree of good communication between internal customer and internal vendor • Manages inventory by sight

reduce the average inventory level, batch size must also be reduced. In order to achieve this objective, setup cost must be reduced as well. This can only be accomplished by reducing setup time. Smaller setup times allow for smaller batches and less work in processing inventory.

Setup time reduction can be achieved by implementing methods improvement. It can be applied to almost any batch production scenario. Machine setup is comprised of two distinct types of work elements as follows:

1. *Internal Elements*: Internal work elements can only be performed when the machine is stopped.
2. *External Elements*: External work elements are external to the machine and can be performed even when the machine is in operation. External elements do not require the machine to be stopped. Thus, external elements can be performed while the machine is still processing the previous job.

Setup time can be reduced in a particular change over by designing the setup tooling (such as fixtures, molds, dies, etc.) and planning the changeover procedure in such a way that it consists of as many external elements as possible. However, it should also be remembered that the internal elements need to be prioritized. This is because the nonproductive time of the machine during changeover is largely dictated by the internal elements. Table 8.2 presents some examples of internal work elements and external work elements.

Internal elements in a setup can be shortened in time duration by resorting to some of the following approaches:

- Time studies, motion studies, and methods improvements should be utilized in order to minimize the time of internal work elements.

TABLE 8.2

A Comparison of Internal Work Elements and External Work Elements Typically Occurring During a Production Setup

Internal Work Elements	External Work Elements
Removal of dies, fixture, molds, and other tooling used in the previous production operation	Obtain the tooling for the next production job from the storage room
Positioning and affixing the tooling for the next production job in sequence	Components of the tooling for the next job are assembled next to the machine prior to being attached to the machine for the next production run
Fine tuning adjustments and alignments of the tooling	Reading and comprehending engineering drawings for new setup
Manufacturing trial parts and trying out the tooling prior to starting the actual production run	Downloading the part program and reprogramming the machine for the next job

- Internal elements can be considerably shortened by resorting to two workers working in parallel instead of just one worker performing the task alone. In theory, the downtime can be reduced by half since the job can be accomplished twice as fast. Remember, though, that this solution is not universally applicable to all changeover scenarios.
- Adjustments, by their very nature, consume time and should be avoided or minimized as much as possible.
- Fasteners such as nuts, bolts, and screws consume valuable time during the fastening and unfastening process. Setup time can be reduced by resorting to quick acting fasteners instead.
- If the application calls for bolts and nuts to be used and no other solution will do, U-shaped washers should be used instead of O-shaped washers. This is because U-shaped washers allow for easy insertion between parts with only partial disassembly. Contrast this to an O-shaped washer in which the entire nut and bolt assembly needs to be disassembled completely.
- Try to design using modular fixtures as much as possible. They consist of a base plate and insert tooling that can be quickly changed for each new part. Insert tooling is the only component that needs to be changed in such fixtures. The base plate remains attached to the machine.

In general, the following approaches can be used to reduce setup times in production operations:

- Study the different types of problems that typically cause delays in setups. Focus resources on developing permanent solutions to such problems.
- The amount of change required in a setup can be minimized by scheduling batches of similar style of parts sequentially.
- Along this vein, group technology and cellular manufacturing can be used in order to produce similar part styles on the same equipment. This will reduce the amount of work to be performed during changeover operations. A schematic of a group technology layout is depicted in Figure 8.3.

Table 8.3 presents some terms commonly used to describe levels of improvement.

We have discussed the concept of just in time manufacturing and its advantages and operation in detail in the preceding sections. Let us examine the impact of setup time reduction in a manufacturing operation. This can be accomplished by observing its effect on economic order quantity. Consider Example 8.2.

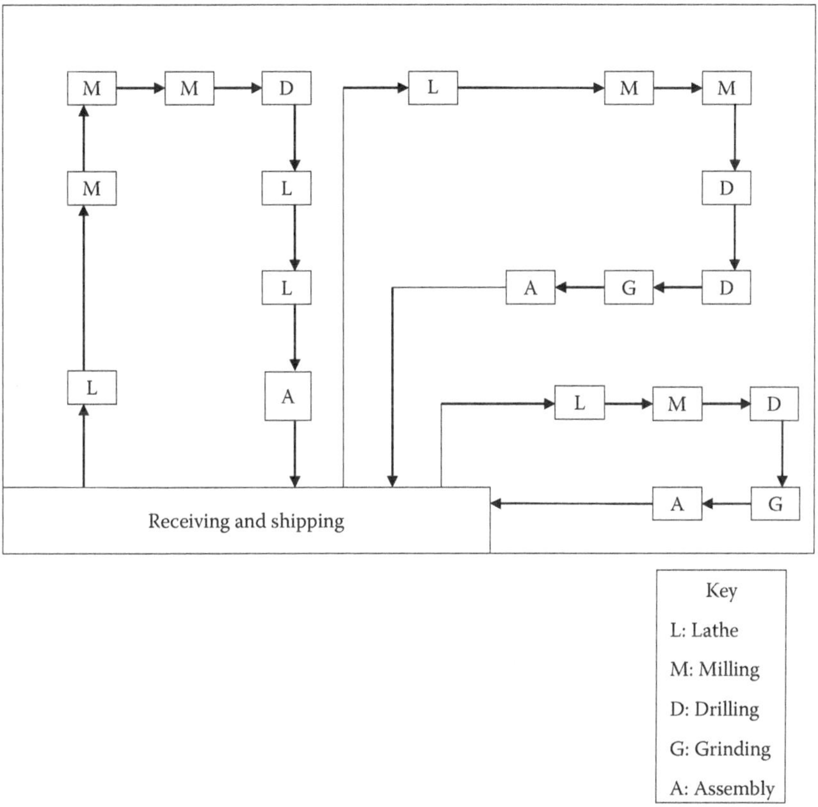

FIGURE 8.3
An example of a group technology layout.

TABLE 8.3

Examples of Terms Commonly Used in Setup Time Reduction

Term	Description
SMED	Single Minute Exchange of Dies. Dies should be changed over in less than 10 minutes.
OTED	One Touch Exchange of Dies. Dies should be setup within one minute.
NOTED	NOn Touch Exchange of Dies. Dies should be exchanged without any human intervention. The changeover process is accomplished automatically. Automatic tool changers are an example of this type of exchange.
RETAD	Rapid Exchange of Tools and Dies. A general term for various procedures designed to accomplish setup time reduction.

EXAMPLE 8.2

Consider the data presented in Example 8.1. This is given as follows:

The annual demand for a certain made-to-stock item is 25,000 pieces per year. Each unit of the item costs \$35. The holding rate is taken to be 20% per year. It takes 7 hours to setup the equipment in order to produce a batch of pieces. The cost of equipment downtime plus labor is \$200/hour.

Now suppose that the setup time was actually reduced from 7 hours to 15 minutes. What is the economic order quantity and total inventory cost?

Solution

Based on the data given in the original problem, we computed the value of EOQ as follows:

Set up cost, C_{su} = 7 hours × \$200/hour = \$1400.

Holding cost per unit = 0.20 × \$35 = \$7.

We can compute the value of economic order quantity by using Equation 6 and substituting values of different variables as follows:

$$\text{EOQ} = \sqrt{\frac{2 \times 25{,}000 \times \$1400}{\$7}} = 3162.27\,\text{units} = 3163\,\text{units}.$$

The total inventory cost corresponding to this EOQ was calculated as follows:

$$\text{TIC} = \frac{Q \cdot C_h}{2} + \frac{C_{su}\,D_a}{Q}$$

Substituting respective variables in the equation, we computed this cost as \$22,135.94.

Now, let us consider the changed scenario, wherein setup time is reduced from 7 hours to 15 minutes. This changes the underlying variables significantly as follows:

Setup cost, C_{su} = (15/60 hours) × (\$200/hour) = \$50.

The holding cost remains unchanged. Substituting the new value into the equation for economic order quantity, we obtain the following:

$$\text{EOQ} = \sqrt{\frac{2 \times 25{,}000 \times \$50}{\$7}} = 597.61\,\text{units} = 598\,\text{units}.$$

Notice the reduction in EOQ, from as high as 3163 units (with a setup time of 7 hours) to as low as 598 units (with a setup time of 15 minutes). This constitutes an 81% reduction in EOQ.

The corresponding total inventory costs are calculated as follows:

$$\text{TIC} = \{(598) \times (\$7) \times (0.5)\} + \{(\$50) \times (25{,}000)/598\} = \$4183.30$$

This constitutes a reduction of 81.1% in total inventory costs compared to the earlier value of \$22,135.94 (with a 7 hour setup time).

8.3.3 Development of Stable and Reliable Production Operations

Stable and reliable production operations can be developed using the following strategies:

1. Production leveling
2. Ensuring on time delivery
3. Defect free components and materials
4. Reliable production equipment
5. Capable, cooperative, and committed workforce and a reliable base of suppliers

In this section, we will briefly address each of the aforementioned components.

1. *Production Leveling*: Production leveling implies a smooth flow of production that will deviate as little as possible from its fixed schedule. It should be borne in mind that any disturbances or changes in downstream operations are magnified in upstream operations. Suppose that the demand in assembly operations (downstream operation) was to change by as little as 5%. This can lead to as much as a 25% change in production of parts (an upstream operation). This ripple effect in upstream operations can occur because of factors such as overtime, variations from regular work procedures, and unscheduled setups that are necessary to meet downstream demand. It is important to maintain a constant master production schedule over time in order to ensure smooth workflow. The only problem with maintaining a smooth workflow is that demand is not constant. It fluctuates. It is imperative for the production system to adjust to these fluctuations. This adjustment can be accomplished using production leveling. It involves distribution of changes in product mix as well as the volume of production over a period of time. The following procedures enable production to be leveled:
    - Production leveling can be achieved if overtime is scheduled during busy periods.
    - Using inventories of finished products to absorb daily fluctuations in demand can help level production.
    - Cycle times of production operations can be adjusted based on the fluctuations in demand, thus helping to level production.
    - Production in small batch sizes enables easy absorption of demand fluctuations. Batch size reduction is facilitated by techniques that reduce setup time. The ideal situation consists of a batch size of one unit.

Production leveling results in greater agility in terms of the degree of responsiveness to demand fluctuations. It can also lead to shorter lead times and reduced WIP inventory as well as inventory of finished goods and a regular workload assigned to workers engaged in production.

2. *Defect Free Components and Materials, On Time Delivery, and Reliable Production Equipment*: Just in time production is enabled by many factors working together. Each factor is a cog in the wheel. Thus, in order to ensure that inventory is received only when required, it is important to ensure defect free components and materials and reliable production equipment in addition to on time delivery. Since this discussion pertains to just in time manufacturing, on time delivery is almost implied. We have already examined the importance of small lot sizes in the context of just in time systems. Due to small lot sizes, it is necessary for parts to be delivered before any stock outs can occur at downstream stations. If stock outs were to occur, it would starve downtime stations for work, thereby putting a halt to any further production.

   The importance of defect free components and materials in the JIT system can hardly be overemphasized. It is obvious that if defective parts were produced at the upstream production station, they would be unfit to be used at the downstream station. This is especially true if the downstream station engaged in parts assembly. Such being the case, production can be severely impacted. In the worst case situation, it could even stop. It is easy to comprehend the importance of defect free products, especially in the light of the small lot sizes used in JIT systems. This makes it imperative to control quality during production instead of relying upon inspectors to inspect quality post production.

   A high degree of reliability of production equipment is necessary for a JIT system to function effectively. JIT is characterized by low work in process. This means that machines cannot breakdown at any time during the shift. It is important that machines and equipment be designed for reliable operation to ensure smooth functioning of JIT systems. Total productive maintenance practices are used to ensure equipment reliability and minimize or even eliminate equipment downtime.

3. *Capable, Cooperative, and Committed Workforce and a Reliable Base of Suppliers*: The role of capable, cooperative, and committed workers in a JIT system is of paramount importance. The system is unable to function without this all important element. Cross training of workers is an important aspect of the JIT system. This is because small lot sizes often require the workers to work on a variety of tasks in order to produce a correspondingly large variety of parts. Inspectors and production workers should work together at each

workstation to ensure the quality of parts being processed during the production process itself. They should also be able to fix minor technical problems that could occur from time to time in order to ensure smooth functioning of the equipment.

Additionally, it is important that raw material suppliers be on board with the JIT philosophy as well. They must be held to the same standards as the company in question. Specific policies must be crafted in order to ensure incoming material quality. Some examples of such policies are presented below:

- The number of suppliers can be reduced. This allows for the chosen suppliers to do more business, thus incentivizing them to be on board with the JIT philosophy.
- A long lasting relationship with suppliers can be established by entering into long term partnerships and agreements with them. This would keep the suppliers from having to competitively bid for each contract.
- Suppliers should be selected based on their capability to meet quality and delivery standards. This implies that they must be able to produce the requisite quality of products and be able to deliver them within the requisite amount of time.
- The company's employees could be stationed in the suppliers' facilities in order to train suppliers to develop JIT systems for their own production.
- Transportation and delivery costs can be minimized by choosing suppliers that are located close to the company's final assembly plant.

8.4 Conclusion

In this chapter, we dealt with concepts related to inventory analysis and costing. The central role played by inventory in a production operation was presented. The costing of inventory was discussed in detail. Ways and means of reducing and minimizing inventory and work in process were analyzed as well. Towards this end, the concept of just in time manufacturing systems was introduced. The important role played by setup time in reducing inventory was discussed and analyzed using actual examples. The two elements of equipment setup: internal elements and external elements were also discussed. Two types of kanbans (signal cards): P-type and T-type kanbans to facilitate JIT production were introduced as well. Also, the key role played by production leveling in the successful operation of a JIT system was examined.

PROBLEMS

1. The annual demand for a certain automotive part is 3000 units per year. The part is produced using a batch model manufacturing system. Annual holding cost per piece is $4.00. It takes 3 hours to set up the machine to produce the part, and the cost of system downtime is $350/hr. Determine (a) the most economical batch quantity for this part and (b) the associated total inventory cost. *(Answer: a. 1255 units; b. $5020/year).*
2. Demand for a certain product is 35,000 units/year. Unit cost is $15.00. Holding cost rate is 35%/yr. Changeover (setup) time between products is 12 hours, and downtime cost during changeover is $250/hr. Determine (a) economic order quantity, (b) total inventory costs, and (c) total inventory cost per year as a proportion of total production costs. *(Answer: 6325 units; b. $33,203.91/year).*
3. A part is produced in batches size of 6000 pieces. Annual demand is 100,000 pieces, and piece cost is $9.00. Setup time to run a batch is 5 hours, cost of downtime on the affected equipment is estimated at $300/hour, and annual holding cost rate is 35%. What would the annual savings be if the product were produced in the economic order quantity? *(Answer: $3709.15/year).*
4. A certain machine tool is used to produce several components for one assembled product. To keep in-process inventories low, a batch size of 200 units is produced for each component. Demand for the product is 5000 units per year. Production downtime costs an estimated at $250/hr. All parts produced on the machine tool are approximately equal in value: $10.00/unit. Holding cost rate is 20% per year. In how many minutes must the changeover between batches be accomplished so that 100 units is the economic order quantity? *(Answer: 1.92 minutes).*
5. Annual demand for a certain part is 30,000 units. At present the setup time on the machine tool that makes this part is 12 hours. Cost of downtime on this machine is $500/hr. Annual holding cost per part is $3.50. Determine (i) EOQ and ii) total inventory costs for this data. Also, determine (iii) EOQ and (iv) total inventory costs if the changeover time could be reduced to six minutes. *(Answer: i. 10,142 units; ii. $35,496.48; iii. 926 units; iv. $3240.75).*
6. A manufacturing plant supplies sheet metal body panels to an automotive final assembly plant. The following data are values representative of the parts made at the plant.

   Annual demand is 200,000 pieces. Average cost per piece is $20 and holding cost is 15% of piece cost. The setup time for the presses is 5 hours and the cost of downtime on any given press is $300/hr. Determine (a) the economic batch size and (b) the total annual inventory cost for the data. If the changeover time for the presses could be reduced to 6 min, determine (c) the economic batch size and (d) the total annual inventory cost.

7. A supplier of parts to an assembly plant is required to make deliveries on a just in time basis daily. The daily demand for one of the parts is 200 parts, five days/week, and 52 weeks/yr. The supplier cannot risk making just 200 parts per day because of the high cost of changing over the production machine. Instead, they must produce in larger batch sizes and maintain an inventory of the parts from which 200 units are withdrawn for shipment every day. Cost per piece is $35 and holding cost is 30% of piece cost. Changeover time for the production machine used to produce the part is 3.5 hours and the cost of downtime on this machine is $270/hr. Compute (a) the economic batch size and (b) the total annual inventory cost for the data. (c) The number of weeks of demand represented by this batch quantity.
8. The following data apply to sheet metal parts produced at a stamping plant that serves a final assembly plant in the automotive industry. Annual demand for each part produced is 250,000 pieces. The average cost per piece is $30. The holding cost is 25% and the setup time for the presses is 7 hours. The cost of downtime on any press is $300/hr. (a) Compute the economic batch size and the total annual inventory cost for the data. (b) If the changeover time could be reduced to 45 min, compute the economic batch size and the total annual inventory cost.
9. A part is produced in batches of size 6300 pieces. Annual demand is 141,000 pieces, while the cost is $9.75 per piece. Setup time to run a batch is 6 hours, the cost of downtime on the relevant equipment is $924 per hour, and annual holding cost rate is 37%. What would the annual savings be if the product were produced in the economic order quantity?

Use the following information to answer the questions below.

Two machines produce the same type of bistro coffee mug. Machine A originally cost $160,000 and Machine B originally cost $320,000. Machine A has a maximum capacity of 30,000 mugs annually, which could be produced in batches of 4000, 6000, and 8000 annually. Machine B produces 50,000 mugs annually that could be produced in batches of 4000, 6000, 8000, and 10000 annually. The fixed cost for Machine A is $24,000 per year, while it is $62,000 per year for Machine B. The following table shows the costs associated with each machine's level of output.

Variable Cost/Unit	0 to 4000 Units	4000 to 6000 Units	6000 to 8000 Units	8000 to 10,000 Units
Machine A	$20/Unit	$22/Unit	$14/Unit	—
Machine B	$23/Unit	$21/Unit	$17/Unit	$10/Unit

10. If 8000 units are produced, plot the profit-volume graph and find the break even point for a price of \$22.75 per unit.
11. If 10,000 units are produced, plot the profit-volume graph and find the break even point for a price of \$18 per unit.
12. If 16,000 mugs are produced, plot the profit-volume graph and find the break even point for a price of \$11.72 per unit.
13. Indicate how the production should be assigned to each machine for the following levels of total output:
    a. 1500 Units
    b. 2000 Units
    c. 4500 Units
    d. 6000 Units
    e. 10,000 Units
    f. 12,500 Units
14. A landscaper has a requirement for potting soil in the amount of 78,000 bags per year. The soil costs \$4.95 per bag and costs \$30 to process a purchase order. The holding cost is \$10 per bag annually. Find the minimum cost purchase quantity.
15. A tool is used to produce several components for a medical device. To keep in process inventories low, a batch size of 300 units is produced for each component. Demand for the product is 4000 units per year. Production downtime costs an estimated at \$450 per hour. All parts produced on the machine tool are approximately equal in value: \$19.22 per unit. Holding cost rate is 45% annually. How long would the changeover be between batches to fulfill an order of 400 units?

9

Administrative and Facilities Costs

9.1 Introduction

In the course of conducting business, certain costs tend to be accrued that are not direct costs. We have dealt with material cost in Chapter 3 of this book and process costing in Chapter 4. We have also dealt with different types of overhead expenses in Chapter 5. In this chapter, we will address costs that support the entire organization: Administration and facilities costs. First, we will consider administrative costs.

Administrative costs can be regarded as general expenses that are related to the operation of the overall administration of an organization. These expenses are accrued in order to achieve common objectives. As such, they cannot be identified with a particular project. These costs may be regarded as both personnel and nonpersonnel costs. Administrative costs are associated with performing administrative and coordinating functions such as the following:

- Accounting, budgeting, performing financial and cash management functions. This consists of the following elements:
  1. *Inventory Valuation*: This represents the full cost of inventory at the end of an accounting period. Different accounting standards require that a correct value be placed on inventory. Inventory can be valued using different methods of valuation such as standard cost, FIFO (First In, First Out) and LIFO (Last In, First Out) methods. It should be noted that inventory valuation is of little use in the day-to-day operation of a manufacturing organization.
  2. *Valuation of Cost of Goods Sold*: This accounting element is closely related to inventory valuation. It consists of ascertaining the cost of particular production jobs or the cost of all units produced. The latter can be computed using process costing. Costs can be tracked using direct costing or it can be calculated by allocating factory overhead costs.
  3. *Analysis of Constraints*: During this step, any bottlenecks in the manufacturing process can be detected. This analysis includes

an examination of the inventory buffer as well as existence of any sprint capacity.

4. *Analysis of Margin*: This element consists of analyzing the profit margin of a product. This is achieved by first compiling all the relevant costs that are accrued during the manufacturing process. The total cost is subtracted from the total revenue in order to compute the profit margin. Margin analysis is also applicable to business units, distribution channels, customers, and product lines. However, many businesses are beginning to realize that incorporation of allocated costs into margin analysis can lead to wrong decisions with regards to sales volume of a product. Thus, this function is beginning to give way to constraint analysis. The most important decision to be made in this regard is to ascertain the most profitable product mix that needs to be produced.
5. *Analysis of Variance*: Analysis of variance consists of comparing actual costs to standard costs or costs that have been budgeted. This is an important step because the underlying budgeted cost itself could be incorrect. In such cases, a favorable variance implies that the underlying standard was set up in a way that was too easy to attain. As such, all variances based on such a standard are bound to be favorable.
6. *Budgeting*: This step is ultimately the responsibility of the production manager. It consists of compiling the information obtained from the preceding five steps and using it for budgeting purposes for manufacturing.

- Costs associated with performing procurement and purchasing functions:

  From the perspective of administrative costs, procurement consists of acquiring goods, services, and work that is essential to an organization. Procurement can be defined as the process of selecting vendors, establishment of terms of payment, strategic scrutiny of vendors, vendor selection, contract negotiation, and the actual purchase of goods.

  Procurement can be closely linked with the very identity of a company. For instance, if an organization is building an identity around the principle of environmental sustainability, then its procurement practices should be reflective of this. In such cases, materials should be sourced from companies having a similar ethic.

  The process of purchasing goods and services is referred to as the "Procure to Pay Cycle." The following steps are involved in the procurement process:

  1. Identification of requirement
  2. Purchase request authorization

3. Purchase request approval
4. Procurement
5. Identification of suppliers
6. Selection of the appropriate vendor
7. Acknowledgement of the purchase order
8. Notice of advance shipment
9. Receipt of goods
10. Recording of the invoice
11. Performing the three-way match
12. Payment to supplier

The process of purchasing can be considered to be a subset of procurement. In most cases, it just implies buying of goods and services. It also involves receiving and payment. The following steps are related to purchasing in particular:

1. Acknowledgement of the purchase order
2. Notice of advance shipment
3. Receipt of goods
4. Recording of invoice
5. Performing the three-way match
6. Payment to supplier

- Costs accrued as a result of managing property:

  In the production context, this can include management of the factory, warehouses, office building, etc. This can also include facilities occupancy cost such as rent, utilities, taxes, and maintenance.

- Costs associated with managing personnel in an organization.
- Costs associated with performance of payroll functions:

  A company that employs people has to account for payroll and fringe benefits. Payroll and benefits include the following items:

  1. *Salaries and Wages*: Salaries are defined as compensation paid to white collar workers such as managers, professionals, office employees, and executives. Gross amount refers to the employee compensation prior to withholdings for taxes, savings plans, etc. Wages, on the other hand, refers to compensation paid to production and manufacturing workers, also referred to as blue collar workers. This compensation is usually stated as a gross hourly figure. Wages are received either weekly or bi-weekly. The total wage earned can be computed by multiplying the employee's hourly rate by the number of hours worked.

2. *Bonuses and Commissions paid to employees*: Bonuses and commissions paid to employees are considered to be a part of salaries.
3. *Overtime pay paid to employees*: The normal working week is considered to be forty hours. Any time worked in excess of forty hours per week is referred to as overtime. Overtime compensation depends on the rate of pay for the employee as well as their job responsibilities. For instance, executives are considered to be exempt from earning overtime pay because their compensation is already high and they can control the number of hours they work. Compare this situation to that of a technician who earns $20,000 per year. Such an employee is not exempt from earning overtime pay. If he works for an executive who decides to work 50 hours per week, he would need to earn overtime compensation of the 10 extra hours he worked for his immediate superior.
4. Payroll Taxes and Associated Costs such as Social Security, Medicare, Federal Income Tax, State Income Tax, State Unemployment Tax, Federal Unemployment Tax, and Worker compensation insurance: Payroll withholdings include the following items:
    i. *Employee portion of Social Security Tax*: Social Security tax is a major component of FICA. The other component is Medicare tax. The employer is required to pay a social security tax in addition to the tax deducted from the employee's salary and wages. As of the writing of this book (2017), the employer is required to withhold 6.2% of the first $127,200 of the employee's annual salary and wages as social security tax. Any amount over that threshold is not subject to social security tax withholding. The amount withheld plus the employer's portion is reported as current liability until they are paid to the government by the employer.
    ii. *Employee portion of Medicare Tax*: Medicare tax is withheld from employees' salaries and wages. The employer is required to pay a Medicare tax as well. Thus, the employer is required to remit to the federal government the employee as well as the employer share of Medicare tax. This means that this particular tax is counted as an employee withholding and an employer expense.

       An employer is required to withhold 1.45% of the annual wages and salary of each employee for Medicare tax. This percentage is applied on the employees' total salary, no matter how large it may be.

       In addition, there is a 0.9% Medicare surtax that is withheld from the employee on wages and salaries that are in excess of certain predetermined amounts.

iii. *Federal Income Tax*: Federal income tax is required to be withheld by the employer in the amount that the employee is expected to owe. This number is based on the wages and salaries of employees. The amount that is withheld is rarely the exact amount that the employee owes the government. The exact amount owed to the government will depend on the employee's income tax at the end of the year. This implies that the employee might end up paying a bit more in taxes or could receive a tax refund.

Federal form W-4 indicates the federal income tax withheld by the employer. This number is based on the employee's salaries or wages, marital status, and the number of dependents claimed as exemptions.

Federal income tax withholdings are reported on the balance sheet of the employer as a current liability. The liability is reduced when the employer pays the amounts to the federal government.

iv. *State Income Tax*: In most states, the employer is required to withhold an amount equal to the state income tax that the employee is expected to owe to the federal government. This tax, just like the federal income tax, is based on the salaries and wages of the employees. Once again, the amount of money withheld is rarely the exact number that the employee is likely to owe to the government. Consequently, at the end of the year some employees might expect to pay a bit more tax while others could expect to get a tax refund based on their individual income situation.

State Income Tax withholdings are reported on the balance sheet of the employer as a current liability. The liability is reduced when the employer pays the amounts to the state government.

v. *State Unemployment Tax*: Unemployment services are administered by individual state governments. They determine the state unemployment tax for each employer. States require that each employer pay the entire unemployment tax. In order to accomplish this, employers build up a large reserve in the unemployment fund of the state in order to have lower unemployment tax rates. On the other hand, employers who have only a small reserve or no reserve at all will have to pay higher unemployment tax rates.

In order to compute the state unemployment tax, the corresponding tax rate is applied to a wage base. This is determined by each state. Consider a state with an unemployment wage base of $14,000. The state unemployment tax rate is applied only to the first $14,000 of an employee's annual salary and wages. If the employer's state unemployment tax rate is 5%, then the maximum unemployment tax cost for the employer would be $700 per year for each employee (This is calculated at 5% of $14,000 = $700).

vi. *Federal Unemployment Tax*: The state unemployment programs are overseen by the federal government. The federal government requires employers to pay a federal unemployment tax of 6%. A credit is given to the employer if they have paid into the state unemployment fund. The maximum credit is 5.4%. In such cases, the federal unemployment tax rate would be 0.6%. This rate is applied to the first $7000 in annual salaries and wages of each employee.

vii. *Worker Compensation Insurance*: Employees who are injured on the job are covered by worker compensation insurance. This is especially relevant in a production and manufacturing environment. Worker compensation insurance rates are a function of at least three variables: Type of industry, type of job being performed, and the employer's history of claims.

Based on industry statistics, a production worker in a meat packing plant is at a greater risk of suffering job related cuts and back injuries. For this reason, worker compensation insurance rates for employees in such industries can be as high as 15%. Contrast this with the situation of an office employee working in the office of the same meat packing plant. Such employees may have rates as low is 1% of salaries and wages. It is to be noted that worker compensation insurance is a significant expense on the part of the employer and thereby a crucial element of payroll accounting.

viii. *Net Pay*: Net pay is defined as the amount of money that remains after all the withholdings have been deducted from the gross pay of an employee. This is also referred to as "take-home pay." This is the amount of cash directly payable to the employee on payday.

5. Employee paid benefits such as holidays, vacations, sick days, insurance (including health insurance, dental insurance, vision insurance, life insurance, and disability insurance), retirement plans, and profit sharing plans.

    i. As a result of escalating costs of group health, dental, vision, life, and disability insurance, most companies require employees to pay a share of the cost of premium. This amount is collected by means of employee directed payroll withholding.

ii. Many companies require employees to contribute company money into an employees' savings program or profit sharing plan. In such cases, the contribution appears as an expense in the period during which the company contribution was earned by the employee. The company could also have the expense and the liability before the amount is actually paid by the company. In such cases, the company is required to record an adjusting entry in order to match the expense to the appropriate accounting period.

iii. Sometimes, companies provide pensions to its employees. This implies that employees will receive monthly payments on an ongoing basis from the company after they retire. The cost of the benefit should be recognized during the years that the employees are actually earning the benefit (by working) and not while the employee has retired.

iv. Post-Retirement Health Insurance: Health insurance is provided by some companies to its employees after they retire. This retiree benefit is part of the compensation package earned by employees while they are working. Based on the matching principle and the principle of accrual accounting, the cost of future insurance coverage is required to be expensed during the productive years of the employees (the years when the employees are working). This is accomplished by debiting an expense and crediting a liability. After the employee retires, the company's payment for insurance will reduce the company's liability as well as its cash.

6. Payroll and fringe benefits are reported on financial statements using the matching principle. According to the matching principle, a company is required to match expenses to the accounting period in which revenues are reported. If such a connection does not exist, the expense should be included on the income statement for the accounting period in which it was incurred. This is important to bear in mind since expenses are usually incurred and paid in different accounting periods. For instance, an expense could be incurred (occurred) in Q3 of 2016 but is only paid in Q4 of that year.

- Costs associated with performance of audits.
- Costs necessary to coordinate the resolution of findings that arise as a result of audits, reviews, and investigations as well as incident reports that are bound arise from time to time during the life of an organization.

- Costs associated with performance of general legal functions.
- Costs accrued in the course of developing systems and procedures: This includes development of information systems required for administrative functions. For instance, costs associated with the purchase, systems development, and operation of systems pertaining to administrative functions such as personnel, procurement, purchasing, property management, accounting, and payroll systems are considered administrative costs.
- Costs accrued as a result of preparation of administrative reports.
- Costs associated with general liability insurance that protects the organization as a whole.
- Costs associated with paying administrative salaries and wages: In addition, applicable fringe benefits may also be categorized as administrative costs. For instance, the salary paid to an executive director can be considered to be an example of administrative salaries and wages. Thus, oversight and monitoring responsibilities related to administrative functions can be considered to be an example of administrative costs.
- The cost of goods and services used for administrative functions is also an example of administrative costs. This includes goods and services such as rental or purchase of equipment, utilities, office supplies, postage, and rental and maintenance of office space.
- Travel costs incurred in the conduct of official business in carrying out administrative activities and the overall management of the organization are also categorized as administrative costs.

It will be obvious from the ongoing discussion that administrative costs are nontechnical charges necessary to operate an organization. They are vital to the overall success of the organization and generally contribute to organizational efficiency. It should be borne in mind that centralized organizations generally have higher administrative expenses that their decentralized counterparts. In general, administrative costs tend to be about 10%–25% of the direct factory labor rate. The responsibility for these costs tend to be distributed within a company. As a result, they do not lend themselves as easily to cost reduction strategies as other categories of cost.

Administrative expenses can be classified as fixed expenses. This is because they would exist irrespective of the level of sales. Thus, they are independent of the sales volume. They are frequently aggregated with general expenses. Administrative expenses are usually listed on the income statement below the cost of goods sold (COGS) and maybe depicted as an

TABLE 9.1
An Example of an Administrative Expense Budget

	Quarter 1	Quarter 2	Quarter 3	Quarter 4
Advertising	$10,000	$10,000	$20,000	$20,000
Insurance	$4000	$4000	$4000	$4000
Payroll taxes	$3000	$3200	$3400	$3600
Rent	$16,000	$16,000	$16,000	$16,000
Salaries	$40,000	$42,000	$44,000	$46,000
Supplies	$3000	$3000	$3000	$3000
Travel and entertainment	$5000	$5000	$5000	$5000
Utilities	$6000	$6000	$6000	$6000
Other expenses	$3000	$3000	$3000	$3000
Total expenses	$90,000	$92,200	$104,400	$106,600

aggregate with other expenses such as general expenses or selling expenses. For tax purposes, administrative expenses that are reasonable, ordinary, and necessary for business operations can be deducted on a company's corporate income tax return. However, they must be deducted in the year they were used. Furthermore, they must have been used in the course of business. An example of an administrative expense budget is depicted in Table 9.1.

9.2 Administrative Cost Ratios

Important ratios from the point of view of administrative costs will be presented in this section.

1. *Administrative Expense Ratio*: The expense ratio is computed to depict the relationship between an individual expense and sales. It is calculated by dividing a particular type of expense by net sales. It is expressed as a percentage. Thus, the administrative expense ratio is calculated by dividing the administrative expense by net sales. The formula for computing this ratio is expressed in Equation 1. Net sales is computed as the difference between gross sales and returns. This ratio allows analysts to measure asset utilization. In other words, it enables analysts to comprehend the level of overhead costs required to support a particular level of revenue. As a company achieves economies of scale, this ratio should decrease in value.

$$\text{Administrative expense ratio} = \frac{\text{Administrative expense}}{\text{Net sales}} \times 100.$$

(Equation 1)

Consider Example 9.1 in order to understand this concept better.

EXAMPLE 9.1

The following information has been extracted from the income statement of Alpha, Inc. Compute the administrative expense ratio based on this information.

Description	Amount ($)
Net sales	1,500,000
Cost of goods sold	900,000
Administrative expenses	75,000
Sales expenses	95,000

Solution

The administrative expense ratio can be computed using Equation 1 as follows:

$$\text{Administrative expense ratio} = \frac{\text{Administrative expense}}{\text{Net sales}} \times 100.$$

(Equation 1)

Substituting the respective values into this equation, we obtain the following:

$$\text{Administrative expense ratio} = \frac{\$75,000}{\$1,500,000} \times 100.$$

Thus, Administrative expense ratio = 4% of net sales. **Solution**

9.3 How to Reduce Administrative Costs

We have observed from the preceding section that the costs of running a company are quite substantial. Administrative cost reduction is almost effective as cutting direct material costs and labor costs. Cutting administrative expenses is likely to result in reducing the value of revenue that needs to be earned in order to break even. This can lead to greater flexibility in long term pricing strategy, improvement in cash flow, and a greater probability of achieving profitability. Administrative costs can be reduced by adopting the following strategies:

- *Renting Facilities instead of Purchasing Them*: Whether production facilities should be purchased or rented is largely based on the scale of the underlying operations. Ownership results in increased fixed costs and higher degree of financial exposure. Monthly rental agreements, on the other hand, cost a bit more in the short term while the ability to end the agreement and relocate to new premises tends to result in significant savings and reduced liability in the long run. Financial exposure is generally limited when facilities are rented.

- *Curbing Travel and Entertainment Expenses*: In order to limit unnecessary travel and entertainment expenses, it is essential to establish an entertainment policy that fits the overall state of the economy.
- *Using Technology to "Telecommute"*: Technology does away with the requirement to have administrative personnel and specialists together in one physical location. Employees who work from home not only eliminate the high cost of office space but the demand for support services as well. As a matter of fact, in some cases, employees who are offered the freedom to work from home will actually accept a lower salary due to the lack of commuting costs.
- *Subleasing Office Space*: Excess, unused office space is defined as office space that is unlikely to be used for over a year. Subleasing this space will afford the opportunity to earn extra income. Some of this income can be used to offset the expense of renting or paying for other facilities. It should be borne in mind before agreeing to a sublet that the space will not be required by the company during the time that such agreement is in place. Explicit language should be included in the contract that allows for early termination just in case the need to use the space in question does arise. This ability to reoccupy said space allows for maximum flexibility, which is reflected in the agreement in the form of reduced rent or payment of a penalty in the case of early termination.
- *Refinancing Existing Debt*: This step is especially important from the point of view of a production and manufacturing enterprise. Long term debt on real estate, equipment, and machinery should be refinanced during times when the interest rates are low. We have already studied the concept of interest rate and time value of money in Chapter 1 of this book.
- *Elimination of "Paper"*: Most office environments are replete with paper. This paper is used for documentation purposes. The process of managing all this paper is time consuming and, therefore, expensive. The process of managing paper involves secretarial functions such as sorting, filing, and finding files. Investment in a scanner and digitization of documents enables them to be stored in well-organized electronic files. This leads to significant savings in space and administrative costs. It is very important to bear in mind that all files need to be backed up every day. This is especially true of any files that contain customers' identification, financial records, and contact information.
- *Sharing Marketing Expenses*: Almost all companies can identify products and services that are complimentary to their own. Manufacturers of such products and services could be contacted

and a marketing arrangement could be negotiated for mutual benefit. This strategy results in expanding a company's marketing "reach" without incurring the additional expense of hiring more sales staff and advertising and marketing personnel.

- *In-House Maintenance of Equipment*: In a production environment, it is likely that equipment will break down periodically. If repair and maintenance is a significant portion to total cost, it is better to hire a mechanic who would perform maintenance of equipment in house. This will result in better equipment that has been regularly maintained, thereby doing away with the need to replace expensive equipment.
- *Keeping Track of Tools and Supplies*: Tools and supplies can be classified as consumable items. In most cases, consumables are an overlooked asset because they are almost invisible. This results in employees misplacing them or even taking them home for their own personal use. In such cases, consumables have to be replaced frequently, thereby leading to increased expenditure and rising costs. In order to avoid this scenario, a consumables policy can be adopted. Such a policy should require that all office supplies be kept in a secure area. They should be available only to designated employees. A similar policy should be enacted in the factory where hand tools and other expensive consumables are commonly used. Tools should be assigned to individual workers. Additionally, each worker should be held personally responsible for the safety of the equipment. They should be held personally responsible for the cost of the equipment or tool if it is found to be lost or missing.

In the next section of this chapter, we will address the topic of facilities cost. In particular, the issues related to facilities construction, facilities maintenance, and water supply will be dealt with.

9.4 Facilities Costs

Facilities are a very important element in the production and manufacturing of goods. Business facilities constitute the physical location and structure where a business is located. The most common types of facilities include office buildings, warehouses, and factories. Physical structures can be new or old. It is obvious that it costs money to maintain and run a facility. This can be defined as facilities cost. According to Generally Accepted Accounting Principles (GAAP), facilities expenses are reported on the income statement of a production operation.

A facility can either be purchased or rented. If the facility is purchased, the mortgage interest that is paid on the money borrowed to pay for the facility is deductible. The depreciation (the loss in value of the property due to a variety of factors, this will be discussed in detail in Chapter 13) is a deductible amount on the income statement as well. On the other hand, if a facility was rented, the actual amounts paid are listed on the income statement. This would include the rent for the land, office buildings, warehouses, and storage facilities as well as factories. If the office furniture, fixtures, and equipment is also rented, those expenses are also listed on the income statement.

In general, facilities operating expenses are classified into three categories as follows:

- *Project costs*: Project costs are a direct consequence of new construction projects as well as renovation projects. Related costs include architectural costs, engineering expenses, construction costs, and cost of furniture and fixtures. Estimates are routinely developed before a major construction project is embarked upon. Construction cost estimates are usually concerned with estimating the cost of roads leading up to the facility, the actual building, and supporting structures. Consequently, a greater deal of emphasis is placed upon variations in size as well as the quantity of construction material that is likely to be required. Examples of such estimates include "cubic yards of concrete," "number of bricks," "amount of earth to be moved," "dimensions of structural member," etc.

  Construction cost estimates are classified as approximate estimates and detailed estimates. Approximate estimates are used to assess the impact a design change can make on the potential costs of a project. Detailed estimates, on the other hand, are obtained by individually estimating the cost of materials, equipment, direct labor, and overhead using accurate numbers pertaining to each of the aforementioned quantities.

  There are two types of detailed estimates:

  1. *Lump Sum Estimate*: In the lump sum estimate, a total cost figure is submitted for the purpose of bid consideration.
  2. *Unit Cost Estimate*: Unit cost estimates are generally made on the basis of cost per unit length, area or volume. Examples of unit cost estimates include cost of pavement, earth work, gutters, etc.

     Estimating construction cost involves estimating equipment and labor costs as a function of the amount of material that needs to be moved and installed. In the field of construction, direct labor times are obtained from historical records. In manufacturing industries, standard time is often used to estimate direct labor cost. There is a dearth of this type of data in the construction industry. Direct labor times are reflective of operator skills and equipment capacities. Direct labor time is converted into direct labor cost by considering the hourly wage of direct labor employees and the rates charged for equipment to recover operating costs, maintenance, and annual depreciation. Construction cost estimates contain the following elements:

1. *Material Cost*: For the purposes of a construction cost estimate, material cost is defined as the cost of all construction material that is used during construction. For instance, cost of concrete, wood, glass, shingles, electrical, and plumbing supplies, etc., are examples of material cost. In order to estimate material cost, estimates and quotes are obtained from prospective vendors. Delivery cost for construction materials are much higher than those for materials used in the manufacturing industry. This necessitates the separate consideration of these costs and the corresponding cost estimate should be obtained from the vendor. Material as a share of total cost is much higher in the construction industry than for the manufacturing industry. Thus, the accuracy of the corresponding estimates is even more important.
2. *Direct Labor Cost*: Direct labor cost is defined as the cost incurred due to payment of salaries and wages to construction workers. Examples of direct labor cost in the construction industry include hourly wages paid to carpenters, bricklayers, plumbers, electricians, etc. The hourly rates include cost of welfare benefits, retirement pensions, vacations, employee insurance, and so on. Estimates of direct labor cost are obtained from historical records of past jobs. The cost is estimated as a function of the material to be installed and moved during the construction operation. Hourly rates vary by skill level. They are higher for jobs requiring a higher skill level and vice versa.
3. *Equipment Cost*: Most construction processes require the use of some equipment to facilitate the process. Equipment costs are estimated in order to recover the cost associated with use, operation, and maintenance of all equipment used during construction. If heavy special purpose equipment such as earthmovers and cranes are rented, the rental fees are expressed as an estimate line item. This practice is commonly adopted by most contractors in the construction industry.

   If, on the other hand, the equipment is owned, the cost of use is denoted as the depreciation cost of the equipment. We will deal with the topic of depreciation at length in Chapter 13. Depreciation costs need to be distributed on an hourly basis. In order to accomplish this, a consideration needs to be given to the number of hours the equipment would be used annually. The amount of depreciation to be recovered during the year is a function of the taxable life of the equipment. Even if the utilization of the equipment is low, the depreciation amount remains fixed. If low utilization is expected, usage costs must be increased accordingly to ensure that annual depreciation will be covered over a period of twelve months.

4. *Overhead Cost*: Overhead costs in the construction industry are defined as any cost that can be charged to a project, which cannot be classified as materials cost, equipment cost or direct labor cost. Overhead costs are further classified into two categories: Main office cost and job site cost. The main office overhead cost is indicative of the cost of operating a centrally located office facility. The central office is typically away from the job site. Examples of such costs include telephone expenses, rent, utilities, office supplies, etc. Main office overhead generally ranges from 6% to 15% of the business volume of a construction contractor.

   Job site overhead consists of a variety of costs such as labor cost for field personnel, cost of temporary job facilities, trucking, bonds and insurance, etc. Labor costs for superintendents, foremen, engineers, and watchmen are considered to be field personnel costs. These costs are estimated to be about 20%–40% of direct payroll costs. The cost of tool shed, job site offices, and portable toilets are examples of temporary facilities costs. They also include rental or depreciation charged plus the costs to install and dismantle the aforementioned facilities, cost of temporary water, heat, electricity, roads, etc.

   For the purpose of taxes and insurance at the job site, a variety of overhead costs are included such as social security tax, unemployment compensation tax, workmen's compensation and employer's liability insurance, public liability and property damage insurance, and fringe benefits.

   - *Infrastructure costs*: Infrastructure costs are incurred as a result of increased demand. For instance, increased demand could result in the need for an expanded parking lot. Costs incurred in the expansion project can be considered to be infrastructure costs.
   - *Ongoing Costs*: Ongoing facilities costs include the cost of facilities maintenance and repair, utilities, landscaping of the grounds, and insurance premiums as well as property taxes.

Individual facilities expenses and their total amounts are transferred to the income statement. They are listed along with the other operating expenses.

In order to better understand the concepts related to facilities cost, consider Example 9.2.

EXAMPLE 9.2

An ironworker works 9 hours/day, 5 days per week. A base wage of $22 per hour is paid for all work performed during an eight hour day, five days per week. An overtime rate of time and one-half is paid for all work performed over eight hours per day. Double overtime is paid for work on Saturdays. The social security tax is 7.65% and the unemployment tax is 3% of actual wages. The rate of worker's compensation insurance is $12.50 per $100 of base wage. Public liability and property damage insurance rate is $3.25 per $100 of base wages. Fringe benefits are $3.25 per hour. Compute the average cost per hour to hire the ironworker.

Solution

Number of actual hours worked = 9 hours/day × 5 days/week = 45 hours.
Pay hours = weekly straight time + weekly overtime + Saturday overtime.
Thus, pay hours = (5 days × 8 hours/day @1) + (5 days × 1 hour/day @1.5).
This implies that pay hours = 40 hours + 7.5 hours = 47.5 hours.
Base wage rate is given as $22 per hour.

$$\begin{aligned}\text{Average hourly pay} &= (\text{pay hours}/\text{actual hours}) \times \text{base wage} \\ &= (47.5\,\text{hours}/40\,\text{hours}) \times \$22/\text{hour} = \$26.125/\text{hour}\end{aligned}$$

CALCULATION OF BENEFITS

Average hourly pay:	$26.125
Social security tax: 7.65% of $26.125	$1998
Unemployment tax: 3% of $26.125	$0.78375
Worker's compensation: ($12.50/$100) × $22	$2.75
Public liability/property damage: ($3.25/$100) × $22	$0.715
Fringe benefits are $3.25 per hour	$3.25
Total average cost per hour	$35.62175

Thus, it would cost $35.62175 per hour to hire the ironworker. **Solution**

9.5 Conclusion

Important concepts related to administrative and facilities costs were introduced in this chapter. Administrative costs are crucial to the smooth functioning of the organization as a whole. Facilities costs have to be incurred in order to construct, manage, and maintain a facility such as a factory, assembly plant, warehouse or an office building. These costs have to be incurred in the course of conducting business. As such, they are indispensable. However, there are ways and means to reduce administrative costs. Some of these solutions were discussed in this chapter. The goal of any production manager should be to reduce administrative costs in such a way that said reduction does not affect the production process adversely.

10

Establishing the Selling Price

10.1 The Relationship between Selling Price and Profitability

Profitability is the lifeblood of any business enterprise. In mathematical terms, profit is defined as the difference between the selling price of a product and the total cost incurred in manufacturing, marketing, and selling that product. This implies that the revenue generated by business activities must exceed all the costs that were incurred in undertaking those activities. Cost estimation as well as profitability is necessary for determining the economic advantage of any business. The economic advantage of a business determines the ability of a company to be competitive and retain its competitive edge in the future. A squeeze on profit, from inaccurate cost estimation, escalating costs, increased market competition or excessive supply, sets the stage for financial failure. It can result in withdrawing new products from the market a short time after they are introduced or abandoning projects regardless of inspiration.

The role of a product's profitability in determining its success cannot be overly exaggerated. It should be noted that outside of recovering the cost of business activities, profits must be available in sufficient size in order to pay taxes (local, state, and national), dividends to stockholders, interest on borrowed capital, research and development funding, reinvestment in upkeep and modernization, and investment in exploring other options.

Profitability of a product is a function of the cost structure. The development of realistic estimates of the costs of an activity will result in appropriate allocation of resources and improved chances of project completion. It will also result in fewer partially completed activities that will be cancelled for lack of funds. Furthermore, activities that have already been started will be more likely to be seen through to successful completion. The profitability and most efficient work output are two major requirements of accurate cost estimation. A business that produces high cost estimates and bases its price on those estimates is more likely to fail, as it becomes less competitive in the marketplace. By the same token, a business that produces low cost estimates will become unprofitable and fail as well.

10.2 Cost and Price Structure

The basic cost and price structure is depicted in Figure 10.1. The sum of direct labor cost and the direct materials cost is known as the prime cost. The direct labor cost is defined as the cost of actual labor used to actually produce the product. The direct materials cost include the cost of raw materials or semi-finished materials that can be directly attributed to the product. We have already examined the role played by direct material costs and direct labor costs in Chapter 2. Direct material cost was examined in further detail in Chapter 3. Process costing was discussed in detail in Chapter 4.

The cost of goods manufactured is obtained by adding factory overhead to the prime cost. This is also referred to as the manufacturing cost. The overhead cost includes items such as indirect materials cost (factory supplies and lubricants), indirect labor costs (cost of supervision, inspection cost, and salaries of factory clerks), and fixed and miscellaneous costs such as rent, insurance, taxes, depreciation, maintenance and repair, utilities, and small tools. The role played by overhead expenses in a costing structure has been examined in detail in Chapter 5.

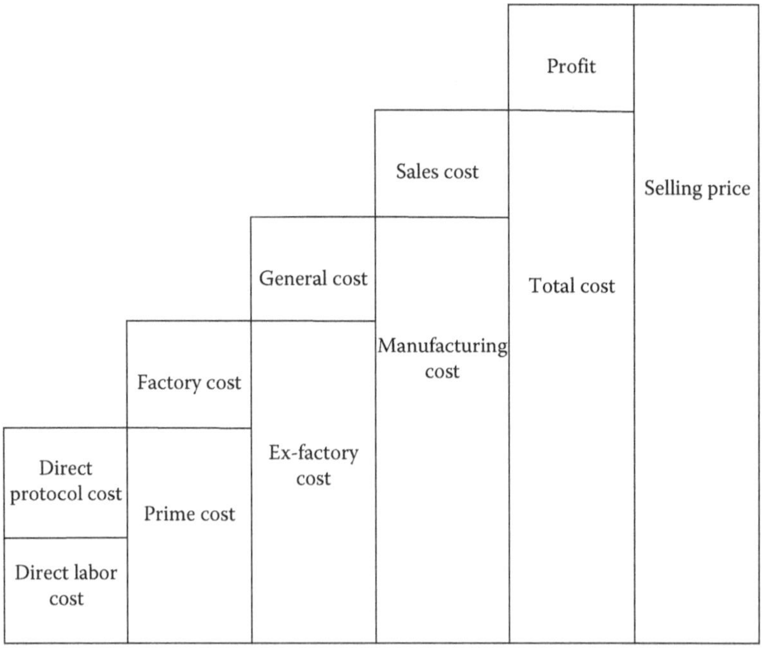

FIGURE 10.1
Interrelationship between product and cost components.

When the cost of distribution, administration, and sales are added to manufacturing costs, we get the cost of goods sold, commonly called COGS. Addition of the profit to COGS determines the selling price.

The distribution costs typically include the following items:

- Advertising costs
- Entertainment and travel costs
- Expenses incurred towards freight and transportation
- Rent and insurance expenses
- Office expenses (stationary, postage, etc.)
- Miscellaneous selling expenses

The administrative expenses include:

- Administrative and office salaries
- Engineering expenses
- Rent and insurance expenses
- Accounting and legal expenses
- Office expenses
- Miscellaneous administrative expenses

The cost of sales staff salaries and commissions to dealers and warranty costs are collectively referred to as selling expenses.

Table 10.1 provides a succinct definition of the various types of costs. This list has been adapted from Ostwald (1974). The following are examples of different kinds of fixed costs:

- Insurance premiums (premiums paid to safeguard against fire, theft, flooding, and occupational hazards)
- Depreciation on buildings, machinery, and equipment
- Property taxes (sometimes states and communities give tax breaks to industrial corporations to attract them, for example, South Dakota)
- Factory indirect labor costs (examples include wages of security personnel, secretarial, clerical, janitorial, accounting staff, etc.)
- Interest on investment capital (cost of borrowing money)
- Cost of general supplies (supplies used by indirect labor—copying, stationary, forms, etc.)
- Management and administrative expenses (salaries and wages paid to legal and corporate staff)
- Engineering cost (design and other engineering personnel, R&D, etc.)
- Cost of rentals or lease (building, equipment, etc.)

TABLE 10.1

Nomenclature of Different Types of Cost (Ostwald, 1974)

Capital cost	The cost of obtaining capital and is expressed as interest rate.
Depreciated cost	A noncash tax expense deduction for recovery of fixed capital from investments whose economic value is gradually consumed in the business operation.
Detailed cost	The value of the detailed estimate obtained with almost complete disclosure of engineering design data using various methods of estimating.
Direct cost	Cost that is traceable to a unit of output such as direct labor costs or direct materials costs.
Direct labor cost	The labor cost of actually producing goods or services.
Direct materials cost	The cost of raw materials or semi-finished materials that can be traced directly to an operation, product, project or system design.
Engineering cost	The total of all costs incurred in a design to produce complete drawings and specifications or reports; included are the costs, salaries, and overhead for engineering administration, drafting, reproductions, cost engineering, purchasing and construction, costs of prototype, and design costs.
Estimated costs	Predetermined value of cost using rational methods.
Fixed cost	Costs that are independent of output.
Future cost	Costs to be incurred at a future date.
Historical cost	A tabulated cost of actual cash payments consistently recorded.
Indirect cost	That cost not clearly traceable to a unit of output or a segment of business operation, such as indirect labor costs and indirect materials costs.
Joint cost	Exists whenever, from a single source or material or process, there are produced units of goods having varying unit values.
Manufacturing overhead	Includes all production costs, except direct labor and direct materials.
Marginal/ incremental cost	The added cost of making one additional unit for an operation or product without additional fixed cost.
Measured cost	A cost based on time relationships to dollars using mathematical rules.
Operating cost	Comprises two distinct cost elements, direct labor and direct materials.
Operation cost	Includes labor, materials, asset value consumed, and appropriate overhead cost pursuant to the operation design.
Opportunity cost	The estimated dollar advantage foregone by undertaking one alternative instead of another.
Optimum cost	An operation, product, project or system economic value for which a minimum (or maximum, as appropriate) is uncovered for specified design variables using variational methods.
Period cost	The cost associated with a time period.
Policy cost	A cost based on the action of others; considered fixed for the purpose of estimating.
Preliminary cost	The value of a preliminary operation, product, project or system design estimate. It is usually obtained quickly with a shortage of information.

(Continued)

TABLE 10.1 (*Continued*)

Nomenclature of Different Types of Cost (Ostwald, 1974)

Prime cost	The total of labor and materials directly traceable to a unit of output.
Product cost	Includes operation costs, purchase materials, overhead, general and administrative expenses, and appropriate design and selling costs.
Project cost	The investment or capital cost proposed for approval in a single evaluation of an engineered project.
Replacement cost	A present cost of the design equipment or facility intended to take the place of an existing design of equipment or facility.
Standard cost	Normal predetermined cost computed on the basis of past performance, estimates or work measurement.
Sunk cost	The past or continuing cost related to past decisions that are unrecoverable by current or future decisions.
System cost	Usually a hypothetical cost for the evaluation of complex alternatives. It may include elements of operation, product or project costs.
Unit cost	Implies in manufacturing the sum of total material, labor, and manufacturing overhead divided by the quantity produced; for an investment, it is the installed cost of the producing unit in convenient units of production.
Variable cost	The element of cost that varies in proportion to the rate of output.

- Marketing and sales expenses (salaries and wages paid to marketing and sales people, communication services, transportation and delivery expenses, warehouse rentals, etc.)

The different variable costs often encountered in a production environment are listed below.

- Materials cost
- Labor costs (including production supervision)
- Cost of maintaining production equipment
- Energy cost (power, gas, oil, etc.)
- Utilities cost (water, sewer, etc.)

10.3 Product Pricing Methods

Setting price is a complex task. It invites reaction from the consumers as well as the competition. Often, pricing has to conform to some legal requirement. While a lack of competition encourages hiking prices, competition puts pressure to lower prices. Pricing is also a function of well-established practices, such as "sales," "coupons," "introductory offers," "holiday sales," etc. Dean (1951) has suggested some basic ways to determine prices:

- Prices that are proportional to full cost, that is, that produce the same percentage net profit for all products.
- Prices that are proportional to incremental costs, that is, that produce the same percentage contribution margin over incremental costs for all products.
- Prices with profit margins that are proportional to conversion cost, that is, that do not consider purchase material costs. Conversion cost corresponds to the value-added concept.
- Prices that produce contribution margins that depend on elasticity of demand.
- Prices that are systematically related to the stage of market and competitive development of an individual member of the product line.

The first three rules emphasize costs.

In general, the following four methods of product pricing are used:

1. Conference and comparison method
2. Investment method
3. Full cost method
4. Direct costing or contribution method

10.3.1 Conference and Comparison Method

As the name implies, this method involves people knowledgeable about the product and the market. They meet and discuss competitor's prices, effect of volume, past pricing history, and price changes. They also evaluate the response of sales people and consumers and the share of the market secured or expected to secure. Product costs and profitability requirements are also considered. All inputs are considered and the final decision is made by committee.

10.3.2 Investment Method

The basic model is as follows:

Sales dollars, S = [Rate of return i, in percentage × Investment I in dollars]/(1 − tax rate t) + C, the total cost excluding investment

Dividing the total sales dollars by the expected number of units to be sold determines the selling price.

10.3.3 Full Cost Method

The method is also known as the *cost-plus* or *markup* method. It is given as follows:

$$\text{Selling price} = (\text{Total cost less investment})/(1 - \text{markup }\%) \times [\text{Number of units sold}]$$

10.3.4 Direct Costing or Contribution Method

This method is used when the costs vary closely with volume. All expenses are divided into fixed and variable components and only the variable costs move with volume changes. As the fixed costs get prorated over the larger volumes, profits increase (increase on increments). With increased profits, one can manipulate the markup in the full cost method (keep it the same or reduce it with increasing volume) to determine pricing.

10.4 Equations for Determining Selling Price

In general, the various relationships between the different elements of cost are presented in the following equations:

$$\text{Prime cost (direct cost)} = \text{Direct material cost} + \text{Direct labor cost} + \text{Direct engineering cost} + \text{Direct expenses.} \quad \text{(Equation 1)}$$

$$\text{Factory cost (manufacturing cost)} = \text{Prime cost} + \text{Factory expenses.} \quad \text{(Equation 2)}$$

$$\text{Production cost} = \text{Factory cost} + \text{Administrative cost.} \quad \text{(Equation 3)}$$

$$\text{Total cost} = \text{Production dost} + \text{Marketing, distribution, and selling expenses.} \quad \text{(Equation 4)}$$

$$\text{Selling price} = \text{Total cost} + \text{Profit (also referred to as markup).} \quad \text{(Equation 5)}$$

In order to better understand the concepts outlined in this chapter, let us consider a few illustrative examples.

EXAMPLE 10.1

For the data presented in the following table, compute the prime cost, factory cost, total cost, and selling price per item for the year 2016.

Item Description	Amount ($)
Inventory as of 12/31/2015	50,000
Inventory purchased during 2016	80,000
Inventory as of 12/31/2016	30,000
Direct labor cost	17,000
Direct engineering cost	13,000
Direct burden	2500
Factory expenses	20,000
Administrative expenses	14,000
Selling and distribution expense	7500
Markup rate	15%
Number of units manufactured	1000 units

Solution

We use Equations 1–5 in order to solve this problem.

From Equation 1, we have the following:

$$\begin{aligned}\text{Prime cost (direct cost)} &= \text{Direct material cost}\\ &\quad + \text{Direct labor cost}\\ &\quad + \text{Direct engineering cost}\\ &\quad + \text{Direct expenses.} \qquad \text{(Equation 1)}\end{aligned}$$

The value of direct labor, direct engineering cost, and direct expenses can be read from the data.

The direct material cost is computed as follows:

$$\begin{aligned}\text{Direct material used during 2016} &= \text{Inventory on hand as of 12/31/2015}\\ &\quad + \text{Inventory purchased during 2016}\\ &\quad - \text{Inventory on hand on 12/31/2016.}\end{aligned}$$

Thus,

$$\begin{aligned}\text{Direct material cost during 2016} &= \$50{,}000 + \$80{,}000 - \$30{,}000\\ &= \$100{,}000\end{aligned}$$

In order to compute prime cost, we use Equation 1.

From the ongoing discussion, we have the following values;

Direct material cost = \$100,000
Direct labor cost – \$17,000
Direct engineering cost = \$13,000
Direct expenses (expressed as direct burden in this case) = \$2500

Adding these elements of cost, the prime cost can be calculated as follows:

Prime cost = \$100,000 + \$17,000 + \$13,000 + \$2500 = \$132,500.
Solution 1.1

Next, we compute factory cost using Equation 2 as follows:

Factory cost (manufacturing cost) = Prime cost + Factory expenses.
(Equation 2)

From solution 1.1, we have the value of prime cost as \$132,500.

From the data presented in the table, factory expenses = \$ 20,000.

Substituting these values into Equation 2, we obtain the value of factory cost as follows:

Factory cost = \$132,500 + \$20,000 = \$152,500. **Solution 1.2**

Next, we compute production cost using Equation 3 as follows:

Production cost = Factory cost + Administrative cost. (Equation 3)

From solution 1.2, we have the value of factory cost as \$152,500.

From the data presented in the table, administrative expenses = \$ 14,000.

Substituting these values into Equation 3, we obtain the value of production cost as follows:

Production cost = \$152,500 + \$14,000 = \$166,500. **Solution 1.3**

Next, we compute total cost using Equation 4 as follows:

Total cost = Production cost + Marketing,
distribution, and selling expenses. (Equation 4)

From solution 1.3, we have the value of production cost as \$166,500.

From the data presented in the table, distribution and selling expenses = \$7500.

Substituting these values into Equation 4, we obtain the value of production cost as follows:

Total cost = \$166,500 + \$7500 = \$174,000. **Solution 1.4**

The total number of units manufactured = 1000.

Thus,

The total cost per unit = \$174,000/1000 units = \$174/unit.

Finally, we compute the selling price by using Equation 5 as follows:

Selling price = Total cost + Profit (also referred to as markup). (Equation 5)

From solution 1.4, we have the value of total cost as $174,000.

From the data presented in the table, the value of markup = 15%.

Substituting these values into Equation 5, we obtain the value of selling price as follows:

Selling price = $174,000 + 15% of $174,000.

Thus,

Selling price = (1.15) × $174,000 = $200,100. **Solution 1.5**

The total number of units manufactured = 1000.

Thus,

The selling price per unit = $200,100/1,000 units = $200.10/unit. **Solution 1.6**

EXAMPLE 10.2

For the data presented in the following table, compute the prime cost, factory cost, total cost, and profit per item for the year 2016 for an electronics manufacturer. The company produced 100 items last year.

Item Description	Amount ($)
Inventory as of 12/31/2015	55,000
Inventory purchased during 2016	550,000
Inventory as of 12/31/2016	75,000
Direct labor cost	250,000
Direct engineering cost	35,000
Depreciation charges: plant and equipment	40,000
Depreciation charges: office equipment	5000
Rent, taxes, and insurance	15,000
Administrative expenses	7500
Office utilities	10,000
Factory utilities	15,000
Direct expenses	3500
Sales commission	20,000
Manager's salary	130,000
Salary paid to office staff	55,000
Income from sales	1,200,000

Factory Overhead	Amount ($)
Rent, taxes, and insurance paid for the factory	13,000
Depreciation of plant and equipment	40,000
Factory utilities	15,000
Salary paid to plant manager	130,000
Total factory overhead	198,000

Administrative Expenses	Amount ($)
Depreciation of office equipment	5000
General administrative expenses	7500
Office utilities	10,000
Salary paid to office staff	55,000
Total administrative expenses	77,500

Sales Expenses	Amount ($)
Sales commission	20,000
Total sales expenses	20,000

Solution

Once again, we use Equations 1 through 5 in order to solve this problem.

From Equation 1, we have the following:

$$\begin{aligned}\text{Prime cost (direct cost)} = {} & \text{Direct material cost} \\ & + \text{Direct labor cost} \\ & + \text{Direct engineering cost} \\ & + \text{Direct expenses.} \qquad \text{(Equation 1)}\end{aligned}$$

The value of direct labor, direct engineering cost, and direct expenses can be read from the data.

The direct material cost is computed as follows:

$$\begin{aligned}\text{Direct material used during 2016} = {} & \text{Inventory on hand as of} \\ & \text{12/31/2015} + \text{Inventory purchased} \\ & \text{during 2016} - \text{Inventory on hand} \\ & \text{on 12/31/2016.}\end{aligned}$$

Thus,

$$\begin{aligned}\text{Direct material cost during 2016} = {} & \$55{,}000 + \$550{,}000 \\ & - \$75{,}000 = \$530{,}000.\end{aligned}$$

In order to compute prime cost, we use Equation 1.

From the ongoing discussion, we have the following values;

Direct material cost = \$530,000
Direct labor cost = \$250,000
Direct engineering cost = \$35,000
Direct expenses = \$3500

Adding these elements of cost, the prime cost can be calculated as follows:

Prime cost = \$530,000 + \$250,000 + \$35,000 + \$3500 = \$818,500.
Solution 2.1

Next, we compute factory cost using Equation 2 as follows:

Factory cost (manufacturing cost) = Prime cost + Factory expenses.
(Equation 2)

From solution 2.1, we have the value of prime cost as \$818,500.

From the data presented in the table, the total value of factory expenses = \$ 198,000.

Substituting these values into Equation 2, we obtain the value of factory cost as follows:

Factory cost = \$818,500 + \$198,000 = \$1,016,500 **Solution 2.2**

Next, we compute production cost using Equation 3 as follows:

Production cost = Factory cost + Administrative cost. (Equation 3)

From solution 2.2, we have the value of factory cost as \$1,016,500.

From the data presented in the table, administrative expenses = \$ 77,500.

Substituting these values into Equation 3, we obtain the value of production cost as follows:

Production cost = \$1,016,5,500 + \$77,500 = \$1,094,000. **Solution 2.3**

Next, we compute total cost using Equation 4 as follows:

Total cost = Production cost + Marketing, distribution, and selling expenses (Equation 4)

From solution 2.3, we have the value of production cost as \$1,094,000.

From the data presented in the table, distribution and selling expenses = \$ 20,000.

Substituting these values in Equation 4, we obtain the value of production cost as follows:

Total cost = \$1,094,000 + \$20,000 = \$1,114,000. **Solution 2.4**

Total profit = Total income from sales − Total cost.

From the table, we know that income from sales = $1,200,000.

Thus,

Profit = $1,200,000 − $1,114,000 = $86,000.

We also know that the company manufactured 100 items last year.

Thus,

Total profit per item = $86,000/100 = $860/item. **Solution 2.5**

EXAMPLE 10.3

A manufacturing company produces 500 pumps. The percent markup is 25%. All other relevant cost information is provided in the following table. Compute the production cost and selling price per pump based on this data.

Item Description	Amount ($)
Cost of raw material	3,500,000
Direct labor cost	1,200,000
Direct expenses	100,000
Salaries paid to plant manager and staff	300,000
Plant utilities	25,000
Plant and equipment depreciation	65,000
Warehouse expenses	25,000
Office utilities	3750
Office depreciation	7500
Salary paid to plant engineer	15,000
Engineering expenses incurred on the plant	6500
Administrative staff salaries	35,000
Sales staff salaries and commissions	27,500
Percent markup	25%
Number of pumps manufactured	500 NOS

Solution

We use Equations 1 through 5 in order to compute the production cost per pump as well as the selling price per pump. Given the fact that there are only two unknowns to be computed, some consolidation of equations will need to be done. This is demonstrated as follows:

From Equation 3, we know the following:

Production cost = Factory cost + Administrative cost. (Equation 3)

Working our way back to Equation 1, Equation 3 can be rewritten as follows:

$$\text{Production cost} = \text{Direct material cost} + \text{Direct labor cost} \\ + \text{Direct engineering cost} + \text{Direct expenses} \\ + \text{Factory expenses} + \text{Administrative cost.}$$

From the data presented in the above table, we know the values of the following entities:

Direct material cost = Cost of raw material = \$3,500,000
Direct labor cost = \$1,200,000
Direct expenses = \$100,000

Factory Expenses

Salaries paid to plant manager and staff = \$300,000
Plant utilities = \$25,000
Plant and equipment depreciation = \$65,000
Warehouse expenses = \$25,000
Salary paid to plant engineer = \$15,000

Engineering Expenses Incurred on the Plant = \$6500

Total factory expenses = \$436,500.

Administrative Expenses

Office utilities = \$3750
Office depreciation = \$7500

Administrative Staff Salaries = \$35,000

Total administrative expenses = \$46,250

Substituting the respective values into the equation for production cost, we obtain the following:

$$\text{Production cost} = \$3{,}500{,}000 + \$1{,}200{,}000 + \$100{,}000 \\ + \$436{,}500 + \$46{,}250$$

Thus,

$$\text{Production cost} = \$5{,}282{,}750$$

Production cost per unit = \$5,282,750/500 units = \$10,565.5/unit.

Solution 3.1

In order to compute the selling price per pump, we need to first compute the total cost.

The total cost is obtained by adding sales overhead to the production cost. This is done as follows:

Sales staff salaries and commissions = \$27,500.

Thus,

Total cost = \$5,282,750 + \$27,500 = \$5,310,250.

The markup rate is 25%.

Thus,

Total revenue = \$5,310,250 × 1.25 = \$6,637,812.50

Selling price per unit = Total revenue/Number of units manufactured

Thus,

Selling price per unit = \$6,637,812.50/500 = \$13,275.63. **Solution 3.2**

EXAMPLE 10.4

Compute the selling price per unit of product if the fixed cost is \$100,000 per year. The variable cost of the product is \$35/unit and the markup rate is 10%. A total number of 1000 units of the product are manufactured.

Solution

In order to solve this problem, we first need to compute the total manufacturing cost. This is achieved by using the basic formula delineated in Chapter 12 on breakeven analysis.

We know that total cost = fixed cost + variable cost.

From the data presented in the problem, we know that fixed cost = \$100,000.

Total variable cost = Variable cost per item × Number of items manufactured.

Thus,

Total variable cost = \$35/unit × 1000 units = \$35,000.

Adding the two quantities together, we obtain the value of total manufacturing cost as follows:

Total cost = Fixed cost + Variable cost.

Substituting the individual values, we obtain the following:

$$\text{Total cost} = \$100{,}000 + \$35{,}000 = \$135{,}000.$$

$$\text{Selling price of the product} = \text{Total cost} \times \text{Markup}.$$

We know from the data given in the problem that the markup rate = 10%. Thus,

$$\text{Revenue} = \text{Total cost} \times (1.10).$$

This implies that revenue = \$135,000 × 1.10 = \$148,500.

$$\text{The selling price per unit} = \text{Total revenue/Number of units manufactured.}$$

Thus,

$$\text{Selling price per unit} = \$148{,}500/1000.$$

This implies that the selling price per unit = \$148.50. **Solution**

10.5 Conclusion

This chapter discussed the various elements of cost that are crucial in determining the selling price of a product. We have already dealt with various methods of cost estimation in this book. The purpose of this chapter was to demonstrate the role played by these elements of cost in deciding the selling price of the product. It should also be emphasized at this juncture that determination of selling price requires a diverse background on the part of the estimator. At the very least, the person should have backgrounds in business, economics, finance, and engineering. It is very unlikely that one would have all the necessary tools for this purpose and, as a result, estimating cost and selling price is often a committee decision.

PROBLEMS

1. Compute the total prime cost, production cost per unit, factory cost per unit, and selling price per unit for a manufacturer of electric pumps for pumps manufactured during 2014. The manufacturer makes 2000 pumps and the markup rate is 15%. Use the data presented in the table below in order to compute the solution.

Item Description	Amount ($)
Inventory as of 12/31/2013	150,000
Inventory purchased during 2014	180,000
Inventory as of 12/31/2014	60,000
Direct labor cost	37,000
Direct engineering cost	33,000
Direct burden	12,500
Factory expenses	30,000
Administrative expenses	24,000
Selling and distribution expense	7500
Markup rate	15%
Number of units manufactured	2000 units

2. An electronics manufacturer is trying to determine the selling price for some electronic equipment. The fixed cost for manufacturing the equipment is $50,000 per year. The variable cost is $25/unit. Compute the selling price per unit if the desired markup rate is 35% and 1200 units are manufactured. *(Answer: $90/unit)*
3. The selling price of an item is $50/unit. The manufacturer produces 100 items at a variable cost of $15/unit. Compute the total fixed cost per year if the markup rate is 25%.
4. An engineering manager is trying to estimate the fixed cost per unit for a household cleaning gadget. The gadget sells for $75 per unit. The markup rate is 15% and the variable cost per unit is $40. The manufacturer produces 1500 units of product. *(Answer: $25.217/unit)*
5. Compute the selling and distribution expense per unit for a manufacturing operation using the data presented in the following table. *(Answer: $25/unit)*

Item Description	Amount ($)
Direct material	250,000
Direct labor cost	47,000
Direct engineering cost	29,000
Direct burden	32,700
Factory expenses	50,000
Administrative expenses	74,000
Selling price per unit	232.375
Markup rate	25%
Number of units manufactured	3000 units

6. A manufacturing process has fixed costs of $60,000 per year with variable costs of $30 per unit. If the company in question sells each unit for $35, how many units must be sold each year in order to reach breakdown?

7. The selling price of a lamp is $78 per unit. The manufacturer produces 500 lamps at a variable cost of $20 per unit. What is the total fixed cost per year if the markup rate is 22.8%?
8. Determine the selling and distribution expense per unit for a manufacturing operation using the data below:

Item Description	Amount ($)
Direct material	500,000
Direct labor cost	90,000
Direct engineering cost	76,000
Direct burden	57,000
Factory expenses	90,000
Administrative expenses	120,000
Selling price per unit	560.500
Markup rate	41%
Number of units manufactured	9700 units

9. The Tyra Company has fixed costs of $500,000 per year. Its primary import, diamonds, sells at a revenue of $3000 per diamond and has variable costs of $1000 per unit. How many diamonds does the company need to sell each year to breakeven?
10. Goddess Wear has created a designer pump. Each pair is sold for $700. If the company produces 1000 pairs with a variable cost of $350 per pair, what would be the fixed cost per year if the markup is 40%?
11. Fixed costs for manufacturing a product is $500,000 per year. Variable costs are $38 per unit. The product can be purchased from another manufacturer for a flat rate of $20 per unit. How many units should be used each year in order for the original manufacture? What is the alternative to breakeven?
12. The manufacturing costs of Products H and M are $30 per unit. With a selling price of $32 per unit and wanted profit of 11%, answer the following:
    a. What is the profit if 1000 units of Product H and 3000 units of Product M are sold?
    b. What if 2500 units of Product H and 900 units of Product M are sold?
    c. If the manufacturing costs of Product H are $11 and those of Product M are $22, what is the profit?
13. The engineering consultant at Stardust, Inc. is trying to determine the fixed cost per unit for an automatic kitchen utensil. The utensil sells for $62 per unit. The markup rate is 13% and the variable cost per unit is $20. The manufacturer produces 7200 units.

14. The price of an item is $100 per unit. The manufacturer produces 2000 items at a variable cost of $90 per unit. What is the total fixed cost per year if the markup rate is 29%?
15. The manufacturing costs of electronics 1 and 2 are $70 per unit. With a selling price of $150 per unit and wanted profit of 40%, answer the following:
    a. What is the profit if 3000 units of Product 1 and 2000 units of Product 2 are sold?
    b. What if 900 units of Product 1 and 1300 units of Product 1 are sold?
    c. If the manufacturing costs of Product 1 are $210 and those of Product 2 are $300, what is the profit?
16. The Thompson Toy Company has the ability to create 850,000 toys annually. Their fixed cost of production is $550,000 with a variable cost of $10 per unit and revenue of $25 per unit. What percent of the company's ability needs to be used for the company to breakeven?
17. Pseudo Line Cosmetics is determining the fixed cost per bottle of their new body lotion. The lotion sells for $20 per bottle. The markup rate is 11% and the variable cost per unit is $8. The manufacturer produces its first 7000 bottles.
18. Dumbbells are being sold at Power Up Athletics for $50 per set. The manufacturer produces 500 items at a variable cost of $27 per set. Determine the total fixed cost per year if the markup rate is 15%.
19. Maxim Boating is considering two types of coatings for a storage tank. Material S will have a first cost of $30,000 and a 12-year life if repaired at the end of year six. Material U will cost $20,000 initially and $3000 per year through its 9-year life. At an interest rate of 17% per year, how much would be spent for repairing Material S that would make the two breakeven?
20. Determine the selling and distribution expense per unit for a manufacturing operation using the data below:

Item Description	Amount ($)
Direct material	200,000
Direct labor cost	52,000
Direct engineering cost	36,000
Direct burden	37,000
Factory expenses	70,000
Administrative expenses	91,000
Selling price per unit	387.500
Markup rate	31%
Number of units manufactured	5400 units

Part II

Decision Making

11

Comparison of Alternatives

11.1 Comparing Alternatives Using Rate of Return as a Measure of Performance

A production manager is often confronted with situations that require them to choose between different mutually exclusive alternatives. This is especially true when it comes to making financial decisions, such as loans and investments. When it comes to production equipment, the engineering manager is often required to choose based on financial constraints such as initial cost, annual benefits offered by the equipment, etc. This is in addition to technical considerations such as equipment functionality and capability. Eventually, questions arise, "How can I make a well informed decision? What metric should I use to evaluate the propriety of my decision?" In Chapter 1 we have already considered how to use different metrics, such as present worth analysis, equivalent uniform cash flow analysis, and future worth analysis in order to help us make well informed decisions. This chapter intends to help answer the aforementioned questions by introducing readers to the concept of rate of return analysis. Additionally, we shall also deal with related concepts such as minimum attractive rate of return (MARR) and incremental analysis.

11.2 What Is Rate of Return?

Rate of return is the most common metric used to judge investment performance in the financial world. There are a variety of metrics used in making financial decisions such as present worth, future worth, equivalent uniform annual cost or benefit, and so on. Yet the rate of return on an investment is the single most important metric that is used to make financial decisions. It is keenly important that it can be used to scale up a portfolio of investments. When compared against its counterparts like that of present worth or equivalent uniform annual cost (or benefit), rate of return offers a more thorough and coherent picture of the nature of the investment. Consider statements such as the following:

- The net present worth of this project is $75,000.
- The equivalent uniform annual benefit accrued as a result of investing in this project is $5000/year.
- The project is likely to produce a rate of return equal to 35%.

Notice that the statement describing the net present worth and equivalent uniform annual benefit of the project only paint a part of the picture. These metrics provide incomplete information, especially when the interest rate is unknown. The statement describing the rate of return is a far better portrayal of the project's likely performance. It takes into account monies invested, along with the benefits likely to be accrued as a result of that investment. Return can only be earned on a specific amount of money invested into a project. Return earned on an investment makes itself apparent through the appearance of a positive benefit. Note that if an investment only produces more expenses, it would be deemed to have failed. The rate of return on such an investment will be a negative number. Similarly, one can project going forward that if a certain amount of money invested in project "X" can produce a rate of return of 23%, what is such an investment likely to produce if it were to be scaled up (more money were to be invested into it)? In order to make a final decision, the rate of return is compared against a minimum attractive rate of return (MARR). The value of MARR is the same value of "i" that is used to compute present worth and equivalent uniform annual cash flow.

In financial parlance, *internal rate of return (IRR)* is defined as the rate at which the present worth is equal to zero. This implies that the equivalent uniform annual worth should also be equal to zero at this rate. Thus, the present worth of cost is equal to the present worth of benefit when the rate of return is equal to the IRR.

11.3 Calculating Rate of Return

The rate of return on an investment is computed by equating the present worth of benefits with the present worth of costs. However, the question is bound to arise: "How can the present worth of any cash flow, whether it is cost or benefit, be computed if the rate of return is not known?" This is a valid question. The answer lies in the fact that the equation needs to be solved for the unknown value of "i." This procedure is explained as follows:

$$\text{Present worth of cost} - \text{Present worth of benefits} = 0.$$

This implies that:

Present worth of benefits = Present worth of cost.

Thus, we have the following expression:

$$\frac{\text{Present worth of benefits}}{\text{Present worth of cost}} = 1. \qquad \text{(Equation 1)}$$

The same concept can be extended to include equivalent uniform annual cash flows as follows:

$$\frac{\text{Equivalent uniform annual benefit (EUAB)}}{\text{Equivalent uniform annual cost (EUAC)}} = 1. \qquad \text{(Equation 2)}$$

When solving for the rate of return, it is to be noted that the internal rate of return (IRR) is the only unknown. The remainder of the values, such as cash flows (including present worth, time periods, equivalent uniform annual cash flow, etc.) is known.

Consider the following example in order to better understand this concept.

EXAMPLE 11.1

$10,000 is invested into an investment account at the end of every year. The objective is to make this sum of money grow to $3,000,000 at the end of 35 years. What rate of return should be earned by the investment in order to achieve the objective?

Solution

Using the concepts learned in Chapter 1, we know that the series of cash flows amounting to $10,000 every year constitutes a uniform series. The future value of this series is known to be $3,000,000. Thus, the following quantities are known to us:

$$A = \$10{,}000;\ F = \$3{,}000{,}000;\ n = 35 \text{ years}.$$

We need to solve for the unknown quantity, namely the rate of return that must be earned on the investment in order to fulfill its intended objective. Thus, the quantity is "i" = ?

Using formulas derived from Chapter 1, we can solve the problem as follows:

$$F = A\ (F/A, i, n).$$

The value of "i" is unknown.

Thus, we have the following expression:

$$\$3{,}000{,}000 = \$10{,}000\ (F/A, i, 35\ \text{years}).$$

Solving the above equation, we get the following:

$$\$3{,}000{,}000/\$10{,}000 = (F/A, i, 35).$$

This implies that 300 = (F/A, i, 35).

We can compute the value of "i" by using the compound interest tables and linear interpolation.

From the interest rate tables, we know the following values to be true for the F/A factor:

1. (F/A, 10%, 35 years) = 271.025
2. (F/A, 12%, 35 years) = 431.663

Thus, our internal rate of return must lie in between these two values. We will use linear interpolation in order to compute the actual value as follows:

$$\frac{300 - 271.025}{i - 0.10} = \frac{431.663 - 271.025}{0.12 - 0.10}$$

This implies that:

$$\frac{28.975}{i - 0.10} = \frac{160.638}{0.02}$$

Thus, we have the following:

$$28.975 = (i - 0.10)\ (8031.9)$$

Dividing by 8031.9 on both sides, we obtain the following:

$$i - 0.10 = 0.0036$$

Adding 0.10 to both sides, we get:

$$i = 0.1.36 = 10.36\%.$$

Thus, the internal rate of return on the investment should be 10.36% per year in order to achieve the intended goal. **Solution**

EXAMPLE 11.2

The following figure represents returns on an investment of $1500. Compute the internal rate of return.

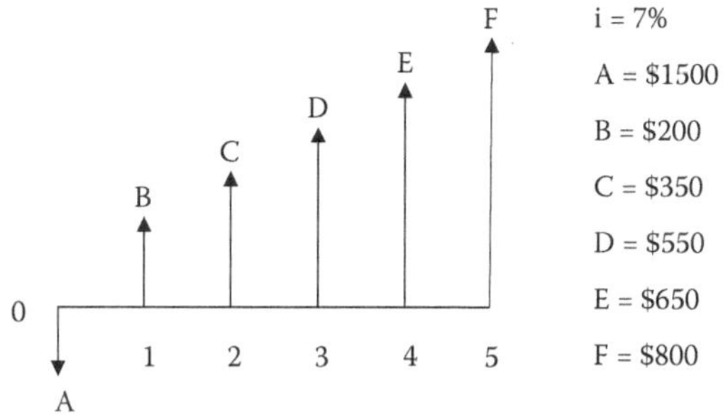

Solution

Looking at the cash flow diagram, it is easy to recognize that it constitutes an arithmetic gradient. Using the concepts learned in Chapter 1, we have the following value for the gradient:

$$A = \$200;\ G = \$150;\ n = 5 \text{ years};\ P = \$1500;\ i = ?$$

The interest rate is not known. This is the value that needs to be computed.

We can solve this problem either by using the equivalent uniform cash flow approach or the present worth approach. Let us solve it using the present worth approach. This is accomplished as follows:

We know from Equation 1 that in order to compute "i," PW (cost) = PW (benefit).

We know from the cash flow diagram that PW (cost) = 1500.

Using arithmetic gradient formula from Chapter 1, we can compute the present worth of benefits as follows:

$$\text{PW (benefit)} = A\ (P/A, i, n) + G\ (P/G, i, n).$$

Substituting respective values into the above equation, we obtain the following:

$$\text{PW (benefit)} = \$200\ (P/A, i, 5) + \$150\ (P/G, i, 5).$$

It is not possible to compute the values of the individual factors without first knowing the value of "i."

We do know that PW (cost) = PW (benefit).

Thus, \$1500 = 200 (P/A, i, 5) + \$150 (P/G, i, 5).

The only way to solve this problem is to try different values of (P/A) and (P/G) in the right hand side of the equation until it closely matches the left hand side of the equation.

From the interest rate tables, we know the following values:

Interest Rate	(P/A, 5 Years)	(P/G, 5 Years)
6%	4.212	7.934
7%	4.100	7.647
8%	3.993	7.372
9%	3.890	7.111
10%	3.791	6.862
15%	3.352	5.775

Substituting values into the equation we get the following:

For $i = 5\%$; \$200 (4.212) + \$150 (7.934) = \$2032.5 > \$1500
For $i = 7\%$; \$200 (4.100) + \$150 (7.647) = \$1967.05 > \$1500
For $i = 10\%$; \$200 (3.791) + \$150 (6.862) = \$1787.5 > \$1500
For $i = 15\%$; \$200 (3.352) + \$150 (5.775) = \$1536.65 ≥ \$1500

At 15%, both sides of the equation are a close match. Thus, we can conclude that the internal rate of return on the investment is about 15% per year. **Solution**

EXAMPLE 11.3

A corporate bond with a coupon rate of 10% was purchased for \$1000. The company promised to pay the purchaser of the bond an interest of \$50 on the \$1000 face value of the bond every six months. The company also promised to repay the \$1000 that it borrowed from the bondholder at the end of 10 years. The bondholder sold the bond for \$900 at the end of one year. What was the rate of return received by the bondholder?

Solution

From the problem description, the following facts are obvious:

- Purchase price of the bond was \$1000. Thus, the present value of the cash flow is a negative \$1000.
- The bond was held for a period of one year since its purchase. This encompasses two 6-month periods. Thus, the bondholder received two payments of \$50 each at the end of the first 6 months and 12 months, respectively.
- The bond was sold for \$900 at the end of one year.

In order to compute the rate of return earned on the transaction, we equate the present worth of cost to the present worth of all benefits received.

In the present case, the only cost was the amount of money spent in purchasing the bond. Thus, the present worth of cost is –\$1000.

In terms of benefits, the bondholder received two payment of \$50 each at the end of 6 months and 12 months, respectively. He also received a lump sum payment of \$900 at the end of 12 months when the bond was sold.

We will consider the cash flow on a semiannual basis since that is when the cash flows actually occur. Since the transaction occurred over the period of one year, the number of semiannual time periods in this transaction is equal to 2.

Thus, the present worth of benefits can be written as follows:

$$\text{PW (Benefits)} = \$900\ (\text{P/F, i, 2}) + \$50\ (\text{P/A, i, 2}).$$

Notice that the two \$50 payments and the end of six months and 12 months constitute a uniform series. Alternatively, the following expression can also be written to compute the present worth of benefits:

$$\text{PW (Benefits)} = \$900\ (\text{P/F, i, 2}) + \$50\ (\text{P/F, i, 1}) + \$50\ (\text{P/F, i, 2}).$$

We can try different rates of interest in order to solve this problem.

We know the following values from the compound interest rate tables to be true:

Interest Rate	(P/A, 2)	(P/F, 2)
0.5%	1.985	0.9901
0.75%	1.978	0.9852
1%	1.970	0.9803
1.5%	1.963	0.9755
2%	1.949	0.9659

Substituting values into the equation we get the following:

For $i = 1\%$; $\$900\ (0.9803) + \$50\ (1.970) = \$980.77 < 1000$
For $i = 0.75\%$; $\$900\ (0.9852) + \$50\ (1.978) = \$985.58 < 1000$
For $i = 0.5\%$; $\$900\ (0.9901) + \$50\ (1.985) = 990.34 \leq 1000$

Thus, the interest rate is approximately 0.5% per six months. This translates to a nominal interest rate of $2 \times 0.5\% = 1\%$ per year. The effective interest rate or the internal rate of return (IRR) is given as $(1 + 0.005)^2 - 1 = 0.010025 = 1.0025\%$. **Solution**

EXAMPLE 11.4

Consider the bond presented in Example 11.3. From the data presented in that example, we know that a total of \$100 was paid to the bond holder by the end of the year. This implies that the nominal interest rate on the bond was actually 10% per year. Assume that the underwriter was paid a fee of 2% to sell the bond, then compute the effective annual interest rate paid by the company.

Solution

In order to solve this problem, we need to consider the cash flow from the perspective of the company offering the bond. If the commission paid to the underwriter to sell the bond was \$0, the company would receive a sum equal to the full face value of the bond, that is, $1000 when the bond was sold. This, however, is not the case. We know from the problem description that the underwriter had a commission of 2% of the face value of the bond. Thus, the amount of money received by the company is reduced by 2% and is now equal to $980.

Also, in order to fulfill their obligations to the bondholder, the company needs to accomplish the following:

- Pay an amount equal to $50 every six months in the form of interest. This implies that the cash flow occurred over 20 semi-annual time periods during an interval of ten years.
- Repay the $1000 that it borrowed from the bondholder at the beginning of the term.

Thus, the cash flow equation can be written as follows:

Present worth (benefit) = $980. From the point of view of the company, this is a positive cash flow and will be counted as a benefit.

$$\text{Present worth (cost)} = \$1000\ (P/F, i, 20) + 50\ (P/A, i, 20).$$

Once again, from the company's perspective, the payment of $50 every six months over a period of 10 years constitutes a uniform series, which is designated as $A = \$50$, $n = 20$ semi-annual time periods.

In order to compute the rate of return, we write the following equation:

$$\$1000\ (P/F, i, 20) + 50\ (P/A, i, 20) = \$980.$$

Trying different interest rates to solve the equation, we know the following values of P/F and P/A for different interest rates to be true:

Interest Rate	(P/A, 20)	(P/F, 20)
4%	13.590	0.4564
5%	12.462	0.3769
6%	11.470	0.3118
7%	10.594	0.2584

Substituting values into the equation we get the following:

For $i = 4\%$; $1000 (0.4564) + $50 (13.590) = $1135.9 > $980
For $i = 5\%$; $1000 (0.3769) + $50 (12.462) = $1000 > $980
For $i = 6\%$; $1000 (0.3118) + $50 (11.470) = $885.3 < $980

It is clear from the above calculation that the target interest rate must lie somewhere between 5% and 6%. We use linear interpolation in order to compute the exact interest rate. This is done as follows:

$$\frac{980-1000}{i-0.05} = \frac{885.3-1000}{0.06-0.05}$$

This implies the following:

$$\frac{-20}{i-0.05}=\frac{-114.7}{0.01}$$

Performing cross multiplication, we get the following:

$$-0.2=-114.7i+5.735$$

Subtracting the term 5.735 from both sides, we get:

$$-5.935=-114.7i.$$

Dividing by −114.7 on both sides, we get the following:

$$i=0.05174=5.174\%.$$

Remember that this interest rate is obtained over a semiannual time frame. Thus, the nominal interest rate is $5.174\% \times 2 = 10.348\%$ per year.

The effective interest rate is given by the following expression:
$i_a = 1.05174^2 - 1 = 0.106157 = 10.6157\%$ per year. **Solution**

EXAMPLE 11.5

A company is contemplating whether it should buy a piece of property so it can expand its operations. The seller has offered to finance the property with 20% down and the balance will be due in seven equal annual payments at 10% interest charged annually. The price of the property would be $450,000. The other option is to pay cash for the property, in which case, the seller would be willing to take $400,000 in an all cash transaction. The company has decided to borrow the $400,000 from their financial institution so that they can pay the $400,000 instead of financing the property with 20% down and accepting the seller's terms. Compute the internal rate of return for the loan the seller offered to finance.

Solution

As mentioned in the problem statement we have two options as follows:

Option 1

Pay cash worth $400,000 in order to buy the property.

Option 2

Make a 20% down payment on a purchase price of $450,000 and pay the balance in seven equal annual payments. The down payment, being 20% of $450,000, amounts to $90,000. The remaining payments are calculated by using the following formula:

$$A = P\,(A/P, 10\%, 7\text{ years})$$

The value of P in this case is 80% of the purchase price of \$450,000 = \$360,000.

Thus, A = \$360,000 (A/P, 10%, 7) = \$360,000 (0.2054) = \$73,944 per year

The cash flow for this option is presented in the following table:

Notice that there is a difference in the cash flow per year if one option was chosen over the other. For instance, if option 1 was chose over its counterpart, the company would end up spending an additional amount of money equal to (\$400,000–\$90,000) at year 0. This equates to an additional expense of \$310,000 at year 0. However, they would end up saving \$73,944 per year for the next seven years. This is due to the fact that the property was bought in a full cash transaction, thus doing away with the need for annual payments.

Year	Payment (\$)
0	90,000
1	73,944
2	73,944
3	73,944
4	73,944
5	73,944
6	73,944
7	73,944

The incremental difference in cash flows between the two options is depicted in the following table.

Year	Option 1: All Cash Transaction (\$)	Option 2: Seller Financing (\$)	Difference (\$)
0	−400,000	−90,000	−310,000
1		−73,944	73,944
2		−73,944	73,944
3		−73,944	73,944
4		−73,944	73,944
5		−73,944	73,944
6		−73,944	73,944
7		−73,944	73,944

Consider the column designated as "difference" in the above table. This is the crux of the problem and therein lies the key to solving the problem.

We already know that the rate of return is obtained by setting the present worth of cost equal to the present worth of benefit. From the above table, it is clear that the present worth of cost is equal to \$310,000. This item is clearly a cost item due to the negative sign. Also, it implies that an additional sum equal to \$310,000 is paid by the company in order to acquire the property. The recurring cash flow that occurs over seven year in the amount of \$73,944 clearly constitutes a uniform series. This is the amount of money saved every year by purchasing the property for cash. Therefore, it constitutes a series of benefits. The internal rate of return can be obtained by using the following equation:

$$\text{PW (cost)} = \text{PW (benefit)}.$$

From the above discussion, it is clear that PW (Cost) = $310,000.

$$\text{PW (benefit)} = A\,(P/A, i, 7\text{ years}).$$

Substituting the respective values in the above equation, we obtain the following:

$$\text{PW (benefit)} = \$73{,}944\,(P/A, i, 7\text{ years})$$

The value of i is the internal rate of return, also designated by the notation IRR.

Thus, we have the following equation:

$$\$310{,}000 = \$73{,}944\,(P/A, i, 7).$$

Solving this equation, we obtain the following:

$$(P/A, i, 7) = 4.192.$$

A look at the interest rate tables reveals that this value corresponds to interest rates between 12% and 15%.

For IRR = 12%, (P/A, 7) = 4.564
For IRR = 15%, (P/A, 7) = 4.160

Thus, using linear interpolation, we can obtain the true value of the IRR as follows:

$$\frac{4.192 - 4.564}{i - 0.12} = \frac{4.16 - 4.564}{0.15 - 0.12}$$

Solving the above equation, we obtain:

$$\frac{-0.372}{i - 0.12} = \frac{-0.404}{0.03}$$

Cross multiplying, we get the following equation:

$$-0.404\,(i - 0.12) = 0.03\,(-0.372)$$

Thus, $(i - 0.12) = 0.0276$.

Adding 0.12 to both sides of the equation, we get the following:

$$\text{IRR} = 0.1476 = 14.76\%.$$

The internal rate of return charged by the seller is 14.76% per year.

Solution

In the following section of the chapter, we will consider the concept of Incremental Analysis.

11.4 Incremental Analysis

Incremental analysis deals with computing the internal rate of return (IRR) on the difference between two alternatives. In order to accomplish this objective, we compute the difference in cash flow by subtracting the higher cost alternative from the lower cost alternative. The incremental difference in the internal rate of return is designated by the notation: ΔIRR. The two alternative decisions are tabulated as follows:

Relationship between Rates of Return	Outcome
$\Delta\text{IRR} \geq \text{MARR}$	Higher cost alternative is selected
$\Delta\text{IRR} < \text{MARR}$	Lower cost alternative is selected

It is best to demonstrate the practical utility of incremental analysis by way of solving actual examples.

EXAMPLE 11.6

The cash flows of two mutually exclusive alternatives are presented in the following table. The minimum attractive rate of return (MARR) is 9%. Select the best alternative using the internal rate of return method.

Year	Alternative A	Alternative B
0	−$2500	−$4500
1	$3250	$6000

Solution

We select the higher cost alternative only if the additional cost of such an alternative produces sufficient additional benefits to compensate for the additional amount of money necessary to buy it.

This can be expressed by way of a simple equation as follows:

$$\text{Higher cost alternative} = \text{Lower cost alternative} + \text{Difference}$$

In our current example, alternative B is the higher cost alternative since it costs \$4500. Alternative A costs \$2500. Thus, the difference in cost between the two alternatives if \$2000. The differences between the two alternatives are tabulated as follows.

Year	Alternative A	Alternative B	Alternative B–Alternative A
0	−\$2500	−\$4500	−\$2000
1	\$3250	\$6000	\$6000–\$3250 = \$2750

We equate the present worth of cost with the present worth of benefit in order to compute the internal rate of return. Thus, we have the following equation:

$$\text{PW (cost)} = \text{PW (benefit)}.$$

Thus, \$2000 = \$2750 (P/F, IRR, 1).

But we know that $F = P\,(1 + i)^n$

Thus, in order to compute the present worth we can rewrite the above equation as follows:

$$P = F/\ (1 + i)^n$$

In the present case, the value of n is equal to 1 as can be observed from the data.

Thus, $P = F/\ (1 + i)$.

Substituting this is into the equation for IRR, we obtain the following expression:

$$\$2000 = \$2750/(1 + i).$$

Multiplying both sides of the equation by the term $(1 + i)$, we get

$$\$2000\ (1 + i) = \$2750.$$

Dividing both sides of the equation by \$2000, we obtain the following:

$$(1 + i) = \$2750/\$2000 = 1.375.$$

Solving the equation, we obtain the value of $i = 1.375 - 1 = 0.375$.

Thus, the value of the internal rate of return, IRR = 37.5%.

Comparing the IRR with the MARR, we observe that 37.5% > 9%. Thus, investment in the higher cost alternative is advisable. **Solution**

EXAMPLE 11.7

In this example, we will resolve Example 11.6, except the increment will be in the form of Alternative A–Alternative B instead of the other way round. All pertinent data remains the same.

Solution

The incremental differences between the two alternatives are presented below.

Year	Alternative A	Alternative B	Alternative A–Alternative B
0	–$2500	–$4500	$2000
1	$3250	$6000	–$6000 + $3250 = –$2750

Once again, as before, we equate the present worth of cost with the present worth of benefit in order to compute the internal rate of return. Thus, we have the following equation:

$$\text{PW (cost)} = \text{PW (benefit)}.$$

Thus, $2000 = $2750 (P/F, IRR, 1).

Using the same logic that was used in Example 11.6, we obtain the following:

$$\$2000 = \$2750/(1 + i).$$

Solving this equation similar to the previous example, the value of the internal rate of return, IRR, is obtained as 37.5%.

Note that there is a key difference where the Alternative A–Alternative B represents a positive $2000 difference between the two alternatives. Being a positive cash flow, this does not constitute an investment. Given the fact that funds in the amount of $2000 are being deposited, this sum represents a loan that the entity would obtain. The sum of –$2750 constitutes the payment made on the loan. As such, the IRR of 37.5% represents interest being paid on the loan. An interest rate of 37.5% is quite an exorbitant rate to be paid on a loan. With such being the case, this particular increment is not feasible.

EXAMPLE 11.8

A company is considering investment in two machines. Both machines have useful lives of seven years and no salvage value. The minimum attractive rate of return MARR is 10%. Machine X costs $25,000 and results in a uniform annual benefit amounting to $6500 over seven years. Machine Y costs $29,000 and produces a series of cash flows that constitute an arithmetic gradient as presented in the following table. In light of the cost and benefit accruing as a result of investment in either machine, which machine should be selected?

Year	Machine X	Machine Y
0	–$25,000	–$29,000
1	$6500	$7000
2	$6500	$7500

Continued

Year	Machine X	Machine Y
3	\$6500	\$8000
4	\$6500	\$8500
5	\$6500	\$9000
6	\$6500	\$9500
7	\$6500	\$10,000

Solution

The incremental cost and benefit accruing as a result of choosing machine Y over machine X are depicted in the following table.

Year	Machine X	Machine Y	Machine Y–Machine X
0	−\$25,000	−\$29,000	−\$4000
1	\$6500	\$7000	\$500
2	\$6500	\$7500	\$1000
3	\$6500	\$8000	\$1500
4	\$6500	\$8500	\$2000
5	\$6500	\$9000	\$2500
6	\$6500	\$9500	\$3000
7	\$6500	\$10,000	\$3500

In order to compute the internal rate of return, we equate the present worth of costs to the present worth of benefits. This is expressed as follows:

$$\text{PW (cost)} = \text{PW (benefit)}.$$

Thus, $\$4000 = \$500\,(P/A, \text{IRR}, 7) + \$500\,(P/G, \text{IRR}, 7)$.

Observe from the column entitled "Machine Y–Machine X" that the series of benefits beginning at the end of the first year with \$500 and ending with \$3500 (at the end of year 7) constitute an arithmetic gradient with the value of $A = \$500$ and the value of $G = \$500$.

Dividing both sides of the above equation by \$500, we obtain the following:

$$8 = (P/A, \text{IRR}, 7) + (P/G, \text{IRR}, 7).$$

From the interest rate tables, we can observe the following:

IRR	(P/A, 7)	(P/G, 7)
20%	3.605	8.255
25%	3.161	6.773
30%	2.802	5.622
35%	2.508	4.717

The constraint is that the sum of (P/A, IRR, 7) and (P/G, IRR, 7) should be equal to 8.

We notice from the above table that at IRR = 30%, (P/A, 7) + (P/G, 7) = 2.802 + 5.622 = 8.424.

Also, at IRR = 35%, (P/A, 7) + (P/G, 7) = 7.225.

Thus, the actual value of the IRR must lie between 30% and 35%. We can compute it using linear interpolation as follows:

$$\frac{8 - 8.424}{\text{IRR} - 0.30} = \frac{7.225 - 8.424}{0.35 - 0.30}$$

Solving the above equation, we obtain the following:

$$\frac{-0.424}{\text{IRR} - 0.30} = \frac{-1.199}{0.05}$$

Cross multiplying the terms, we obtain the following:

$$-0.0212 = 0.3597 - 1.199\ (\text{IRR}).$$

Thus, IRR = −0.3809/(−1.199) = 0.3177 = 31.77%.

Comparing the IRR of 31.77% to the MARR of 10%, we conclude that the incremental investment in machine Y is desirable due to the fact that IRR > MARR. **Solution**

Next, we will consider an application of incremental analysis to a problem involving alternatives with different useful lives.

EXAMPLE 11.9

An engineering manager is considering two mutually exclusive machines for purchase. The various costs and benefits associated with each machine are tabulated below. If the minimum attractive rate of return (MARR) is 15%, which machine should the manager choose based on IRR calculation?

Item	Machine Y	Machine Z
Purchase price	$450,000	$1,500,000
Uniform annual benefit	$200,000	$250,000
Salvage value	$110,000	$400,000
Useful life in years	5	10

Solution

One of the first things that should be noticed about the data is the fact that the two machines have different lives. Machine Y is expected to have a useful life of 5 years, whereas machine Z is expected to "live" for 10 years. Thus, the analysis has to be done considering the least common multiple of the two useful lives. The least common multiple of 5 and 10 is 10. For this reason, our analysis will be based on a 10-year span of time.

In order to consider machine Y on a 10-year timespan, we assume that a new machine will be purchased and installed at the end of the term of the first machine (5 years). The purchase price and uniform annual benefit for the replacement machine are assumed to be identical to the first machine. Thus, the cash flow diagram for machine Y over 10 years is represented in the following figure:

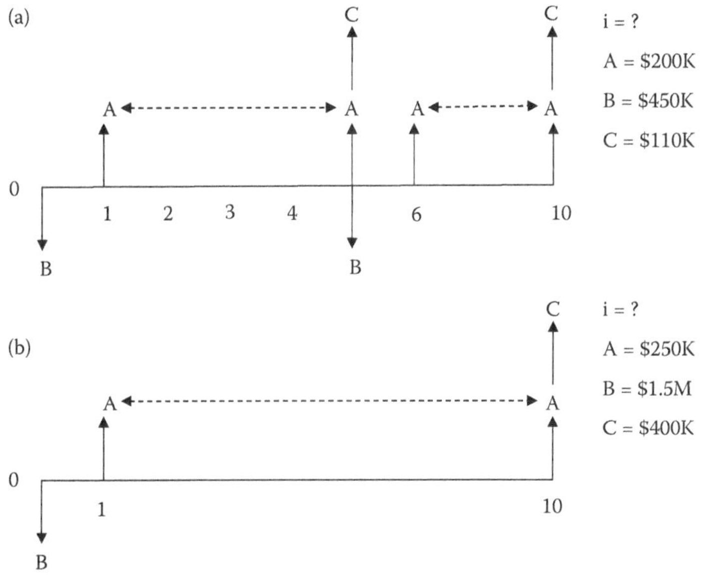

Machine Z costs more than Machine Y. The incremental analysis will be performed for the Z–Y increment. It is depicted in the following table.

Year	Machine Y	Machine Z	Machine Z–Machine Y
0	−$450,000	−$1,500,000	−$1,050,000
1	$200,000	$250,000	$50,000
2	$200,000	$250,000	$50,000
3	$200,000	$250,000	$50,000
4	$200,000	$250,000	$50,000
5	−$140,000	$250,000	$390,000 = $50,000 + $340,000
6	$200,000	$250,000	$50,000
7	$200,000	$250,000	$50,000
8	$200,000	$250,000	$50,000
9	$200,000	$250,000	$50,000
10	$200,000 + $110,000	$250,000 + $400,000	$50,000 + $290,000

Notice that the cash flow for machine Y for year 5 is not in accordance with the uniform annual benefit for all other years. This is due to the fact that at the end of the fifth year, the following cash flow accrues as a result of purchasing machine Y:

- Annual benefit worth \$200,000: Positive cash flow
- Salvage value worth: \$110,000: Positive cash flow
- Purchase price for identical machine when the first machine expires: −\$450,000: Negative cash flow

This amounts to a net cash flow as follows:

$$+\$200{,}000 + \$110{,}000 - \$450{,}000 = \$ - \$140{,}000 \text{ at the end of the fifth year.}$$

This makes the increment = \$250,000 − (−\$140,000) = \$390,000.

We will write this amount as: \$340,000 + \$50,000 so as to not break the uniform series that started at the end of year 1.

We write the following equation for the increment (Z–Y) in order to compute the internal rate of return (IRR):

$$\text{PW (cost)} = \text{PW (benefit)}.$$

Substituting the respective values into the above equation, we have the following

$$\$1{,}050{,}000 = \$50{,}000\,(P/A, IRR, 10) + \$340{,}000\,(P/F, IRR, 5) + \$290{,}000\,(P/F, IRR, 10).$$

Dividing throughout by 50,000 we obtain the following:

$$21 = (P/A, IRR, 10) + 6.8\,(P/F, IRR, 5) + 5.8\,(P/F, IRR, 10).$$

Note that it is not necessary to divide throughout by 50,000, but doing so makes the equation more manageable.

In order to solve this equation, we try substituting values of different factors for various interest rates in the above equation until the two sides of the equation match.

IRR	(P/A, 10)	(P/F, 5)	(P/F, 10)
0.5%	9.730	0.9754	0.9513
0.75%	9.600	0.9633	0.9280
1.00%	9.471	0.9515	0.9053
1.25%	9.346	0.9398	0.8832

Substituting the above values into the equation, we obtain the following:

For IRR = 0.5%; (9.730) + 6.8 (0.9754) + 5.8 (0.9513) = 21.880, which is slightly higher than 21.

For IRR = 0.75%; (9.600) + 6.8 (0.9633) + 5.8 (0.9280) = 21.53, indicating that we are getting closer to 21.

For IRR = 1%; (9.471) + 6.8 (0.9515) + 5.8 (0.9053) = 21.19
For IRR = 1.25%, (9.346) + 6.8 (0.9398) + 5.8 (0.8832) = 20.85

It is obvious that the actual IRR must lie between 1% and 1.25%. As before, we compute the actual value by using linear interpolation as follows:

$$\frac{21-21.19}{\text{IRR}-0.01}=\frac{20.85-21.19}{0.0125-0.01}$$

$$\frac{-0.19}{\text{IRR}-0.01}=\frac{-0.34}{0.0025}$$

Cross multiplying, we get:

$$-0.000475 = -0.34\ \text{IRR} + 0.0034$$

Thus, $-0.003875 = -0.34$ IRR
Dividing both sides of the equation by -0.34, we get:

$$\text{IRR} = 0.011397 = 1.1397\%.$$

We observe from the above calculation that the internal rate of return (IRR) is 1.1397%. Comparing this with the MARR (10%), we notice that the IRR is way below the minimum attractive rate of return. In this situation, machine Z is not preferable over machine Y and money should not be invested in it at the expense of machine Y. **Solution**

11.5 Conclusion

This chapter dealt with the all-important concept of rate of return analysis, internal rate of return, and the minimum attractive rate of return. Numerous problems were solved in order to clearly demonstrate the importance of each of these concepts. It should be reiterated at this point that rate of return is perhaps the single most important metric of financial performance in existence. One of the principal reasons for this is that it imparts a clear picture of the performance of financial loans and investments. Payback period of initial investment can be quickly ascertained on this basis. Also, it provides a highly useful basis for scaling up an investment portfolio and thus helps to determine other metrics such future value.

PROBLEMS

1. Ann invests $6500 at the end of every year over a period of 25 years into a savings account. She desires to retire at the end of 25 years with

an amount equal to $2,000,000. What is the rate of return she should earn on her investments? *(Answer: 17.193% per year)*

2. Carl invested $50,000 into a dot com stock. At the end of four years, his account was worth $300,000. What rate of return did he earn on his investment? *(Answer: 56.508% per year).*
3. Abby borrowed $15,000 from the bank. In order to repay her loan, she agreed to make monthly payments of $800 at the end of every month for the next 48 months. Compute the nominal annual rate of return for the bank. What is the effective rate of return she agreed to pay?
4. Given the following cash flows, compute the rate of return for the following investment.

Year	Cash Flow ($)
0	−13,500
1	0
2	2500
3	5000
4	10,000
5	15,000
6	20,000

5. Compute the rate of return on the following investment.

Year	Cash Flow ($)
0	−25,000
1	0
2	7000
3	8000
4	9000
5	8000
6	7000

6. Jeff obtained a loan from his bank. Compute the internal rate of return he paid based on the following cash flow.

Year	Cash Flow ($)
0	27,000
1	−2500
2	−5000
3	−7500
4	−10,000
5	−12,500
6	−15,000
7	−17,500

7. An investor bought a corporate bond with a face value of $1000 for $750. The bond pays a 5% coupon interest rate quarterly. The investor held the bond to maturity for a period of 10 years. Compute the nominal and effective interest rate he received.
8. An investor bough a rental property for $75,000 cash. Each year, he paid property taxes of $1200. He also received an income of $10,000 per year. He kept the property for ten years. What rate of return was achieved on the investment? *(Answer: 3% per year).*
9. An investor wishes to buy a tract of timber for $1,000,000. He believes that it will produce an income of $300,000 the first year and decline $50,000 every successive year. The investor intends to hold the investment for five years. Compute the rate of return he stands to earn on his investment. *(Answer: 1.73% per year).*
10. An engineering manager is considering which of two alternative machines he should buy. If the MARR is 10%, which machine should he select based on the following information? Use incremental analysis.

Year	Machine X ($)	Machine Y ($)
0	−45,000	−62,000
1	18,000	25,000
2	18,000	25,000
3	18,000	25,000
4	18,000	25,000

11. Using incremental analysis, choose the better alternative subject to an MARR of 10%.

Year	Machine X ($)	Machine Y ($)
0	−15,000	−15,000
1	−7500	6000
2	−5000	6000
3	5000	6000
4	5000	6000
5	5000	6000

12. Using incremental analysis, choose the better alternative subject to an MARR of 15%.

Year	U ($)	V ($)
Purchase price	12,000	20,000
Uniform annual benefit	4000	4500
Useful life (years)	6	12

13. Using incremental analysis, choose the better alternative subject to an MARR of 15%.

Year	A	B	C	D
0	–$2200	–$1200	–2500	–$3200
1	$800	$450	$950	$1200
2	$800	$450	$950	$1200
3	$800	$450	$950	$1200
4	$800	$450	$950	$1200
5	$800	$450	$950	$1200

14. Susan is considering two investment opportunities as presented in the table below. If the MARR is 15%, which alternative should she select?

Cash Flow Item	X ($)	Y ($)
Purchase price	3000	2000
Uniform annual benefit	500	450
Salvage value	400	0
Useful life (years)	3	2

15. Three mutually exclusive alternatives are being considered for installation in a new factory. Each alternative has a 15 year useful life. If the MARR is 10% which alternative should be selected?

Cash Flow Item	A ($)	B ($)	C ($)
Purchase price	60,000	75,000	83,000
Uniform annual benefit	10,500	12,500	13,700

16. You're considering an investment opportunity. Using your MARR to bring all the cash flows associated with the investment to time zero, you discover that the net present value is zero. What is the rate of return this investment yields?

17. Using incremental analysis, consider projects X, Y, and Z for the last decade. Which is the best alternative? The MARR is 11%.

	X ($)	Y ($)	Z ($)
Initial	–360K	–820K	–630K
Annual	69K	165K	131.5K

18. You need to install an air compressor. You have 4 choices for 5 years. The MARR is 22%. The expenses are as follows:

Project	Price ($)	Annual Cost ($)	Resale Value ($)
A	−150K	−57K	17K
B	−190K	−28K	21K
C	−198.5K	−24.8K	35K
D	−115K	−22K	19K

19. If you were to purchase a stock for $300 now, what is the rate of return on your investment if the stock is worth $647.50 at the end of 3 years?
20. Consider the following investment projects. Determine the range of MARR where Project 2 would be preferred over Project 1.

	Flow Line	Manufacturing Cell
Installment cost	$25,000	$20,000
Expected life	7 years	7 years
Salvage value	$0	$0
Variable cost per rail	$8.00	$11.00

Year	Project 1 (Net Cash Flow)	Project 2 (Net Cash Flow)
0	−$1400	−$2220
1	700	1700
2	1500	1500
IRR	21.75%	16.54%

21. Chad's Manufacturing is working on producing a new product. The price point is $12.95 per unit. The cost of the production equipment is $250,000. Operating, maintenance, and management costs are $5500 annually. With a 6-year plan and a MARR of 12%, how many units must be sold annually for the company to break even?
22. Al decides to spend $2400 more on a hybrid engine rather than a standard gasoline engine. It will result in an annual fuel savings of $500. How many years must this savings continue in order for Al to justify the investment if money is worth 12% per year, compounded annually?

For problems 23–24 consider the data below:

23. Based on MARR of 11%, what is the annual rate of production where the alternatives are equally economical?
24. If an estimated 400 rails were to be produced every year, which alternative would be best? What will be the total annual cost?
25. Pristine Snow Ski Resort & Lodge has been paying for snow removal from its parking lots at $700 per day. The proper machinery costs

$35,000. It has a life of 8 years with no salvage value. Operating and maintenance costs are $9000. What is the break even value for the number of days per year that snow removal is required in order to rationalize purchasing the machinery? MARR is 11%.

For problems 26–28, consider the following.

A needed service can be purchased for $400 per unit. The same service can be provided by equipment that costs $300,000 and has a salvage value of $40,000 at the end of 15 years. Annual operating expenses will be $6500 annually, plus $55 per unit.

26. What will be the incremental IRR on the investment if 500 units are provided per year?
27. If estimates are correct, what will be the incremental IRR on the investment if 340 units are provided per year?
28. If the company requiring this service has a MARR of 17%, what would be the alternative to select for the service levels in 11 and 12?
29. Steve borrowed $30,000 from his parents. In order to repay their loan, he agreed to make monthly payments of $1000 at the end of every month for the next 60 months. Compute the nominal annual rate of return for his parents. What is the effective rate of return he agreed to pay? *(Answer: 2.64% per month; 31.68% per year).*
30. A venture capitalist purchased a bond with a face value of $4000 for $600. The bond pays a 9% interest rate quarterly. The investor held the bond for 8 years. What is the nominal and effective interest rate received?

12

Break Even and Benefit Cost Analysis

12.1 What Is Break Even Analysis?

Break even analysis is a very important tool that is commonly used in economic decision making. The essence of break even analysis is that in every business scenario, there will be a situation wherein the total cost incurred in producing a product or offering a service will invariably be offset by the total revenue obtained from selling it. This scenario is referred to as break even. "Breaking even" occurs when the total cost is equal to total revenue implying that neither any profit nor loss is incurred. It can also be used to compare process costing between different alternative processes. Break even analysis is used for a variety of economic evaluations such as:

- To compare different processes of manufacturing in order to determine the most economical option at various production levels.
- Once the production run (level) has been determined, it is necessary to establish the minimum selling price of the product. Break even analysis is helpful in this evaluation. If revenue is equal to total production cost, then the minimum selling price of the product will have to be equal to total production cost (which is equal to revenue at break even production levels) divided by the number of units produced (production run).
- To determine the actual number of units that has to be produced so that total revenue will be equal to total cost. This production run is referred to as break even point.

It will be obvious from the preceding discussion that in order to use break even analysis, it is essential to be able to correctly evaluate total cost of production. We have already dealt with this concept in previous chapters. In Chapter 2, we dealt with cost estimation, in Chapter 3, we dealt with material costing, Chapter 4 considered process costing and overhead costs and capital expenses were dealt with in Chapters 5 and 6. For the purpose of using break even analysis effectively, we need to consider the total cost of production. This analysis is presented in the following section.

12.2 Two Main Components of Cost

There are two main components of cost: fixed cost and variable cost. Fixed costs, as the term implies, remain unchanged over time. No matter what the production quantity, this component of cost is always unchanged. Fixed costs are independent of the number of units produced. Thus, if a company were to manufacture 10,000 widgets or 1,000,000 widgets, there are certain costs that would remain unchanged. Examples of such costs include the following:

- Depreciation charges taken on a plant, building, and equipment
- Salaries (such as salaries paid to clerical and administrative staff, janitorial services, technical personnel such as engineering staff, etc.)
- Property taxes

Depreciation charges are a function of variables, such as the asset price, useful life of the asset, and salvage value. Production quantity usually does not play any role in determining the depreciation charge unless unit of production methods of depreciation are used.

Salaries, it will be appreciated, are fixed in nature, no matter what the employee output. Thus, a highly productive employee is paid about the same salary as someone who is not as productive.

Property taxes are a function of appraised value of the property and are independent of production quantity. A company that owns a production plant would be obligated to pay property taxes even if it produced zero units of output. On the other hand, no matter how wildly productive the company was, its property taxes would remain the same and would have no correlation with the production output.

Variable costs, however, are directly related to the production quantity. Thus, a greater number of units produced results in a higher variable cost and vice versa. Examples of variable cost include direct material cost, direct labor cost, utility cost used for production, inventory carrying costs, etc. Direct material cost is defined as the cost of all material that is a direct component of the product. Hence, it is easy to recognize that direct material cost rises in a linear fashion with respect to the number of units produced. A similar reasoning can be extended to the definition of direct labor cost. Utility costs like water and electricity related expenses also rise in tandem with the production quantity.

There are instances wherein the costs do not rise linearly with respect to production quantity. Instead, they rise in a stepwise pattern. For instance, when production quantity exceeds the maximum limit, wages must be paid for overtime work or to retain additional labor. In such cases, labor costs will neither be fixed nor variable. Instead they will increase in stepwise increments. These costs are referred to as semi-variable costs. Inclusion of semi-variable costs results in a changing slope of the variable cost curve.

12.3 Mathematical Derivation of Total Cost as a Function of Fixed Cost and Variable Cost

The total cost of any production operation can be ascertained by using the following equation:

$$\text{Total cost (TC)} = \text{Fixed cost} + \text{Variable cost.}$$

Fixed costs are always expressed in term of units of currency such as \$. Variable costs, given their correlation with number of units produced, are expressed in terms of amount of money per unit of output (\$/unit). The total cost is expressed as \$. For both sides of the equation to be consistent, we arrive at a modified version of the above equation as follows:

$$\text{Total cost (\$)} = \text{Fixed cost (\$)} + \text{Variable cost (\$/unit)} \times (\text{“X” number of units}). \quad \text{(Equation 1)}$$

It will be observed that on the right hand side of the equation, when the term (\$/unit) is multiplied by the number of units produced, we arrive at the total variable cost, expressed in terms of (\$). It is very important to clearly draw this distinction.

Total cost is expressed as the cost necessary to produce "X" number of units. It should be obvious by now that it is inclusive of fixed cost and variable cost.

Fixed cost is expressed as a sum of all the fixed costs incurred during the period of time over which "X" number of units were produced.

Variable costs (\$/unit) are expressed in terms of the sum total of all variable costs that are incurred in order to produce "X" units of output.

12.3.1 Break Even Analysis to (Compare Two Methods of Production)

Various methods of production can be compared from the perspective of total production cost using break even analysis. Different production methods have different fixed costs based on the type of equipment, facilities used for production, and other process characteristics. The same is true for variable costs. For instance, the fixed cost and variable cost for a process like forging could be different from that of a process such as casting. The question for a production manager is which of the different processes should be chosen. This question can be answered using break even analysis.

Consider a situation in which a production manager needs to choose between two production processes: Y and Z. The two processes can be compared from the point of view of total cost. The break even quantity corresponds to the production quantity when total cost for process Y is equal to that of process Z.

We have the following equation:

$$\text{Total cost (Y)} = \text{Total cost (Z)}.$$

We know that total cost = fixed cost + variable cost.

Substituting this relationship into the above equation, we obtain the following:

$$\text{Fixed cost (Y)} + \text{Variable cost (Y)} = \text{Fixed cost (Z)} + \text{Variable cost (Z)}.$$

But, total variable cost = cost/unit × number of units produced. Let "X" be the number of units produced.

Thus, the total cost equation can be rewritten as follows:

$$\text{Fixed cost (Y)} + \text{Variable cost (Y)} \times (\text{X}) = \text{Fixed cost (Z)} + \text{Variable cost (Z)} \times (\text{X}).$$

In this case, "X" is the number of break even units. In other words, it is the number of units that must be produced so that the total cost for process Y is equal to the total cost for process Z.

Solving the above equation for the unknown value of "X", we obtain the following:

$$\text{X}\,[\text{VC (Y)} - \text{VC (Z)}] = \text{FC (Z)} - \text{FC (Y)}.$$

VC: Variable cost
FC: Fixed cost

Dividing both sides of the equation by X, we obtain the following expression:

$$\text{Break even quantity} : \text{X} = \frac{\text{FC (Z)} - \text{FC (Y)}}{\text{VC (Y)} - \text{VC (Z)}}. \qquad \text{(Equation 2)}$$

If the costs were to be plotted on a graph, the following principle will be noticed:

PRINCIPLE

- The alternative with the lower fixed cost will be preferable over the alternative with the higher fixed cost as long as $X <$ break even value.
- Above that value, that is, if $X >$ break even quantity, the option with the lower variable cost is preferred over the option with the higher variable cost.

Let us consider an example to better understand the concepts presented in this section.

EXAMPLE 12.1

A production manager needs to fulfill on order of 500 units before the end of the quarter. They are considering which of two alternative processes will help them do so. Both processes are equally efficient in terms of production capability. The manager has access to cost information for both processes. This information is presented in the following table.

Expense	Type of Process Cost	Process Y	Process Z
Maintenance cost	Fixed ($)	1000	400
Depreciation	Fixed ($)	1000	600
Direct material cost	Variable ($/unit)	4.0	4.0
Direct labor cost	Variable ($/unit)	2.50	4.70
Utilities cost	Variable ($/unit)	1.50	3.30
Total fixed cost	Fixed ($)	2000	1000
Total variable cost	Variable ($/unit)	8.00	12.00

He asks you to help him make a decision.

Solution

In order to solve this problem, we use the formula presented in Equation 2. As engineers, we need to be objective and reach an objective solution. No guesses.

There are two methods to solve this problem: the algebraic method and the geometric method. We will consider the algebraic method first.

Method 1: Algebraic Method

Using Equation 2, we have the following:

$$\text{Break even quantity}: X = \frac{FC\,(Z) - FC\,(Y)}{VC\,(Y) - VC\,(Z)}. \quad \text{(Equation 2)}$$

Substituting the respective values into the equation, we get the following result:

$$X = \frac{1000 - 2000}{8.00 - 12.00} = \frac{-\$1000}{-\$4.00} = 250\,\text{units}.$$

Notice that the negative sign in the numerator cancels the negative sign in the denominator.

In order to pick the ideal alternative, we compute the total cost for each process for one unit of production below the break even quantity and one unit of production above the break even quantity. It would be futile to test for a value equal to break even quantity because we know that by definition, the total cost for each process at break even is identical.

Following the above principle, we have the following:

Total cost (Y) 249 units = \$2000 + \$8 (249) = \$3992. (Lower total cost)

Total cost (Z) 249 units = \$1000 + \$12 (249) = \$3988.

This implies that alternative Z should be chosen for $X < 250$ units, the break even quantity. Notice that alternative Z has a lower fixed cost (\$1000) compared to alternative Z (\$2000). Thus, the lower fixed cost alternative is chosen for a production quantity lower than break even.

Testing for $X = 251$ units, we have the following:

Total cost (Y) 251 units = \$2000 + \$8 (251) = \$4008. (Lower total cost)

Total cost (Z) 251 units = \$1000 + \$12 (251) = \$4012.

This implies that alternative Y should be chosen for $X > 250$ units, the break even quantity. Notice that alternative Y (\$8/unit) has a lower variable cost that alternative Z (\$12/unit).

The solution is in agreement with the principle stated before.

Thus, the final solution is as follows:

If $0 \leq X \leq 249$: Choose alternative Z
If $251 \geq X \geq 500$: Choose alternative Y

Method 2: Geometric Method

This problem can also be solved geometrically, wherein the total costs are plotted on a graph. The geometric solution is actually easier in that it enables quick visualization of the problem and its solution.

In order to plot the total cost of each process, we use the basic formula for total cost:

$$T.C = F.C + (V.C) \times (\text{number of units produced}).$$

Notice that this formula looks eerily similar to that of a straight line: y = mx + C, wherein, 'm' is the slope of the straight line and 'C' is the Y coordinate when plotted in the XY space.

Thus, returning to the problem at hand, we obtain the total cost for each process as follows:

$$T.C\ (Y) = \$2000 + 8X, \text{ where X is the break even quantity.}$$

$$T.C\ (Z) = \$1000 + 12X.$$

Instead of solving the equations algebraically, we plot them. The geometric plot is presented in Figure 12.1.

It will be noticed from the figure above that the fixed costs for both processes are positive even when the number of units produced is zero. This property is in keeping with the definition of fixed cost as explained previously in the chapter.

It is easier to solve the problem using the graphical method. The solution is as follows:

Choose the lower line since it represents lower cost. Thus, from the above figure, it will be observed that the line corresponding to process Z is physically lower compared to the line representing process Y as long as the number of units produced are less than the break even quantity ($X = 250$). For values of $X > 250$, the situation is reversed. In that case, the line corresponding to process Y is lower than the one corresponding to process Z. Thus, process Y must be chosen if $X > 250$ in this case.

Notice that we obtain the same result regardless of the method used.

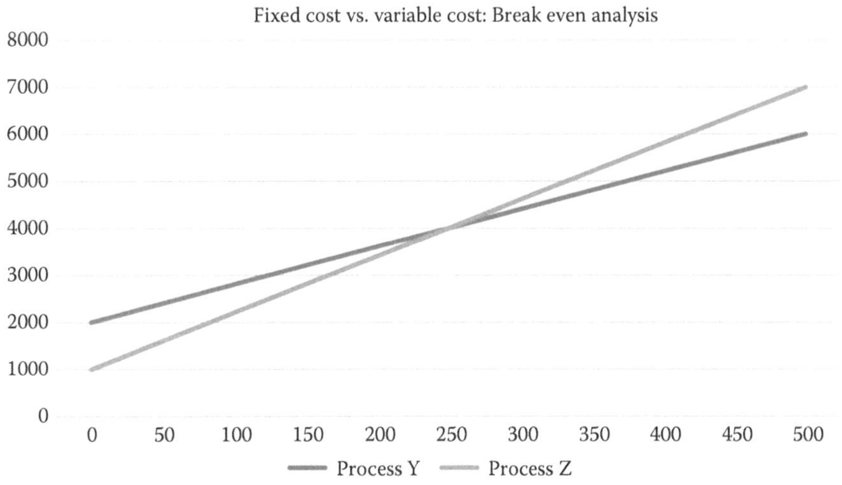

FIGURE 12.1
Break even analysis by comparing two processes.

12.3.2 Break Even Analysis (Compare More Than Two Alternatives)

In the previous section, we dealt with a situation where two different alternatives were compared based on total cost. In real life, there are instances where we are confronted with several equally capable alternatives from which to choose. The concept of break even analysis is equally applicable to such scenarios.

In general, if there is 'n' number of alternatives under consideration, then the total number of possible break even points is computed using the following equation:

$$\text{Number of break even points} = \frac{n!}{2! \times (n-2)!}$$

Thus, if five alternatives are under consideration, the above equation can be used to compute the total number of break even points as follows:

$$\text{Number of break even points } (n = 5)$$
$$= \frac{5!}{2! \times (5-2)!} = \frac{120}{2 \times (3)!} = \frac{120}{2 \times 6} = 10 \text{ break even points.}$$

Both algebraic and geometric methods may be used to figure out the solution in this case as well. Let us consider an example in order to understand this application of break even analysis.

EXAMPLE 12.2

A production manager may choose from three different manufacturing processes that can be implemented in a new factory. He shows you the costing data he has obtained for each of the three manufacturing processes in terms of fixed costs and variable costs for each process and asks you for your advice. What would you advise him?

			Alternatives		
Item	**Fixed Cost**	**Variable Cost**	**A**	**B**	**C**
Maintenance cost	Fixed ($)		1000	400	800
Depreciation charges	Fixed ($)		1000	600	400
Direct material cost		Variable ($/unit)	4.00	4.00	4.00
Direct labor cost		Variable ($/unit)	2.50	4.70	2.90
Energy cost		Variable ($/unit)	1.50	3.30	3.10
Total fixed cost	Fixed ($)		2000	1000	1200
Total variable cost		Variable ($/unit)	8.00	12.00	10.00

Solution

In this example, the production manager can choose from three different processes. Thus, the total number of break even points possible in this scenario can be ascertained by using the following equation:

$$\text{Number of break even points (n = number of alternatives)} = \frac{n!}{2! \times (n-2)!}$$

Thus, if n = 3, substituting that value into the above equation, we obtain the following:

$$\text{Number of break even points (n = 3)} = \frac{3!}{2! \times (3-2)!} = \frac{6}{2 \times 1} = 3 \text{ break even points.}$$

The three break even points are obtained from the following combinations: A vs B, B vs C, and A vs C.

As before, we can solve this problem using both algebraic as well as geometric methods.

Method 1: Algebraic Method

In order to compute the break even point between processes A and B, we use the total cost equation as follows:

$$\text{T.C (A)} = \text{T.C (B)}.$$

Let X be the break even quantity.

Using the data given in the problem statement, we obtain the following expression:

$$\$2000 + 8X = \$1000 + 12X.$$

Solving this equation, we obtain the following:

$$\$1000 = 4X, \text{ thus } X = 250 \text{ units.}$$

Similarly, the break even point between alternatives B and C are computed as follows:

$$\text{T.C (B)} = \text{T.C (C)}.$$

Let X be the break even quantity.

Using the data given in the problem statement, we obtain the following expression:

$$\$1000 + 12X = \$1200 + 10X.$$

Solving this equation, we obtain the following:

$$\$200 = 2X, \text{ thus } X = 100 \text{ units.}$$

The break even point between alternatives A and C is computed as follows:

$$\text{T.C (A)} = \text{T.C (C)}.$$

Let X be the break even quantity.

Using the data given in the problem statement, we obtain the following expression:

$$\$2000 + 8X = \$1200 + 10X.$$

Solving this equation, we obtain the following:

$$\$800 = 2X, \text{ thus } X = 400 \text{ units.}$$

The decision as to which process should be used when can be arrived at by using the technique described in the previous problem. It will be appreciated that in the case of multiple (more than two) alternatives, the algebraic process can become quite cumbersome.

The cost for each process based on the number of units produced is presented in the following table.

	Processing Costs ($)		
Number of Units	**Process A**	**Process B**	**Process C**
0	2000	1000	1200
50	2400	1600	1700
100	2800	2200	2200
150	3200	2800	2700
200	3600	3400	3200
250	4000	4000	3700
300	4400	4600	4200
350	4800	5200	4700
400	5200	5800	5200
450	5600	6400	5700
500	6000	7000	6200
550	6400	7600	6700
600	6800	8200	7200

The geometric process is especially useful in visualizing and solving such problems. We solve the example using the geometric method as follows.

Method 2: Geometric Method

In order to plot the total cost of each process, we use the basic formula for total cost:

$$\text{T.C} = \text{F.C} + (\text{V.C}) \times (\text{number of units produced}).$$

As stated before, this formula represents a straight line: $y = mx + C$, wherein, 'm' is the slope of the straight line and 'C' is the Y coordinate when plotted in the XY space.

Thus, returning to the problem at hand, we obtain the total cost for each process and break even points between pairs of processes as follows:

Process pairs A and B

$\$2000 + 8X = \$1000 + 12X$; X is the break even point. X = 250 units.

Process pairs B and C

$\$1000 + 12X = \$1200 + 10X$; X is the break even point. X = 100 units.

Process pairs A and C

$\$2000 + 8X = \$1200 + 10X$; X is the break even point. X = 400 units.

All three processes are plotted on a chart as represented in the following Figure 12.2.

It will be noticed from the above figure that for $0 \leq X \leq 100$, process B is the lower cost alternative and should be preferred.

For $101 \leq X \leq 400$, process C is the lower cost alternative and should be preferred.

For $401 \leq X \leq \infty$, process A is the lower cost alternative and should be preferred.

The same conclusion can be drawn from the table presented in the algebraic method solution. From the graphical solution, note that the break even point between process A and process B is immaterial to the solution.

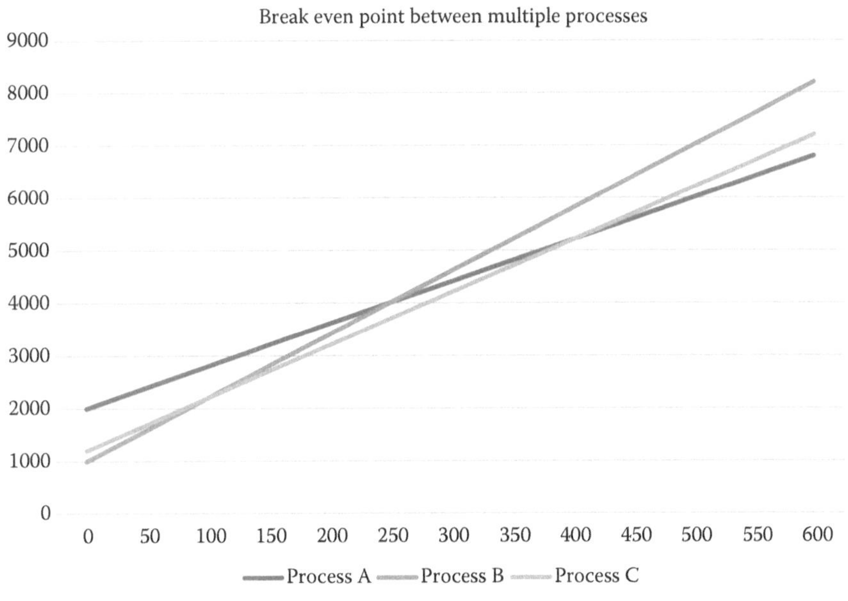

FIGURE 12.2
Break even point between multiple processes.

12.3.3 Break Even Analysis Used to Compute Product Profitability

Break even analysis is frequently used to compute sales volume necessary to achieve break even. In this case, the total revenue is compared with total cost in order to arrive at the break even sales volume.

Consider a situation in which a sales manager is trying to figure out the volume of sales that must be achieved so as to recover the cost of production. In this case, we have the following equation:

$$\text{Total cost} = \text{Total revenue}.$$

Thus, substituting the equation for total cost into the above equation, we get the following:

$$F.C + V.C\ (X) = \text{Total revenue}.$$

If 'S' is the selling price per unit, the above equation can be rewritten as follows:

$$F.C + V.C\ (X) = S\ (X).$$

F.C = fixed cost incurred in production
V.C = variable cost incurred in production
X = total number of units produced
S = selling price per unit

Solving the above equation with the objective of finding the value of X, we obtain the following:

$$X\ (S\text{-}V.C) = F.C.$$

Solving the above equation for the value of X, we obtain:

$$X = \frac{F.C}{S - V.C}. \qquad \text{(Equation 3)}$$

PRINCIPLE

- If production quantity < break even point, loss is incurred
- If production quantity = break even point, neither loss nor profit is obtained
- If production quantity > break even point, profit is obtained

Consider Example 12.3 in order to better understand the concepts presented in this section.

EXAMPLE 12.3

A sales manager has been given the task of studying sales data in order to determine the break even selling price for a product. The pertinent data has been presented in the following table. The manager enlists your help to try and figure out the break even selling price. What is your advice?

Item	Fixed Cost or Revenue	Variable Cost or Revenue	Numeric Value
Administration overhead	Fixed($)		1500.0
Sales and distribution overhead	Fixed($)		1000.0
Depreciation charges	Fixed($)		3000.0
Direct material cost		Variable($/unit)	5.0
Direct labor cost		Variable($/unit)	1.00
Total fixed cost	Fixed($)		5500.0
Total variable cost		Variable ($/unit)	6.0
Selling price per unit		Variable($/unit)	16.00

Solution

This problem can be solved both algebraically as well as geometrically. We will consider the algebraic method first.

Method 1: Algebraic Method

In order to ascertain the break even sales volume, we will use Equation3 as follows:

$$X = \frac{F.C}{S - V.C}. \qquad \text{(Equation 3)}$$

wherein X is the break even sales volume

F.C: Total fixed costs

V.C: Total variable costs

S: Selling price per unit

Substituting the respective values from the table into the equation, we obtain the following:

$$X = \frac{\$5500}{16.00 - 6.00} = \frac{\$5500}{10.00} = 550 \text{ units of product.}$$

Note that if $X < 550$ units, the company stands to incur a loss. Conversely, if $X > 550$ units, the company will make a profit. This is proven as follows:

If $X < 550$, consider a case wherein $X = 540$. In such a case, the total variable cost would be the product of variable cost per unit and the number of units sold.

Thus,

$$\text{Total variable cost} = (\$6/\text{unit}) \times (540 \text{ units}) = \$3240.$$

Total fixed cost = \$5500. Remember that this cost is fixed regardless of the output.

Thus,

$$\text{The total cost} = \text{fixed cost} + \text{variable cost} = \$5500 + \$3240 = \$8740.00$$

$$\text{The total revenue} = \text{selling price per unit} \times \text{number of units sold.}$$

Thus,

$$\text{Total revenue} = (\$16/\text{unit}) \times (540 \text{ units}) = \$8640.$$

Comparing total cost with total revenue, we notice that \$8640 < \$8740 implying a loss.

On the other hand if $X > 550$, consider a case wherein $X = 560$ units.

Thus,

$$\text{Total variable cost} = (\$6/\text{unit}) \times (550 \text{ units}) = \$3300.$$

Total fixed cost = \$5500. Remember that this cost is fixed regardless of the output.

Thus,

$$\text{The total cost} = \text{fixed cost} + \text{variable cost} = \$5500 + \$3300 = \$8800.00$$

$$\text{The total revenue} = \text{selling price per unit} \times \text{number of units sold.}$$

Thus,

$$\text{Total revenue} = (\$16/\text{unit}) \times (560 \text{ units}) = \$8960.$$

Comparing total cost with total revenue, we notice that \$8960 > \$8800 implying a profit.

Thus, our final answer is verified to be correct.

Method 2: Geometric Method

As before, this problem can also be solved geometrically wherein the total costs and revenue are plotted on a graph. Given the obvious ease of visualization, the geometric solution is easier to read and compute.

In order to plot the total cost of each process, we use the basic formula for total cost:

$$\text{T.C} = \text{F.C} + (\text{V.C}) \times (\text{number of units produced}).$$

As mentioned before in Example 12.1, this formula is that of a straight line in the form of $y = mx + C$.

The total cost of production is obtained as follows:

$$\text{T.C} = \$5500 + 6X,$$

where X is the break even quantity.

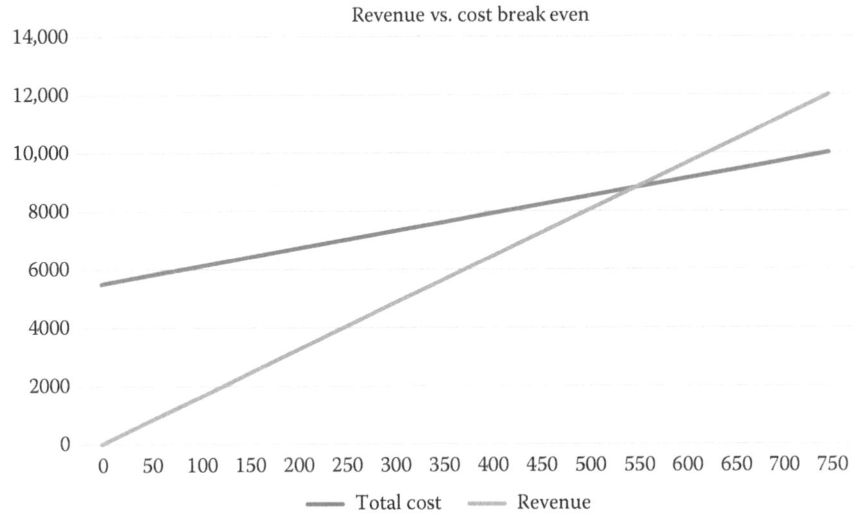

FIGURE 12.3
Comparing total revenue with cost: break even analysis.

Total revenue, R = selling price per unit × number of units sold

Thus, R = (\$16/unit) × (X).

Instead of solving the equations algebraically, we plot them. The geometric plot is presented in Figure 12.3.

Notice that there are no fixed components for the revenue curve. Revenue is strictly a function of number of units sold. Thus, if no units of the product are sold, the revenue is equal to zero.

When comprehending the geometric solution, as before, the lower line represents lower cost or revenue depending on the situation. Thus, from the above figure, it will be observed that the line corresponding to revenue is physically lower compared to the line representing total cost as long as the number of units produced are less than the break even quantity ($X = 550$). For values of $X > 550$, the situation is reversed. In that case, the line corresponding to total cost is lower than the one corresponding to revenue. Thus, the company stands to make a profit of $X > 550$ in this case.

Notice that we obtain the same result regardless of the method used.

12.3.4 Break Even Analysis Using Semi-Variable Costs

Sometimes, the classification of costs is not as straightforward as fixed cost or variable costs. Costs could increase in steps. For instance, hiring new personnel in order to fulfill a contract is an example of such a case. Stepwise increase in costs is referred to as type 1 semi-variable cost. These types of changes generally occur when production reaches a certain level. It is a discrete change, thus altering the slope of the cost curve such that it ceases to have a constant slope.

Sometimes costs could increase in such a way that their slope changes. Slope can change either positively or negatively. The former case implies an increasing rate of change and vice versa. This type of semi-variable cost is referred to as type 2 costs. For instance, taxes play an important role in costing. Taxes cannot be levied unless a company is profitable. By definition, as we have already seen, profit occurs after the break even point has been reached and production has surpassed that level.

Consider Example 12.4 in order to better understand this concept.

EXAMPLE 12.4

A production manager is concerned with changing costs at his factory. He has been required to make some changes such as hire additional labor, etc., to help him with production. He shows you data pertinent to his production process and process parameters and asks you to interpret it. What is your interpretation?

Item	Fixed Cost ($)	Variable Cost ($/unit)	Range of Production	Price
Depreciation charge	Fixed ($)		Entire range	1000
Factory overhead	Fixed ($)		0–1000	600
	Fixed ($)		1001–2000	1800
Direct material cost		Variable ($/unit)	Entire range	4.00
Direct labor cost		Variable ($/unit)	0–800	1.00
		Variable ($/unit)	801–1600	2.00
Sales price		Variable ($/unit)	Entire range	10.00
Total fixed costs	Fixed ($)		0–1000	1600
	Fixed ($)		1001–2000	2800
Total variable costs		Variable ($/unit)	0–800	5.00
		Variable ($/unit)	801–1600	6.00

Solution

It should be obvious that the application of the concept of break even analysis in this example is with respect to cost versus profit.

Notice from the above table that cost increments occur when the production quantity reaches 800 units and 1000 units. When production quantity surpasses 800 units, a different labor rate is applied. Similarly, when production quantity surpasses 1000 units, a different factory overhead is applied.

Using Equation 3, we obtain the first break even point for the range (0–800) as follows:

$$X = \frac{F.C}{S - V.C}. \qquad \text{(Equation 3)}$$

From the table, it will be observed that when 800 units are produced, the total fixed cost is $1600 since 800 is within the 0–1000 units range.

The sales price is variable for the entire production and is listed at \$10/unit. From the table, the total variable cost corresponding to a production quantity of 800 units is \$5/unit.

Substituting the respective values into Equation 3, we obtain the following:

$$X = \frac{\$1600}{\$10.00 - \$5.00/\text{unit}} = 320 \text{ units.}$$

This implies that for $0 \leq X \leq 320$ units, cost > revenue.

Also, when $321 \leq X \leq 400$, revenue > cost.

The second change occurs at $X = 1000$ units as mentioned before. At this stage, the variable rate has changed and must be accounted for. Equation 3 will have to be modified in order to account for this change. This modification is written as follows:

$$X = \frac{F.C + L1 \times (V.C1 - V.C2)}{S - V.C2}. \qquad \text{(Equation 4)}$$

wherein:

F.C represents the fixed cost
S represents the selling price per unit
L_1 represents the production limit for the first variable cost rate
$V.C_1$ represents the first variable unit cost
$V.C_2$ represents the second variable unit cost

Substituting the respective values into Equation 4, we get the following:

$$X = \frac{\$1600 + 800 \text{ units} \times \left(\frac{\$5}{\text{unit}} - \frac{\$6}{\text{unit}}\right)}{\left(\frac{\$10.00}{\text{unit}} - \frac{\$6.00}{\text{unit}}\right)} = \frac{\$1600 - \$800}{(\$4.00/\text{unit})} = 200 \text{ units.}$$

The production range is 800–1000 units. 200 units is below the lower threshold of that range implying that this production quantity will be profitable.

The different costs and revenues over different production ranges are depicted below.

Number of Units	Fixed Cost (\$)	Variable Cost (\$)	Total Cost (\$)	Revenue (\$)
0	1600	0	1600	0
500	1600	2500	4100	5000
550	1600	2750	4350	5500
600	1600	3000	4600	6000
650	1600	3250	4850	6500
700	1600	3500	5100	7000

Continued

Number of Units	Fixed Cost ($)	Variable Cost ($)	Total Cost ($)	Revenue ($)
750	1600	3750	5350	7500
800	1600	4000	5600	8000
850	1600	5100	6700	8500
900	1600	5400	7000	9000
950	1600	5700	7300	9500
1000	1600	6000	7600	10,000
1050	2800	6300	9100	10,500
1100	2800	6600	9400	11,000
1150	2800	6900	9700	11,500
1200	2800	7200	10,000	12,000
1250	2800	7500	10,300	12,500
1300	2800	7800	10,600	13,000
1350	2800	8100	10,900	13,500
1400	2800	8400	11,200	14,000
1450	2800	8700	11,500	14,500
1500	2800	9000	11,800	15,000
1550	2800	9300	12,100	15,500
1600	2800	9600	12,400	16,000
1650	2800	9900	12,700	16,500
1700	2800	10,200	13,000	17,000
1750	2800	10,500	13,300	17,500
1800	2800	10,800	13,600	18,000
1850	2800	11,100	13,900	18,500
1900	2800	11,400	14,200	19,000
1950	2800	11,700	14,500	19,500
2000	2800	12,000	14,800	20,000

The graphical solution to the above problem is depicted in the following Figure 12.4.

12.4 Benefit Cost Analysis

In simple terms, benefit cost ratio is the ratio of the present worth of benefits to the present worth of costs. Alternatively, it can also be expressed as the ratio of equivalent uniform annual benefit to equivalent uniform annual costs. Both the present worth, as well as the equivalent uniform annual benefits and costs, are computed using a minimum attractive rate of return (MARR). Mathematically, this ratio is expressed as follows:

$$\text{Benefit cost ratio} = \frac{\text{Present worth of benefit}}{\text{Present worth of cost}} = \frac{\text{Equivalent uniform annual benefit}}{\text{Equivalent uniform annual cost}}$$

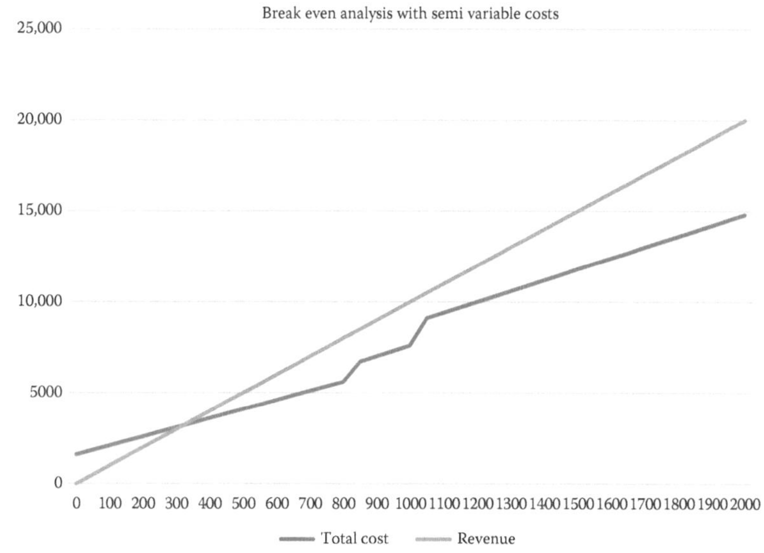

FIGURE 12.4
Break even analysis with semi-variable costs.

The basic concept of benefit cost analysis is to maximize the benefit for every unit of currency invested (cost) into an asset. Different criteria used in benefit cost analysis are presented in Table 12.1.

It needs to be kept in mind that with benefit cost analysis, the total profit is sought to be maximized and not the rate of profit. In order to better understand the concept of benefit cost analysis, let us solve a problem using the principles outlined in Table 12.1.

EXAMPLE 12.5

As a production manager, you are required to analyze two machines that are being considered for purchase. The interest rate is assumed to be 7% per year. Which machine would you buy in light of the data presented in the following table?

Benefit and Cost Details	Machine Y	Machine Z
Purchase price	$400,000	$1,400,000
Uniform annual benefit	$190,000	$240,000
Salvage value	$100,000	$300,000
Useful life in years	6	12

Solution

Before we proceed with benefit cost analysis, we need to make sure that both alternatives are analyzed using the same useful life. It will be noticed from the above table that machine A has a useful life of 6 years whereas machine B has a useful life of 12 years. In such cases, as

TABLE 12.1

Criteria Used to Solve Benefit Cost Analysis Problems

Input or Output	Condition	Criteria for Solution
Input is variable. Output is variable as well	Input and output are both variable. Neither is fixed	Comparison between two alternatives: • Computation of incremental benefit cost ratio (ΔB/ΔC) on the increment of investment between the alternatives • If comparison is between three or more alternatives, the problem should be solved using benefit cost ratio incremental analysis
Input is fixed	Input resources such as money, etc., are fixed	Maximize the value of B/C
Output is fixed	Output to be accomplished, benefit is fixed	Maximize the value of B/C

described in Chapter 1, we analyze the alternatives based on the least common multiple of useful lives, which in this case happens to be 12 years. (Least common multiple of 6 and 12 is 12). In order to compute cash flows for years 7–12 for machine Y, we assume that it is replaced at the end of year 6 and the replacement has identical cash flows as the original.

A graphical depiction of the cash flows for both machines is presented in the following figure.

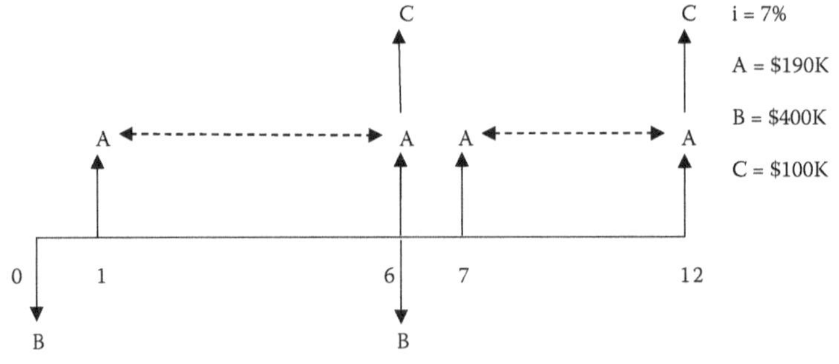

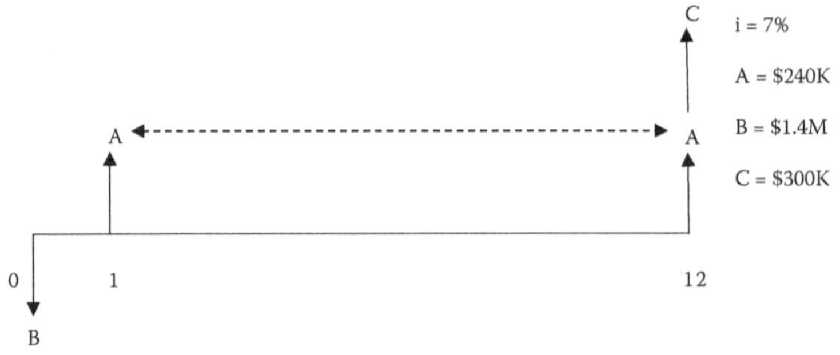

We use the following formula to compute benefit cost ratios:

$$\frac{\text{Benefit}}{\text{Cost}} = \frac{\text{EUAB}}{\text{EUAC}}$$

The equivalent uniform annual benefit is already provided in the data for each machine. Thus, the EUAB for machine Y is $190,000 per year and that for machine Z is $240,000 per year. Thus, the numerator values are already known. In order to compute the equivalent uniform annual costs (EUAC) for both machines, we will convert the purchase price into an annual cash flow. The salvage value can be incorporated into the computation as a decrease in EUAC, given the fact that salvage value represents a positive cash flow.

The EUAC computation for machines Y and Z is presented as follows:

Machine Y

$$\text{EUAC} = \$400{,}000\ (A/P, 7\%, 6) - \$100{,}000\ (A/F, 7\%, 6).$$

But we said we decided to analyze both options based on the least common multiple of useful lives. In this case, since the cash flows for machine Y from years 6 through 12 are identical to those for the first six years, they can just be replicated and the value of EUAC will be the same as the one obtained here.

Thus,

$$\begin{aligned}\text{EUAC} &= \$400{,}000\ (0.2098) - \$100{,}000\ (0.1398) = \$83{,}920 - \$13{,}980 \\ &= \$69{,}940\end{aligned}$$

Thus, the equivalent uniform annual cost (EUAC) for machine Y is $69,940.

The equivalent uniform annual benefit (EUAB) for machine Y is $190,000.

Thus, the benefit cost ratio for machine Y is B/C = 2.7166.

Machine Z

Machine Z has a useful life of 12 years. Thus, no theoretical replacement at the end of year 6 is necessary in this case. The computation for EUAC for machine Z is quite straightforward. It is presented as follows:

$$\text{EUAC} = \$1{,}400{,}000\ (A/P, 7\%, 12) - \$300{,}000\ (A/F, 7\%, 12)$$

$$\text{EUAC} = \$1{,}400{,}000\ (0.1259) - \$300{,}000\ (0.0559).$$

Thus,

The value of EUAC for machine Z = $176,260—$16,770 = $159,490

The benefit cost ratio for machine Z = $240,000/$159,490 = 1.504

Notice from the problem statement that neither input as represented by cost nor the output as represented by the benefit is fixed in this case. Both are variable. In these situations, referring to Table 12.1, we use incremental analysis to find the best alternative.

Incremental Analysis

Here are the final figures prior to beginning incremental analysis:

	Machine Y	Machine Z
EUAC ($/year)	69,940	159,490
EUAB ($/year)	190,000	240,000

Observe that machine Z costs more than Y since $EUAC (Z) > EUAC (Y). Thus, we perform incremental analysis: Z–Y as follows:

$$\frac{\Delta B}{\Delta C} = \frac{\$240,000 - \$190,000}{\$159,490 - \$69,940} = \frac{\$50,000}{\$89,550} = 0.5583 < 1.$$

It is clear from the incremental analysis that the increment is a value <1. Thus, this increment is undesirable. Thus, we select machine Y as the better alternative. In this example, the B/C value for machine Y happened to be greater than that for machine Z. This not always the case. Often, the machine with the lower B/C value is preferable based on incremental analysis. Also, note that incremental analysis should be done only when neither input nor output is fixed.

Next, let us consider an example with more than two alternatives.

EXAMPLE 12.6

An investment company is trying to make a decision as to which of five mutually exclusive investment alternatives it should choose. Each alternative has a fifteen year useful life and no salvage value. Which alternative should be selected if the minimum attractive rate of return in 7%?

The data regarding costs and benefits is presented in the following table.

	X	Y	Z	U	V	W
Cost ($)	8000	4000	12,000	2000	18,000	20,000
Present worth of benefits ($)	15,000	10,000	18,000	3200	18,000	19,500
PW (benefits)/PW (costs)	1.875	2.50	1.5	1.6	1	0.975

Solution

Cursory examination of the problem reveals the fact that neither input as expressed by cost nor output as expressed by benefits is fixed. The present worth of cost and benefits has already been given in the problem and as such there is no need to compute those values separately. Thus, it is necessary to use incremental analysis in order to solve the problem.

The following procedure needs to be followed when solving problems with multiple alternatives using incremental analysis:

1. All alternatives need to be properly identified.
2. The benefit cost ratio of each alternative needs to be computed. If the B/C ratio for any alternative <1, that alternative should be discarded. From the above table, it will be observed that all alternatives have a B/C ratio ≥1 except alternative W. The B/C ratio for alternative W is 0.975, which is less than 1. This alternative will be discarded.
3. The remaining alternatives will then be arranged in increasing order of cost. The accompanying benefits will be listed alongside each alternative as follows:

	U	Y	X	Z	V
Cost ($)	2000	4000	8000	12,000	18,000
Present worth of benefits ($)	3200	10,000	15,000	18,000	18,000
PW (benefits)/PW (costs)	1.6	2.5	1.875	1.5	1.00

Notice from the above arrangement that all alternatives have been arranged in increasing order of cost beginning with $2000 and ending with $18,000. The present worth of benefits for each alternative is listed alongside. The corresponding B/C ratio is listed for each alternative as well.

4. Perform incremental analysis for each alternative pair such as Y–U, X–U, X–Y, etc. Compute the ΔB/ΔC value for each pair. If the ΔB/ΔC value for any increment is <1, that increment will be discarded. This analysis is presented in the following table.

	Y–U	X–U	Z–U	V–U	X–Y	Z–Y	V–Y	Z–X	V–X	V–Z
ΔC ($)	2000	6000	10,000	16,000	4000	8000	14,000	4000	10,000	6000
ΔB ($)	6800	11,800	14,800	14,800	5000	8000	8000	3000	3000	0
ΔB/ΔC	3.40	1.967	1.48	0.925	1.25	1.00	0.571	0.75	0.30	0

It will be observed from the above table that the following increments will be discarded due to the fact that the ΔB/ΔC ratio values are <1.

- Increment V–U will be discarded since the incremental benefit cost value = 0.925 < 1
- Increment V–Y will be discarded since the incremental benefit cost value = 0.571 < 1
- Increment Z–X will be discarded since the incremental benefit cost value = 0.75 < 1
- Increment V–X will be discarded since the incremental benefit cost value = 0.30 < 1
- Increment V–Z will be discarded since the incremental benefit cost value = 0.0 < 1

Thus, we conclude from the above observations that the alternative 'V' is NOT preferable over either of alternatives X, Y, Z or U (the remainder of the alternatives). Due to this fact, alternative V will be discarded as well. This narrows the field down to four alternatives, namely X, Y, Z, and U.

We observe the following among the alternatives Z, Y, Z, and U:

- Increment Y–U has a $\Delta B/\Delta C$ value of 3.40 > 1. Thus, alternative Y is preferred over alternative U.
- Increment X–U has a $\Delta B/\Delta C$ value of 1.967 > 1. Thus, alternative X is preferred over alternative U.
- Increment Z–U has a $\Delta B/\Delta C$ value of 1.48 > 1. Thus, alternative Z is also preferred over alternative U.

This implies that alternative U will be discarded as well since better alternatives (X, Y, and Z) are available.

Narrowing the field down to three alternatives: X, Y, and Z, we make the following observations:

- Increment X–Y has a value of 1.25 > 1. Thus, X is a feasible increment over Y.
- Increment Z–Y has a value of 1.00 = 1. Thus, Z is a feasible increment over Y.

Because feasible increments are available over alternative Y, we discard alternative Y next.

This narrows the field down to two alternatives, namely X and Z.

We notice the increment Z–X has a value of $\Delta B/\Delta C = 0.75 < 1$. Thus, alternative Z is not a feasible increment over X leaving us with one alternative only, namely alternative X. This implies that the investment company should choose alternative X as the best option. **Solution**

12.5 Conclusion

In this chapter, we studied two methods of cost analysis, namely break even analysis and benefit cost analysis. The first half of the chapter dealt with the basic concept of break even analysis. It considered various applications of this method such as profitability analysis, consideration of different alternatives, etc. The concept of fixed cost and variable cost was also discussed in detail.

The second half of the chapter presented the concept of benefit cost analysis. The various scenarios in which this method can be used were presented. The application of incremental analysis to solve problems related to benefit cost analysis in order to maximize total profit was also presented and explained in detail through rigorous problem solving.

PROBLEMS

1. Eagle, Inc. is trying to make a decision as to which production process to choose in order to fulfill an order of 50,000 units. It has identified three alternatives for this purpose. The details of fixed costs and variable costs for each alternative are presented below. Determine the range of production over which the company can choose each alternative.

Production Process	Fixed Cost/Year ($)	Variable Cost/Year ($/unit)
X	200,000	42.00
Y	450,000	12.00
Z	315,000	17.00

2. An assembly operation is performed by a worker at a direct labor cost of $2.00 per unit. A robotic assembler costing $20,000 can perform the same task and reduce the direct labor cost to $0.25 per unit. The assembler will have no value at the end of its useful life of five years and can be disposed of at that time. Determine the minimum number of units that must be assembled each year in order to justify investing in the machine.
3. A company incurs the following costs during its manufacturing operation:
   - Fixed cost of $15,000
   - Variable cost of $25 per unit
   - The relationship between price and demand is given as: $P = 0.50D + 500$, wherein D is the annual demand and P is the sales price per unit of product
   - Total manufacturing cost = fixed cost + variable cost
   - Revenue = demand × price
   - Profit = revenue − total cost

   Based on the information above, compute the following:

   a. What is the break even quantity of units that must be manufactured in order to recover all costs?

   b. If total revenue was maximized, how much profit would be earned by the company?

   c. What is the maximum possible profit that can be earned by the company in light of the aforementioned figures?
4. The sales volume of a product is a direct function of its selling price. This is given by the equation: $S = \$250 - P$, wherein S is the number of units of product sold and P is the selling price per unit.

The manufacturing cost is given by the equation: C = \$2000 + 25S. Determine the following:

a. What is the break even point in terms of the lowest sales volume at which total sale income is equal to the total costs?

b. How many units should the company sell in order to maximize its revenue?

c. How many units should be sold in order to maximize total profit? Is this quantity different from the solution to part b?

5. The costs for three different process designs for an automated welding application have been presented in the following table.

Production Process	Fixed Cost/Year ($)	Variable Cost/Year ($/unit)
U	200,000	40.00
V	700,000	22.00
W	1,250,000	17.00

Compute the range of annual production over which each alternative should be used. The company wishes to manufacture a total of 350,000 units. Chart your answer graphically.

6. Four mutually exclusive alternatives, each with a seven year useful life, are being considered. The minimum attractive rate of return (MARR) is 7%. Which alternative should be chosen? Use benefit cost analysis to solve the problem.

	A	B	C	D
Cost ($)	1200	1000	450	600
Uniform annual benefit ($/year)	320	280	120	290

7. Use benefit cost analysis to determine which of the following three mutually exclusive alternatives should be chosen. Each alternative has a useful life of five years and the MARR is 10%.

	A	B	C
Cost ($)	1000	700	275
Uniform annual benefit ($/year)	250	220	100
Salvage value ($)	90	10	15

8. Using benefit cost analysis, determine which of the following five mutually exclusive alternatives should be chosen. Each alternative has a useful life of ten years and the MARR is 15%.

	A	B	C	D	E
Cost ($)	220	450	650	875	1100
Uniform annual benefit ($/year)	75	145	170	265	315

9. Use benefit cost analysis to determine which of the following alternatives should be chosen. Each alternative has a useful life of 7 years and the MARR is 12%.

	Q	R	S
Cost ($)	6000	430	155
Uniform annual benefit ($/year)	450	390	300
Salvage value ($)	75	28	25

For problems 10–13, consider the following:

The Wheezy Windows Company has the capacity to produce 2,500,000 window units each year. Currently, it is operating at 80% of capacity. The company's annual revenue is $900,000. The annual fixed costs are $400,000 and variable costs are $1 per unit.

10. What is the company's annual profit (or loss)?
11. When does the company break even (sales percentage)?
12. What will be the profit or loss if the plant runs at 92% of capacity (assuming that income per unit and the variable cost per unit remain constant)?
13. What percent of capacity would the company have to run at in order to earn a profit of $100,000?
14. A dam and reservoir provide benefits of $100 million over the next 75 years. The cost of construction, maintenance, and operation will be $50 million. A single flood control dam will have benefits of $30 million and would cost $30.5 million. Modifications can be made to the dam and reservoir to provide the control dam benefits as well, leading to a total of $77 million. If funds are available, what should be the course of action?
15. Lincoln City is considering replacing its police radio. The benefits and costs of the new radio are shown below. What is the replacement's benefit/cost ratio of the annual interest is 7%?

    Purchase cost: $10,000

    Annual savings: $2750

    Life: 12 years

For problems 16–18, consider the following.

Polar Publishing has produced a textbook that is distributed throughout university bookstores. It is estimated that 25,000 copies will be sold per year and the fixed cost of production is $85,000. The average cost of materials is $45.50 and the cost of labor is $17.50.

16. If Polar Publishing sells the textbook for $85, what is:
    a. The annual profit
    b. The break even point in the books produced
    c. The break even point in sales ($)
17. What if the selling price was $75? Repeat parts a–c.
18. What if the selling price was $90? Repeat parts a–c. For problems 19–20, consider the following. The cost of tooling and labor required to set up for a machining job is $700. Once set up, the cost to produce one unit is $6.50 for the material and $2.50 for labor to operate the machinery.
19. If each unit is sold for $11, determine the production quantity required to break even and net profit (or loss) if the size is 1500 units.
20. If each unit is sold for $12.50, determine the production quantity required to break even and net profit (or loss) if the size is 3200 units.

    For problems 21–23, consider the following.

    A firm has the capacity to produce 5,500,000 units each year. Currently, it is operating at 72% of capacity. The company's annual revenue is $3,000,000. The annual fixed costs are $700,000 and variable costs are $4 per unit.
21. What is the company's annual profit (or loss)?
22. When does the company break even (sales percentage)?
23. What will be the profit or loss if the plant runs at 98% of capacity (assuming that income per unit and the variable cost per unit remain constant)?

13

Depreciation

13.1 What Is Depreciation?

In accounting terminology, depreciation can be defined as a decrease in value. Depreciation takes place as follows:

- Decrease in market value of an asset
- Decrease in value of an asset to its owner
- Allocation of the capital cost of an asset over a certain period of time also referred to as "depreciable life"

For all practical purposes, depreciation is an accounting entry that allocates a certain amount of the original cost of capital equipment such as buildings, machines, manufacturing and production equipment, etc., every year to be deducted from revenue for tax purposes. For instance, when a piece of equipment such as a forklift is purchased for a manufacturing material handling operation, it usually lasts many years. However, although it is not consumed directly, it does get "used up" and thus loses value. This loss in value occurs over time because of a variety of factors such as deterioration, obsolescence or just age. Costs also need to be incurred on the maintenance of such equipment in order to keep them in proper working order. This loss in value and concomitant maintenance expense is clearly a business expense. The US government has made provisions that allow companies to deduct business expenses of this nature from their gross income for tax purposes.

Business assets can be depreciated only under the following conditions:

- The asset must be used for the purpose of producing an income.
- The asset must have a determinable useful life which should exceed one year.
- Over the course of time, the asset must get used up, decay, wear out, become obsolete or lose value to the owner through some natural cause.

It is clear from the preceding discussion that depreciation can occur due to two main factors: Deterioration and obsolescence.

13.1.1 Depreciation due to Deterioration

Deterioration can also be characterized as wear that happens over time due to use or a variety of other factors. A machine or piece of equipment that is undergoing deterioration does not function as intended and requires a considerable amount of maintenance and upkeep. In many instances, the machine may work as intended, but the quality of output may suffer a decline due to wear associated deterioration. A car could be an excellent case in point. It will be appreciated that as a machine, a car requires constant maintenance in order to fulfill its intended function. A car that is being used every day deteriorates due to wear. Examples of automotive wear include wear on components such as wheel bearings, brakes, rotors, pistons, steering pumps, etc.

13.1.2 Depreciation due to Obsolescence

A machine could be in excellent working condition and perform its intended function, yet could lose value because of the obsolescence of the underlying technology. Examples of obsolete technology include engineering calculators from the 1970's. These devices used to operate on AA batteries and contained light emitting diodes (LEDs). Needless to say, they consumed a substantial amount of energy and were quite restricted in their computing capabilities. The evolution of the cellular phone has followed a similar pattern. In the 1990s, several high end cars used to feature a "car phone" to enable drivers to make a phone call from their car. With the advent of the cell phone and smart phones, the aforementioned technology quickly became obsolete in the 2000s.

Business assets, regardless of their mode of depreciation, can be classified as either tangible or intangible. The following discussion differentiates between these two classes.

13.1.3 Tangible Property

An asset is deemed as being tangible if it can be seen, felt or touched. Tangible property can be further classified as real property and personal property.

- Real Property: Real property includes assets such as land, buildings, and all things attached to the land such as forests, mines, mineral deposits, etc. Examples include houses, buildings, golf courses, precious metal deposits, fossil fuel deposits, etc.
- Personal Property: Personal property includes equipment, vehicles, furnishings, machinery, etc. that is tangible but cannot otherwise be classified as real property. Examples include cars, trucks, production equipment, material handling devices, etc.

13.1.4 Intangible Property

An asset is deemed as being intangible if it cannot be seen, touched or felt. However, it is valuable to the owner. Examples of intangible property include copyrights, trademarks, franchises, goodwill, etc.

It should be noted that almost all tangible property can be depreciated except land. Land is an asset that is never depreciated. This is because of the fact that land does not wear out, lose value or have a determinable useful life. It is observable that over time, instead of losing value, land actually gains in value. Other tangible assets that cannot be depreciated include factory inventory and leased property. In the case of leased property, only the owner can claim depreciation on the property due to the fact that it is a business asset. The tenant cannot claim depreciation on a leased property. Tangible property such as a car used by a realtor to facilitate real estate transactions (such as driving to and from a property for sale) can be depreciated in proportion to the extent that it is being used for business purposes.

Depreciation is essentially a book keeping entry and as such is a noncash cost. The depreciation is claimed by a business only on paper. The "cost" incurred through depreciation is not carried by any entity. Depreciation, taken as a deduction, reduces the taxable income, thus reducing the amount of taxes paid on business income.

13.2 Terminologies Associated with Computing Depreciation

The following terminologies are commonly used in calculating depreciation regardless of the method of calculation.

- *Purchase Price*: This term is used to refer to the price paid in order to purchase the asset. It is denoted by "B."
- *Depreciation*: This is the amount of "loss of value" that occurs over any particular time frame, and is allowed to be deducted from gross income for tax purposes. It is denoted by the term D_t.
- *Book Value*: Book value is defined as the amount of residual value left after depreciation for any particular year has been accounted for. It is denoted by the term B_t.

$$\text{Book value for year "t"} = \text{cost basis} - \text{depreciation to date}$$

- *Salvage Value*: Salvage value is the amount of value remaining in the asset after all of the depreciation has been accounted for. Thus, by definition, salvage value is usually equal to the purchase price minus sum total of all depreciation. It is denoted by the term S.

$$\text{Salvage value} = \text{book value} - \sum \text{depreciation for all years}$$

It will be clear from the above formulas that over time, book value will approach and eventually equal salvage value.

- *Life of the asset*: Life of the asset is the number of years over which the asset will be depreciated. It is denoted by the term N.

13.3 Straight Line Depreciation

Straight line depreciation is the easiest and most familiar method of computing depreciation. By its nature, the depreciation charge computed using this method is constant every year over which the charge is allocated. The annual depreciation charge using the straight line method can be computed by using the following formula:

$$\text{Annual depreciation charge}\, D_t = \frac{B-S}{N} \qquad \text{(Equation 1)}$$

From the above equation, it will be obvious that the depreciation charge per year will be constant, no matter which year is in question.

Consider the following example:

EXAMPLE 13.1

A building is purchased for a price of $1,000,000. The depreciable life of the as set is five years and its salvage value is assumed to be $250,000. Compute the straight line depreciation schedule for this asset. What is the book value at the end of year 2?

Solution

When dealing with word problems, it is always a good idea to write down the data first. This makes it easier to visualize what has already been presented in the problem and what needs to be computed.

In the present situation, the following data has been provided:

- Purchase price (B) = $1,000,000
- Depreciable life (N) = 5 years
- Salvage value (S) = $250,000

The following are the unknowns that need to be computed using the straight line method:

- Depreciation per year.
- Book value at the end of each year.
- Book value at the end of year 2. This will be computed as a matter of course while solving the problem.

We compute the depreciation per year using Equation 1. For straight line depreciation, the value of depreciation per year is constant regardless of the year under consideration. It is computed by using the data provided above.

$$D_t = \frac{B - S}{N}$$

Thus, the annual depreciation charge $= D_t = \frac{\$1{,}000{,}000 - \$250{,}000}{5}$ $= \$150{,}000$ per year.

Next, the book value at the end of the first year is computed as the difference between the purchase price and the depreciation charge for the first year.

Thus, $B_1 = \$1{,}000{,}000 - \$150{,}000 = \$850{,}000$.

Recursive iteration of the aforementioned principles provides the final depreciation schedule as follows:

Year (t)	Depreciation for Year "t" (D_t)	Sum of Depreciation Charges	Book Value at the End of Year "t" (B_t) = B − Σ Depreciation Charges to Date
1	150,000	150,000	850,000
2	150,000	300,000	*700,000 = Book value at end of year 2*
3	150,000	450,000	550,000
4	150,000	600,000	400,000
5	150,000	750,000	*250,000 = S*

Note that the book value at the end of year five by definition is the salvage value, which was $250,000 as given in the problem.

13.4 Sum of Years' Digits Depreciation (SOYD)

The sum of years' digits deprecation method, (henceforth referred to by its acronym SOYD) is a depreciation method that results in higher depreciation charges upfront which progressively diminish charges over time. Each year, the depreciation charge is equal to a fraction of the total amount of depreciation. This method gets its name (SOYD) from the fact that the denominator of the aforementioned fraction is equal to the sum total of all years over which the asset will be depreciated.

For instance, if the life of the asset is seven (7) years, the value of SOYD will be equal to $7 + 6 + 5 + 4 + 3 + 2 + 1 = 28$. The depreciation charge takes the form of an arithmetic gradient.

The following equations are used when computing the depreciation charge using the SOYD method.

$$D_t = \frac{(N-t+1)(B-S)}{SOYD} \qquad \text{(Equation 2)}$$

$$SOYD = \frac{(N)(N+1)}{2} \qquad \text{(Equation 3)}$$

The following terminologies apply to Equations 2 and 3

D_t = Depreciation charge in year "t"
B = Cost of the asset
S = Salvage value of the asset
N = Number of years over which the asset will be depreciated
SOYD = Sum of years' digits as computed from Equation 3

Consider the following example.

EXAMPLE 13.2

A crane is purchased for a price of $750,000. The depreciable life of the asset is seven years and its salvage value is assumed to be $200,000. Compute the depreciation schedule for this asset using the SOYD method. What is the book value at the end of year 4?

Solution

In the present situation, the following data has been provided:

- Purchase price (B) = $750,000
- Depreciable life (N) = 7 years
- Salvage value (S) = $200,000

The following are the unknowns that need to be computed using the straight line method:

- Depreciation per year.
- Book value at the end of each year.
- Book value at the end of year 4. This will once again be computed as a matter of course while solving the problem.

We use Equations 2 and 3 to solve this problem.

$$D_t = \frac{(N-t+1)(B-S)}{SOYD} \qquad \text{(Equation 2)}$$

$$SOYD = \frac{(N)(N+1)}{2} \qquad \text{(Equation 3)}$$

In order to compute the depreciation charge per year, it is first necessary to compute the value of SOYD. We can either use Equation 3 to accomplish this or simply add up the values of all years from 1 through 7. Thus, $1 + 2 + 3 + 4 + 5 + 6 + 7 = 28$.

Conversely, using Equation 3 in the present case, SOYD = $(7)(7+1)/2 = 28$.

The value of the depreciation charge per year is computed by using Equation 2. It will be noted that unlike straight line deprecation, the charge using the SOYD method will be different for different years. Thus, we have the following depreciation charges per year using SOYD:

For year 1, the value of $t = 1$. For year 2, the value of $t = 2$ and so on. The value of SOYD = 28 as we have already established. This value will remain unchanged for the duration of the problem.

$$D_1 = \frac{(7-1+1)(\$750{,}000 - \$200{,}000)}{28} = \$137{,}500$$

Similarly, by applying the above logic, we have the values for depreciation charge for the remaining years as follows:

$$D_2 = \frac{(7-2+1)(\$750{,}000 - \$200{,}000)}{28} = \$117{,}857.14$$

$$D_3 = \frac{(7-3+1)(\$750{,}000 - \$200{,}000)}{28} = \$98{,}214.28$$

$$D_4 = \frac{(7-4+1)(\$750{,}000 - \$200{,}000)}{28} = \$78{,}571.42$$

$$D_5 = \frac{(7-5+1)(\$750{,}000 - \$200{,}000)}{28} = \$58{,}928.57$$

$$D_6 = \frac{(7-6+1)(\$750{,}000 - \$200{,}000)}{28} = \$39{,}285.71$$

$$D_7 = \frac{(7-7+1)(\$750{,}000 - \$200{,}000)}{28} = \$19{,}642.85$$

The depreciation schedule is presented in the following table:

From the following depreciation schedule, it will be noticed that the depreciation charge per year reduces by the same amount. For instance, the difference between the depreciation charge for years 1 and 2 is \$137,500 – \$117,857 = \$19,643. This is the same difference between depreciation charges for years 2 and 3, 3 and 4, and so on. Thus, it can be proved that the depreciation charges using the SOYD method follow an arithmetic gradient.

Year (t)	Depreciation for Year "t" (D_t)	Sum of Depreciation Charges	Book Value at the End of Year "t" (B_t) = B − $\sum$ Depreciation Charges to Date
1	137,500	137,500	612,500
2	117,857	255,357	494,643
3	98,214	353,571	396,429
4	78,571	432,143	*$317,857 = Book value at the end of year 4*
5	58,929	491,071	258,929
6	39,286	530,357	219,643
7	19,643	550,000	*200,000 = S*

13.5 Double Declining Balance

The Double Declining Balance Method also referred to as DDB applies a constant depreciation rate to an asset's declining book value. Prior to the 1981 and 1986 tax revisions, two rates were commonly in use: the 150% and the 200% of the straight line rate. The 200% is twice the straight line rate. This is the reason it is referred to as the double declining balance.

The depreciation charge per year using double declining balance is computed using Equation 4 with all depreciation notations used in their standard forms:

$$D_t = \frac{(2)(\text{Cost} - \text{Depreciation charges to date})}{N} \qquad \text{(Equation 4)}$$

Consider the following example:

EXAMPLE 13.3

An asset is purchased for a price of $1,250,000. The depreciable life of the asset is five years and its salvage value is assumed to be $100,000. Compute the depreciation schedule for this asset using the double declining balance method. What is the book value at the end of year 3?

Solution

In the present situation, the following data has been provided:

- Purchase price (B) = $1,250,000
- Depreciable life (N) = 5 years
- Salvage value (S) = $100,000

The following are the unknowns that need to be computed using the straight line method:

- Depreciation per year.
- Book value at the end of each year.
- Book value at the end of year 3. This will once again be computed as a matter of course while solving the problem.

We use Equation 4 to solve this problem.

$$D_t = \frac{(2)(\text{Cost} - \text{Depreciation charges to date})}{N}$$

We have the following depreciation charges per year:

For year 1, the depreciation charges to date are equal to \$0, this being the first year.

Thus, $D_1 = \frac{(2)(1,250,000 - 0)}{5} = \$500,000.$

For the second year, depreciation to date amounts to \$500,000.

Thus, $D_2 = \frac{(2)(1,250,000 - 500,000)}{5} = \$300,000.$

For the third year, depreciation to date will be equal to the sum total for the depreciation charges for the first two years. This will be equal to \$500,000 + \$300,000. This amounts to \$800,000.

Thus, $D_3 = \frac{(2)(1,250,000 - 800,000)}{5} = \$180,000.$

For the fourth year, depreciation to date will be equal to the sum total of the depreciation charges for the first three years. This will be equal to \$500,000 + \$300,000 + \$180,000 = \$980,000.

Thus, $D_4 = \frac{(2)(1,250,000 - 980,000)}{5} = \$108,000.$

Using the same principle as elucidated above, we have:

$$D_5 = \frac{(2)(1,250,000 - 1,088,000)}{5} = \$64,800.$$

Year (t)	Depreciation for Year "t" (D_t)	Sum of Depreciation Charges	Book Value at the End of Year "t" (B_t) = B − $\sum$ Depreciation Charges to Date (\$)
1	500,000	500,000	750,000
2	300,000	800,000	450,000
3	180,000	980,000	*\$270,000 = Book value at end of year 3*
4	108,000	1,088,000	162,000
5	64,800	1,152,800	*97,200* $\neq S$

From the above table, it will be noticed that the salvage value does not match the value given in the problem. This is primarily due to the fact that the formula for computing depreciation charges using DDB does not take the salvage value into account. In this particular situation, the depreciation is stopped once the book value becomes equal to the salvage value, thus preventing any company from taking too much depreciation. Conversely, the method can switch back to straight line depreciation from DDB, thus ensuring that sufficient depreciation is taken.

13.6 Modified Accelerated Cost Recovery System (MACRS)

The Modified Accelerated Cost Recovery System (MACRS) method of depreciation is a simplified method of allocating depreciation and is characterized by three major features:

- Salvage lives for all assets are assumed to be zero.
- Tables of annual depreciation percentages are precalculated for simplification of the process.
- Property class lives are less than the actual useful lives of the assets.

The MACRS revolves around the general depreciation system (GDS). This system is based on the declining balance method of computation with a switch to straight line method. This can be contrasted with the alternative depreciation system (ADS) which uses straight line depreciation and provides for a longer recovery period. The ADS system is used according to US law in the following circumstances:

- Any tangible property outside the US
- Any property that is tax exempt or financed by tax exempt bonds
- Farming property placed in service when uniform capitalization rates are not applied

The ADS system can used to depreciate a property that is depreciable using the GDS system. The disadvantage is that it is not possible to switch back to the GDS system once the ADS has been elected. In general, the ADS makes the depreciation deductions less valuable.

13.6.1 How to Use the MACRS GDS System

The following steps need to be followed in order to use the MACRS GDS system:

- Determine the cost basis of the asset (B).

- Determine the property class and recovery period of the asset.
- Determine the date at which the asset was placed in service.

The cost of the asset (B) is defined as the amount of money required to obtain the asset and place it in service. This can include fees and charges paid by the buyer in order to facilitate the purchase. This can include legal and recording fees, abstract fees, transfer taxes, title insurance, back taxes, etc.

Property class and recovery period can be obtained by placing each asset into a MACRS property class that defines the recovery period. Annual depreciation percentages are precalculated for each property class for the sake of simplification. Table 13.1 presents the class lives for several types of GDS

TABLE 13.1

Class Lives and MACRS Property Classes

IRS Asset Class	Asset Description	Class Life (Years) ADR	MACRS Property Class	
			GDS	ADS
00.11	Office furniture, fixtures, and equipment	10	7	10
00.12	Information systems, computers/peripherals	6	5	6
00.22	Automobiles, taxis	3	5	5
00.241	Light general purpose trucks	4	5	5
00.25	Railroad cars and locomotives	15	7	15
00.40	Industrial steam and electric distribution	22	15	22
01.11	Cotton gin assets	12	7	12
01.21	Cattle, breeding or dairy	7	5	7
13.00	Offshore drilling assets	7.5	5	7.5
13.30	Petroleum refining assets	16	10	16
15.00	Construction assets	6	5	6
20.10	Manufacture of grain and grain mill products	17	10	17
22.2	Manufacture of yarn, thread, and woven fabric	11	7	11
24.10	Cutting of timber	6	5	6
32.20	Manufacture of cement	20	15	20
37.11	Manufacture of motor vehicles	12	7	12
48.11	Telephone communications assets and buildings	24	15	24
48.2	Radio and television broadcasting equipment	6	5	6
49.12	Electric utility nuclear production plant	20	15	20
49.13	Electric utility steam production plant	28	20	28
49.23	Natural gas production plant	14	7	14
50.00	Municipal waste water treatment plant	24	15	24
80.00	Theme and amusement park assets	12.5	7	12.5

TABLE 13.2

Property Classes for the MACRS-GDS System

Property Class	Personal Property (All Property Except Real Estate)
3 Year Property	Special handling devices for food and beverage manufacture Special tools for the manufacture of finished plastic products, fabricated metal products, and motor vehicles Property with ADR class life of 4 years or less
5 Year property	Automobiles and trucks Aircraft of nonair transport companies Equipment used in research and experimentation Computers Petroleum drilling equipment Property with ADR class life of more than 4 years and less than 10 years
7 Year property	All other property not assigned to another class Office furniture, fixtures, and equipment Property with ADR class life of 10 years or more and less than 16 years
10 Year Property	Assets used in petroleum refining and certain food products Vessels and water transportation equipment Property with ADR class life of 16 years or more and less than 20 years
15 Year property	Telephone distribution plants Municipal sewage treatment plants Property with ADR class life of 20 years or more and less than 25 years
20 Year property	Municipal sewers Property with ADR class life of 25 years or more
Property Class	**Real Property (Real Estate)**
27.5 Years	Residential rental property not including hotels and motels
39 Years	Nonresidential rental property

and ADS property classes whereas Table 13.2 is a tabulation of MACRS-GDS property classes. The MACRS class that is a most appropriate fit to the asset in question can be ascertained by a variety of methods as delineated below:

- Use the property class as given in the problem.
- The asset description is located in Table 13.2.
- Use IRS tables or Table 13.1 to locate the asset.
- Use class life.
- If it is not possible to locate the asset description, the seven year property class should be used. One of the asset descriptions for the seven year property class is: "All other property not assigned to another class."

Once the MACRS property class, the placed in service date, and cost basis have been ascertained, the annual depreciation deductions can be computed using the following equation:

$$D_t = B \times r_t \qquad \text{(Equation 5)}$$

D_t = Depreciation deduction for year "t"
B = Cost basis of asset being depreciated
r_t = MACRS percentage rate depending on the asset class for year "t"

The following assumptions are made in determining MACRS percentage rates:

- Salvage values for all assets regardless of property class are assumed be zero.
- The first and last years of the recovery period are both assumed to be a half year.
- The declining balance for 3, 5, 7, and 10 year property is 200%, whereas for 15 and 20 year property, it is 150%.

13.6.2 Computation Method

- The following classes use 200% declining balance: 3 year, 5 year, 7 year, 10 year classes.
- The following classes use 150% declining balance: 15 year and 20 year classes.
- All classes convert to straight line depreciation in the optimal year designated by *.
- Half year depreciation is allowed in the first and last recovery years.
- If more than 40% of the year's MACRS property is placed in service in the last three months, a midquarter convention is used.

In order to better understand the concept of MACRS depreciation and distinguish from other methods of depreciation, consider the following example.

EXAMPLE 13.4

Using the MACRS-GDS method, compute the annual depreciation allowances and book values for a material handling device valued at $750,000 for the special handling of food and beverage manufacture. The device is expected to have a salvage value of $100,000 at the end of its useful life. Compute the book value at the end of year 2.

Solution

In the above problem, the following data has been provided:

- Cost basis of the asset (B) is given as $750,000.
- The asset is clearly depreciable and its description can be found in Table 13.2.

- Special handling devices for food and beverage manufacture are listed as a three year property.
- MACRS-GDS applies to the asset under question because of the aforementioned facts.
- The crucial point to bear in mind is that in this problem, the salvage value that is given ($100,000) is extraneous information. Always remember that for the MCRS-GDS system computations, the salvage value is always assumed to be zero. In other words, the entire cost basis of the asset is depreciated over the life of the asset.

Using the depreciation schedule for a three year property (MACRS-GDS system) from Table 13.3, we have the following depreciation computation for the property in question.

Depreciation for the first year is calculated as:

$$D_1 = 33.33\% \text{ of } \$750{,}000 = \$249{,}975.$$

Depreciation for the second year is calculated as:

$$D_2 = 44.45\% \text{ of } \$750{,}000 = \$333{,}375 \text{ and so on.}$$

Using the above figures, the final depreciation schedule for the asset using the MACRS-GDS system is presented in the following table.

Year (t)	MARCS (rt)	Cost Basis	D_t	Cumulative D_t	Book Value at the End of Year "t" (B_t)
1	0.3333	750,000	249,975	249,975	500,025
2	0.4445	750,000	333,375	583,350	166,650
3	0.1481	750,000	111,075	694,425	55,575
4	0.0741	750,000	55,575	750,000	$0 = S = Assumed salvage value

Referring to the table above, it can be inferred that the book value at the end of year 2 is $166,650. Also, notice that the salvage value given in the problem statement ($100,000) was never used in the depreciation computation. As a rule, the MACRS-GDS system depreciates the entire cost basis of the asset. Thus, the salvage value of ALL MACRS-GDS problems is always equal to zero.

Consider another example based on the MACRS-GDS system as follows.

EXAMPLE 13.5

SGP LLC, a limited liability company purchased a resort property for $30,000,000 in May. The cost of the building and structures was $20,000,000 and the land was valued at $10,000,000. The resort was sold in October, 7 years later. Compute the annual depreciations deductions and the book value of the resort when it was sold.

Solution

In order to solve this problem, refer to Table 13.2. The following points need to be noted:

- Resorts and hotels are classified as nonresidential rental property and are depreciated over 39 years.
- The percentage depreciation incurred for each year is presented in Table 13.3.
- Bear in mind that with respect to real estate, only the cost of structures is depreciated. The cost of land is never depreciated.
- In this specific case, the cost of structures ($20,000,000) will be depreciated.
- The cost of land ($10,000,000) will not be depreciated.

Referring to depreciation percentages presented in Table 13.4, we have the following values for annual depreciation:

TABLE 13.3

MACRS Depreciation for Personal Property Using the Half Year Convention

	Applicable Percentage for Property Class					
Recovery Year	3 Year Property	5 Year Property	7 Year Property	10 Year Property	15 Year Property	20 Year Property
1	33.33	20.00	14.29	10.00	5.00	3.750
2	44.45	32.00	24.49	18.00	9.50	7.219
3	14.81*	19.20	17.49	14.40	8.55	6.677
4	7.41	11.52*	12.49	11.52	7.70	6.177
5		11.52	8.93*	9.22	6.93	5.713
6		5.76	8.92	7.37	6.23	5.285
7			8.93	6.55*	5.90*	4.888
8			4.46	6.55	5.90	4.522
9				6.56	5.91	4.462*
10				6.55	5.90	4.461
11				3.28	5.91	4.462
12					5.90	4.461
13					5.91	4.462
14					5.90	4.461
15					5.91	4.462
16					2.95	4.461
17						4.462
18						4.461
19						4.462
20						4.461
21						2.231

*All classes convert to straight line depreciation in the year marked by the asterisk.

TABLE 13.4

MACRS Depreciation for Real Property

Recovery Year	Recovery Percentages for Nonresidential Real Property Based on the Month Placed in Service											
	1	2	3	4	5	6	7	8	9	10	11	12
1	2.461	2.247	2.033	1.819	1.605	1.391	1.177	0.963	0.749	0.535	0.321	0.107
2–39	2.564	2.564	2.564	2.564	2.564	2.564	2.564	2.564	2.564	2.564	2.564	2.564
40	0.107	0.321	0.535	0.749	0.963	1.177	1.391	1.605	1.819	2.033	2.247	2.461

Depreciation in year 1 (property purchased in May):

$D_1 = 1.605\%$ of \$20,000,000 = \$321,000
Depreciation in year 2: $D_2 = 2.564\%$ of \$20,000,000 = \$512,800
Depreciation in year 3: $D_3 = 2.564\%$ of \$20,000,000 = \$512,800
Depreciation in year 4: $D_4 = 2.564\%$ of \$20,000,000 = \$512,800
Depreciation in year 5: $D_5 = 2.564\%$ of \$20,000,000 = \$512,800
Depreciation in year 6: $D_6 = 2.564\%$ of \$20,000,000 = \$512,800
Depreciation in year 7: $D_7 = 2.564\%$ of \$20,000,000 = \$512,800
Depreciation in year 8: $D_8 = 2.564\%$ of \$20,000,000 = \$512,800

This value of depreciation has to be counted since the resort was sold in October in the seventh year.

The book value of the resort when it was sold can be computed by deducting the annual depreciation values from the cost basis of the structures only. The value of the land is not depreciated nor is it accounted for in the book value calculations.

Thus, the book value of the resort would be equal to (\$20,000,000) – (\$3,910,600) = \$16,089,400.

13.7 Unit of Production Depreciation

The unit of production method of depreciation deals with situations in which an asset is depreciated based on the extent to which it is used, not based on time. In these rare situations, the annual allocation for depreciation is expressed by formula given in Equation 6.

$$D_t = \frac{\text{Production for year "t"}}{\text{Total lifetime production for the asset}} \times (B - S) \qquad \text{(Equation 6)}$$

This method is generally used for equipment such as that used for processing natural resources. This can include equipment used to process lumber, sand, gravel, etc. It needs to be pointed out that it is not considered an acceptable method to depreciate general industrial equipment.

In order to fully grasp the concept of this method of depreciation, consider the following example.

EXAMPLE 13.6

Equipment costing \$1,000,000 was purchased for use in talc production. The talc mine will operate for five years. The salvage value of the equipment at the end of its useful life is estimated to be around \$50,000. Compute the unit of production (UOP) depreciation schedule for this equipment. What is the book value at the end of the fourth year of operation? According to the mining engineer, the mine contains about 50,000 cubic meters of commercial quality talc. The rate of harvest is provided as follows:

Year	Talc Production (cubic meters)
1	4000
2	6000
3	10,000
4	20,000
5	10,000

Solution

In the present situation, the following data has been provided:

- Purchase price (B) = \$1,000,000
- Depreciable life (N) = 5 years
- Salvage value (S) = \$50,000

The following are the unknowns that need to be computed using the straight line method:

- Depreciation per year.
- Book value at the end of each year.
- Book value at the end of year 4. This will once again be computed as a matter of course while solving the problem.

Using the formula provided in Equation 6, we can compute the annual depreciation allocation as follows:

$$\text{Depreciation for the first year} = D_1 = \frac{4000}{50{,}000}(1{,}000{,}000 - 50{,}000) = \$76{,}000.$$

$$\text{Depreciation for the second year} = D_2 = \frac{6000}{50{,}000}(1{,}000{,}000 - 50{,}000)$$
$$= \$114{,}000.$$

$$\text{Depreciation for the third year} = D_3 = \frac{10{,}000}{50{,}000}(1{,}000{,}000 - 50{,}000)$$
$$= \$190{,}000.$$

$$\text{Depreciation for the fourth year} = D_4 = \frac{20{,}000}{50{,}000}(1{,}000{,}000 - 50{,}000)$$
$$= \$380{,}000.$$

$$\text{Depreciation for the fifth year} = D_5 = \frac{10{,}000}{50{,}000}(1{,}000{,}000 - 50{,}000)$$
$$= \$190{,}000.$$

The Book value at the end of the fourth year can be computed by subtracting all depreciation to date (until and including the fourth year) and subtracting that value from the cost basis.

Thus, book value at the end of the fourth year is \$1,000,000 – \$760,000 = \$240,000.

The depreciation schedule for this problem is presented below:

Year (t)	Depreciation for Year "t" (D_t)	Sum of Depreciation Charges	Book Value at the End of Year "t" (B_t)
1	76,000	76,000	924,000
2	114,000	190,000	810,000
3	190,000	380,000	620,000
4	380,000	760,000	240,000
5	190,000	950,000	50,000

13.8 Cost Depletion

Depletion can be defined as exhaustion of natural resources due to consumption. Cost depletion is a depreciation method that is very similar to the unit of production method. Examples of depletion include metals, minerals, natural resources such as timber, petroleum reserves, fossil fuels such as coal and natural gas, etc. Since capital investment in an asset is being used up, it is recognized for income tax purposes. Just like its predecessor, it is computed using the cost of the property minus the cost of land. The estimated number of recoverable units is used in conjunction with the salvage value of the asset. The salvage value of the property if any is also taken into account.

EXAMPLE 13.7

Forest Resources LLC obtained a piece of land with standing timber. It was purchased for \$50,000. The land was valued at \$7000. The timber was valued at \$43,000. The amount of standing timber was estimated to be about 2 million board feet. The first year, 200,000 board feet of timber were harvested. Compute the depletion allowance for the year.

Solution

The depletion allowance per 1000 board feet of timber can be calculated by using the following formula:

$$\frac{\text{Value of the asset} - \text{Value of the land}}{\text{Total amount of harvestable resource}}$$

Thus, in the present case, the depletion allowance is computed as follows:

$$\frac{\$50,000-\$7000}{2000\times1000 \text{ board feet}}=\$21.5 \text{ per } 1000 \text{ board feet.}$$

We know that 200,000 board feet of timber was harvested. Thus, the depletion allowance for the year would be 200 × $21.5 per 1000 board feet = 4300 for the year under consideration.

13.8.1 Percentage Depletion

Percentage depletion is an alternate method for calculating depletion and is specifically used in the case of mineral property. It is expressed as an allowance. The allowance is a specific percentage of the gross income obtained from the property annually. It will be appreciated that the concept of percentage depletion is actually quite different from that of depreciation in general. In the depreciation, the asset cost is allocated over the useful life of the asset. In the case of percentage depletion, on the other hand, the allowance is based on the gross income obtained from the property.

It should be borne in mind that the percentage depletion allowance cannot exceed 50% of the taxable income obtained from the asset. The taxable income obtained from the asset should be computed without taking the depletion deduction into account. Table 13.5 presents percentage depletion allowances commonly used for selected deposits.

EXAMPLE 13.8

A gold mine based in the US has a gross income of $2,000,000. The total expenses associated with the mine amount to $1,000,000. Calculate the percentage depletion deduction that is allowable.

Solution

It will be observed from Table 13.5 that gold deposits based in the US have a 15% depletion allowance that is based on gross mining income. Also, the allowable percentage depletion deduction is capped and limited to a maximum of 50% of taxable income. Based on these facts, we have the following computation:

Computed Percentage Depletion

Gross income from gold deposits: $2,000,000
Percentage Depletion: 15%

Thus, computed percentage depletion: $300,000 (15% of $2,000,000).

Applying the taxable income limitation constraint, we have the following:

TABLE 13.5

Percentage Depletion Allowances for Selected Deposits

Deposits	Rate
Sulfur, uranium and if from deposits in the US; asbestos, lead ore, zinc ore, nickel ore and mica.	22%
Gold, silver, copper, iron ore, and certain oil shale from deposits in the US.	15%
Borax, granite, limestone, marble, mollusk shells, potash, slate, soapstone, and carbon dioxide produced from a well.	14%
Coal, lignite, and sodium chloride.	10%
Clay and shale used or sold for use in making sewer pipes or bricks or used or sold for use as sintered or burnt lightweight aggregates.	7.5%
Clay used or sold for use in drainage and roofing tile, flower pots, and kindred products. Gravel, sand, and stone. This excludes stone used or sold for use by a mine owner or operator as dimensional or ornamental stone.	5%

Taxable Income Limitation

Gross income from gold deposits: $2,000,000
Deduct expenses other than depreciation: $1,000,000
Thus, the taxable income from the mine: $1,000,000
Deduction limitation: 50%

Thus, taxable income limitation: $500,000 (50% of $1,000,000).

In this case, the taxable income limitation is $500,000. This is greater than the computed percentage depletion ($300,000). Thus, the allowable percentage depletion deduction is $500,000.

13.9 Conclusion

This chapter described how business assets are claimed as business expenses over time. Several methods of depreciation were discussed in detail and elucidated with the use of accompanying examples.

Though several methods of depreciation are available, the MACRS system is the most commonly used. It assumes a salvage value of zero. The unit of production method of depreciation takes usage into account to compute loss in value. However, this method is used very rarely and is generally not acceptable for most business assets.

It will be appreciated from the solved examples that most depreciation calculations are recursive in nature. Thus using a calculator to compute depreciation values and book values can be quite a tedious experience. Instead, the same task can be facilitated through the use of a spreadsheet such as EXCEL to compute the depreciation values, book values and the depreciation schedule.

PROBLEMS

1. A company purchases some petroleum drilling equipment for \$750,000. This equipment will be depreciated using the MACRS-GDS system. Compute the depreciation schedule as well as the book value of the equipment at the end of the third year.
2. Some equipment used in research and development costs \$250,000 and has an estimated salvage value of about \$20,000. The useful life of the equipment is estimated to be about four years. Compute the depreciation schedule using the SOYD method of depreciation.
3. Some special handling devices for food and beverage manufacture cost \$150,000. At the end of six years, they can be sold for \$25,000. Compute the depreciation schedule for the devices using the following methods:
   - Straight line depreciation
   - Sum of years' digits depreciation
   - Double declining balance method
   - MACRS-GDS depreciation
4. BVR LLC Computers paid \$350,000 for some office furniture, fixtures, and equipment. The company believes the equipment can be sold for salvage for \$10,000 at the end of its useful life of 7 years. Compute the depreciation using the following methods:
   - Straight line depreciation
   - Double declining balance method
   - 150% declining balance
   - MACRS depreciation

   What is the book value of this equipment at the end of the fourth year using each of the above methods?
5. The Midas Gold Assets, Inc company paid \$750,000 for some mining equipment for use in one of its mines. The company believes that the mine contains 30,000 tons of ore and plans are underfoot to mine the entire contents in the next seven years. The rate of production from the mine is estimated to be as follows:

Year	Ore Production (tons)
1	1000
2	5000
3	10,000
4	12,000
5	1000
6	500
7	500

The mine will be exhausted at the end of seven years and the equipment will be worthless. Compute the depreciation schedule for the equipment using the following methods:

- Double declining balance
- MACRS-GDS method
- Unit of production method of depreciation

What is the book value of the equipment at the end of its third year? Contrast the book values using the MACRS and unit of production methods.

6. A new casting machine can be purchased for $175,000. The machine will be used for 15 years and is expected to fetch a salvage value of about $30,000 at the end of its useful life. Calculate the amount of depreciation allowable for this machine during the fifth year using each of the following methods:
    - Straight line method
    - Double declining balance method
    - MACRS depreciation method
    - Sum of years' digits depreciation method
7. Office furnishings have been purchased by a staffing company for $30,000. The company intends to use the furniture for 5 years and expects to obtain a salvage value of $2500 at the end of such time. Compute the depreciation charge for the third year and the book value at the end of the fourth year using each of the following methods of depreciation:
    - Straight line depreciation
    - MACRS depreciation
    - Double declining balance method
8. A CNC machining center costs $200,000 with an estimated life of about 10 years. Compute the depreciation and book value for this machine for the first five years using each of the following methods:
    - 150% declining balance method
    - Double declining balance method
    - MACRS-GDS method
    - Straight line method
9. Black Gold, Inc. purchased some petroleum drilling equipment that is expected to last 9 years. The equipment was purchased for $1,000,000 and has an estimated salvage value of $200,000. The company owns some oil deposits and plans to extract crude oil based on the following schedule:

Year	Crude Oil Production (barrels)
1	15,000
2	35,000
3	50,000
4	65,000
5	40,000
6	30,000
7	20,000
8	15,000
9	5000

Compute the unit of production depreciation schedule for the aforementioned equipment. What is the book value of the equipment at the end of years 3, 7, and 8?

10. A diamond polishing machine can be obtained for $35,000 with an estimated salvage value of $10,000 at the end of its useful life of 7 years. Compute the depreciation schedule using the following methods:
    - MACRS depreciation
    - Double declining balance method
    - Straight line depreciation method
11. A coal mining company bought some extraction equipment for $375,000. Coal will be extracted from the mine at a measured pace over several years. It is estimated that at the end of its useful life, the equipment can be sold for about $50,000. The life of the equipment is estimated to be about 7 years. The schedule of extraction is presented in the following table:

Year	Coal Production (tons)
1	5000
2	15,000
3	17,000
4	22,000
5	15,000
6	5000
7	2000

Compute the depreciation schedule for this equipment using the unit of production (UOP) method. What is the depreciation allowed during the fourth year? Compute the book value at the end of year 5 using the UOP method.

12. Midnight Retreat LLC purchased a small hotel for $1,000,000. The price of the land is estimated to be about $200,000 and the price of the

building is $800,000. The hotel was purchased in April. Compute the MACRS depreciation for the first two calendar years. Assuming that the hotel is sold in October of the third year, compute the MACRS depreciation in that year.

13. A computer was purchased in 2016 for $20,000. It has no salvage value at the end of five years. Compute the MACRS depreciation allowance for years 3 and 4 as well as the book values for those years.
14. Material handling equipment for specialized manufacturing operations costs $450,000 with an estimated life of about 7 years. Compute the depreciation and book value for this machine for the first five years using each of the following methods:
    - 150% declining balance method
    - Double declining balance method
    - MACRS-GDS method
    - Sum of years' digits method
15. The depreciation schedule for some office furniture has been arrived at by using the straight line depreciation method. According to this schedule, the annual depreciation is $2500 for a period of 7 years. The salvage value of the furniture is estimated to be about $7500. Compute the initial cost of the furniture.
16. Atlas Earthmovers purchased an excavator for $250,000. The machine was used for 7 years and disposed of for $75,000 at the end of its useful life. Calculate the allowable depreciation in year 3 for this equipment using the following methods of depreciation:
    - SOYD method of depreciation
    - Double declining balance method
    - MACRS method of depreciation
17. A frontend loader costs $75,000 and has an estimated salvage value of $15,000 at the end of 5 years of useful life. What is:
    - The book value at the end of the fourth year by the straight line method
    - The depreciation charge in the third year if the MACRS method is used
18. An electrostatic precipitator is purchased in 2009 for $160,000. It is expected to last for 10 years and have a $20,000 salvage value. If the operating cost is $8000 per year, determine the depreciation charge 5 years from now according to the MACRS method.
19. A drilling rig was purchased for 1.6 million dollars. It has an estimated life of 12 years and the salvage value at that time is expected to be $650,000. After 7 years, what will be the total depreciation charged for:

- Basic straight line depreciation
- Basic double declining balance depreciation

20. A coal barge was acquired in 2002 for $830,000. It is expected that it will be sold for scrap in 20 years from its purchase date for $45,000. Find the book value of the barge at the end of 2008 for MACRS (GDS).
21. Filters costing $1000 per set are used to separate particles from hot gases being discharged into the air. The filters must be replaced every 3000 hours. If the plant operated 24 hours per day, 30 days per month, what is the maximum depreciation charge per month for the filters?
22. At the beginning of the fiscal year, the aerospace company, Air Fun, purchases a mold for rear pressure bulkheads for $1 million. The estimated salvage value after 8 years is $120,000. Calculate the depreciation deduction and the value for each year using MACRS-GDS allowances.
    - Assume the complete depreciation schedule is used.
    - Assume the asset is sold during the seventh year of use.
23. A sulfur mining operation has a gross income of $700,500 in March from the production of sulfur. All expenses with the exception of depletion expenses for this month amount to $190,000. The percentage depletion rate for sulfur is 15%. What is the depletion allowance for this month?
24. An asset had a first cost of $30,000 with a salvage of $8000 after 9 years. If it is depreciated by the basic declining balance method using a rate of 10%, what will be the:
    a. Depreciation charge in the third year
    b. Depreciation charge in the seventh year
25. An evaporation tower in a petroleum refinery was placed in service in 2009 at a cost of $2 million. This seven year property is expected to be removed at a cost of $150,000 during 2017. Find the depreciation charges over the tower's lifetime using MACRS (GDS).
26. A copper company purchased a mine for $600,000, which contained an estimated 250,000 tons of copper. The company sells copper to wholesalers for $1.10 per pound. In the first year, the company produced and sold 70,000 tons of copper from the mine. If the depletion allowance for copper is 14%, calculate the deductions allowed for one year by the:
    a. The factor depletion method
    b. The depletion allowance method
27. The Ying Company purchased a new computer system that had a first cost of $80,000. The company expects to keep the system for 5 years and then sell it for $12,000. The operating cost of the system is expected to be $27,000 per year. If the company's MARR is 11% per year and the company elects to capital expense $10,000 during the first year:

a. What would the depreciation charge be in the second year according to MACRS?

b. What would be the book value at the end of the third year?

28. A special purpose machine was purchased for $300,000 with an expected life of 11 years. If its salvage value is expected to be zero, what will be the depreciation charge in the first and fifth years for

 a. Basic straight line depreciation

 b. Basic double declining balance depreciation

29. Using depreciation methods, calculate the book value at the end of each year using straight line and double declining balance methods of depreciation for an asset with an initial cost of $75,000 and an estimated salvage value of $15,000 after 6 years. Present the results in tabular form.

30. A truck cost $400,000 and had an estimated useful life of 11 years and estimated salvage value of $4000. How much depreciation should be allocated on the new truck in each year and what is the undepreciated value of the truck at the end of the ninth year, assuming the use of basic straight line depreciation?

31. An electric generating unit is placed in service by a public utility. Present the difference between the percentage depreciation rates each year if the unit is depreciated:

 a. Under MACRS (GDS) using the 15-year property class

 b. Using straight line for 15-year property

 c. Under MACRS (ADS) using the 15-year property class

32. The ABC Company purchased a dump truck for $30,000. The company expects to keep the truck for 5 years and then sell it for $10,000. In the first year the company owned the truck, it took a capital expense deduction of $12,000 as per section 179 of the tax code. If the recovery period for trucks is 3 years, what would be the MACRS depreciation charge in the second?

14

Replacement Analysis

14.1 What Is Replacement Analysis?

Industrial equipment and machinery wear out over time. We have already dealt with the corresponding decline in value that occurs as a result of this wear in Chapter 13. In this chapter, we will consider the different factors that play a role in deciding whether an asset, such as a machine or other equipment, should be replaced. It will be appreciated that this is a unique situation. We have seen other situations, such as present worth analysis, etc., wherein all economic choices dealt with new alternatives. Replacement analysis seeks to answer the question, *"When should the machine be replaced? Now or later?"* If an asset is crucial to the functioning of a system, it is obvious that it will wear out with constant use and will be replaced eventually anyway. Thus, the main objective of a replacement study is to figure out when the asset should be replaced and not whether it should be replaced.

Most replacement studies consist of deciding whether an asset should be replaced at the current moment in time. If management does indeed decide to replace the asset under consideration, then the replacement study is considered to be complete. If on the other hand, the decision is made to not replace at the current moment in time, then the study is considered to be ongoing and the situation will be revisited periodically in the future.

At the heart of it, replacement analysis is an application of equivalent uniform annual cost (EUAC) and annual benefit (EUAB) to alternatives with unequal lives. We have already dealt with the basic concept of EUAB and EUAC in Chapter 1. The distinguishing feature of replacement analysis is that in this method, there is no specific period of study. The economic service life analysis is used to determine the equivalent uniform annual cash flow (whether it is EUAC or EUAB). On the other hand, if the study period has been specific explicitly, the replacement analysis tends to follow a slightly different path.

14.2 The Need for a Replacement Study

A replacement study can be undertaken for three specific reasons as delineated below:

1. *Compromised or Subpar Performance of Equipment*: With constant use, equipment deteriorates over time. This results in decreased reliability as well as productivity. A machine that is less reliable and less productive is bound to result in higher scrap and rework and cost more to operate. This will eventually lead to lower sales, reduced quality, compromised safety, and higher maintenance expenses.
2. *Changed Specifications*: System requirements are often recalibrated. Specifications change for both products and processes. Requirements are updated in keeping with advances in technology and the changing marketplace. This necessitates upgrading equipment in order to keep abreast of such changes. This can be achieved through completely replacing old equipment or by augmenting existing equipment.
3. *Obsolescence*: Obsolescence is a fact of life especially in the technical arena. New technologies are being developed even as we speak. Just think about how far technology has evolved over the past 10–15 years. Remember the "razor phone?" The advent of the iPhone made it obsolete. What about the manual typewriter? Its electronic counterpart and more recently, word processing software have rendered the manual typewriter a relic of a bygone era. Similarly, in terms of industrial machinery, manual lathes and milling machines were rendered obsolete by more precise and accurate computer numerically controlled (CNC) technology. The advent of rapid prototyping or 3D printing is forcing us to rethink our traditional view of manufacturing itself. The relentless forward march of automation is a fact of life. New creative ideas lead to better products with higher productivity and as such should be encouraged. However, this implies that old equipment will eventually be rendered obsolete or "out of sync" and for that reason, must be replaced.

The following basic terminology is used in replacement analysis.

14.3 Replacement Analysis: Terminology

- *Defender*: Defender is defined as the asset that is currently installed. The task of the replacement study is to draw an objective comparison between the current asset and its replacement.

- *Challenger*: Challenger is defined as the asset that is being considered to replace the defender. The task of the replacement study is to compare the two alternatives.
- *Market Value*: How much would the currently installed asset fetch it was sold on the open market today? This is defined as the market value. It is also referred to as trade-in value. It is estimated using the services of appraisers or monitoring sales of similar equipment. Recall from Chapter 1 that salvage value was defined as the value attached to equipment at the end of its useful life. In the case of replacement analysis, the market value at the beginning of any year is considered to be equal to the salvage value at the end the preceding year.
- *Equivalent Uniform Annual Cost (EUAC)*: The EUAC is the single most important metric of comparison used to compare defender and the challenger. In the case of replacement analysis, any benefits produced by the equipment are considered to be equal since the defender is replaced by the challenger, thus restoring function and accompanying benefit. This goes to the very nature and objective of the study itself. Thus, EUAC is used as the primary measure of performance for the sake of this comparison.
- *Economic Service Life (ESL)*: The economic service life for a possible challenger is considered to be the number of years at which the lowest value of EUAC occurs. The numeric value of ESL is established for the best challenger. From the perspective of the defender, it is taken to be the lowest cost life.
- *Defender First Cost*: Defender first cost is defined as the initial investment necessary to procure the defender. The current market value of the defender is used as the appropriate estimate of this figure. It should be borne in mind that trade-in value is not an accurate estimate of first defender cost because it does not represent true market value. The same holds true for depreciated book value as derived from accounting records.

  We have already discussed the possibility that the defender can be upgraded in some way to make it equivalent to the challenger. In such cases, the augmentation cost is added to the market value in order to ascertain the defender first cost.
- *Challenger First Cost*: Challenger first cost is defined as the sum of capital that is at risk when the defender is being replaced by the challenger. Thus, this amount of money must be recovered. In most cases, this sum is equal to the cost of the challenger.

  The trade in value received for the defender also affects the challenger first cost. If this trade in value is higher than expected, the challenger first cost is likely to be reduced significantly. In such a scenario, the challenger first cost would be expressed by the following equation:

$$P - (TV - MV). \quad \text{(Equation 1)}$$

wherein:
P: First cost of the challenger
TV: Trade-in value received for the defender
MV: Market value of the defender

Consider the following example:
Management is considering replacing an existing machine (defender). The machine has a market value (MV) of $75,000. It can be traded in for $90,000. The machine will be replaced by a modern and more capable counterpart (challenger) that costs $200,000. In such a case, the amount of money at risk and which must be recovered can be computed by using the aforementioned equation. This amounts to $200,000 – ($90,000 – $75,000) = $285,000. Note that if the defender was sold at market rates, the capital at risk would amount to $200,000 – $75,000 = $225,000. When the market value is equal to the trade in value, the amount of capita that needs to be recovered is equal to the first cost of the challenger.

It will be obvious from the preceding discussion that the challenger first cost is the amount of money that must be expended in order to purchase and install it. An important point to be noted in this discussion is that the unrecovered capital left in the defendant cannot be used to increase the first cost of the challenger.

- *Sunk Cost*: Sunk cost is defined as the amount of money that already been expended and as such cannot be recovered. Sunk cost, being a real cost, should be treated realistically using appropriate tax laws. It is expressed as the difference between the market value of an asset and its book value. Adding the sunk cost to the purchase price of the challenger will artificially inflate that price. Hence, such practice is not encouraged.
- *Nonowner's Viewpoint*: Nonowner's viewpoint is defined as the perspective that an outsider would have regarding the replacement situation. The biggest advantage of having such a viewpoint is the degree of objectivity and a consequent lack of bias afforded by it.

If the time horizon under consideration in replacement analysis is unlimited in nature or is not specified in the problem, the following assumptions are made:

1. The replacement will be used for an extended period of time.

2. In terms of capability, the challenger has the highest capability when replacing the defender. Eventually, the same cycle of decisions will apply to the challenger. That is to say that in time, the current challenger will assume the role of defender and a new challenger will replace it. This process will be repeated indefinitely over the life cycle of the equipment.
3. Cost estimates for the defender and challenger will be identical across all iterations. Note that this assumption is not entirely correct but can be used as an effective approximation in order to facilitate the problem solving process.

14.4 Economic Service Life (ESL)

Throughout Chapter 1, we have seen that the life over which the asset will be evaluated was known. However, this was just a theoretical exercise, an estimate at best. In real life, the useful life of an asset can only be estimated upfront and not known exactly. In the case of replacement analysis, the service life of an asset is computed using current estimates of cost. This closest estimate of service life of an asset is referred to as economic service life (ESL). Another term for ESL is minimum cost life. It is based solely on the economic aspect of the asset. When the asset is old enough to have reached the ESL, replacement analysis theory suggests that it should be replaced in order to minimize total costs. The ESL of the defender and challenger are determined prior to the replacement analysis study. This is a prerequisite to the study since those are unknown values unless independently determined. The following procedure is adopted in order to compute the ESL of any asset:

- Capital recovery is determined. This is defined as the equivalent uniform annual cash flow (EUAC) as obtained by using the initial investment and any salvage value.
- Total EUAC of operating costs is determined.
- Total EUAC is the sum of capital recovery and the EUAC of annual operating costs.

From the above discussion, we obtain the following equation:

$$\text{Total EUAC} = \text{Capital recovery (CR)} + \text{EUAC of operating costs.} \qquad \text{(Equation 2)}$$

The economic service life (ESL) corresponds to the smallest total equivalent uniform annual costs. The shape of a typical total EUAC curve is depicted in the following Figure 14.1.

Notice that the capital recovery component undergoes a continuous decline over time that is to be expected. Simultaneously, the EUAC of annual operating costs rises steadily over time. This is to be expected as well due to the deterioration that occurs in any machine over time due to wear and tear, friction, and age. For instance, the maintenance expenses on a new car are generally far less than those on a twenty year old car. That is the reason car manufacturers do not offer a twenty year and unlimited mile warranty. Auto warranties are only offered for the first few years and few thousand miles. The manufacturer is aware that operating costs during this period are likely to be minimal at worst. However, after the first few years, said expenses rise steadily as the years roll on. In order to keep operating the vehicle, maintenance expenses need to be borne annually. For this reason, they can be characterized as operating costs. The compound curve denoting total EUAC of costs is concave. This is due to the fact that one of its components is steadily declining while the other is simultaneously rising. These two components are elaborated on as follows:

- *Shrinking Cost of Capital Recovery*: Since capital recovery is the equivalent uniform annual cost of owning the equipment, it decreases with each year of ownership. The same example of car ownership can be cited here in order to understand this phenomenon. The cost of capital recovery shrinks due to depreciation, obsolescence, capital

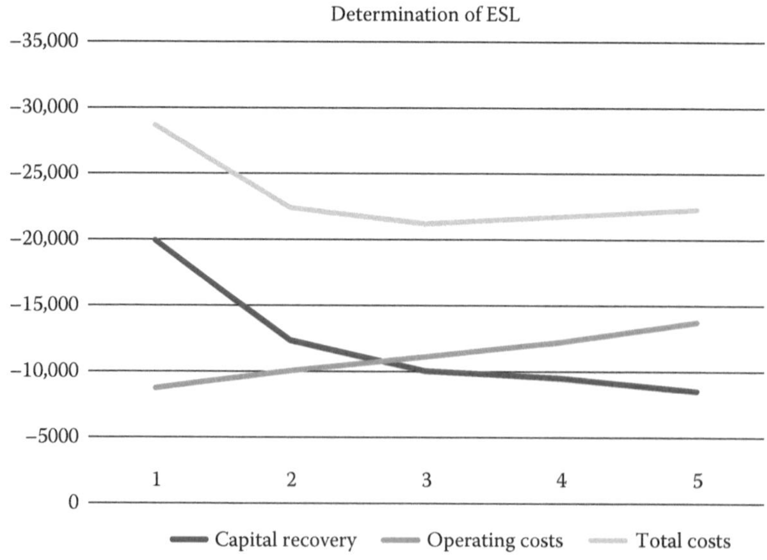

FIGURE 14.1
Figure for Example 14.1.

factors, etc. Salvage value is also included in these calculations. Salvage value itself decreases over time. In most cases, salvage value is an approximation of estimated market value during the year under consideration. Based on the preceding discussion, the capital recovery can be expressed by the following equation:

$$\text{Capital recovery} = -P(A/P, I, n) + S(A/F, I, n) \qquad \text{(Equation 3)}$$

wherein:

P: Purchase price of the asset
S: Salvage value of the asset
I: Interest rate
N: Number of time periods

- *Expanding Equivalent Uniform Annual Operating Cost*: Annual operating costs are usually comprised of maintenance costs and operation costs. In a majority of cases, these costs have a tendency to rise over time. Thus, the equivalent uniform annual cash flow corresponding to these costs rises with the passage of time as well. The equivalent uniform annual cash flow of annual operating costs can be computed as follows:

  1. Compute the present value of each individual cash flow using the P/F factor
  2. Add up all the present values thus obtained
  3. Redistribute the present value by converting it into an annual cash flow using the (A/P) factor

A better understanding of the concepts presented so far can be gained by solving an example. Consider the following example.

EXAMPLE 14.1

A five year old truck is being considered for replacement. Currently, the truck has a market value of $35,000. It is estimated that future values and annual operating costs for the next five years will be in line with the data presented in the following table. If the interest rate is 7% per year, compute the economic service life (ESL) of the truck.

Year	Market Value (MV)	Annual Operating Costs
1	$17,500	–$8750
2	$14,500	–$11,500
3	$10,500	–$13,500
4	$3500	–$16,000
5	$0	–$21,000

Solution

In order to compute the economic service life, it is essential to compute the following two factors:

- Capital recovery
- Equivalent uniform annual worth of operating costs

Computation of Capital Recovery for Year 1

We use Equation 3 to compute the value of capital recovery for each year. Thus, we have the following formula:

$$\text{Capital recovery} = -P(A/P, I, n) + S(A/F, I, n) \qquad \text{(Equation 3)}$$

In this case, the current market value as given in the problem will be considered as the value of "P," which is given in the table as \$35,000. The salvage value of the asset is the market value for year 1 and is given in the table as \$17,500. The value of "n" = 1 since we are interested in computing the capital recovery for the first year. Additionally, we have the following values for compound interest factors:

(A/P, 7%, 1 year) = 1.07
(A/F, 7%, 1 year) = 1.00

Substituting the above values into Equation 3 we obtain the following:

$$\begin{aligned}\text{Capital recovery for year 1} &= 1-\$35{,}000(1.07) + \$17{,}500(1.00)\\ &= -\$19{,}950\end{aligned}$$

Computation of Capital Recovery for Year 2

For year 2, we use P = \$35,000; MV = \$14,500 (from the data given in the problem). Additionally, we have the following compound interest factors:

(A/P, 7%, 2 years) = 0.5531
(A/F, 7%, 2 years) = 0.4831

Substituting the above values into Equation 3 we obtain the following:

$$\begin{aligned}\text{Capital recovery for year 2} &= -\$35{,}000(0.5531) + \$14{,}500(0.4831)\\ &= -\$12{,}353.55\end{aligned}$$

Computation of Capital Recovery for Year 3

For year 3, we use P = \$35,000; MV = \$10,500 (from the data given in the problem). Additionally, we have the following compound interest factors:

(A/P, 7%, 3 years) = 0.3811
(A/F, 7%, 3 years) = 0.3111

Substituting the above values into Equation 3 we obtain the following:

$$\begin{aligned}\text{Capital recovery for year 3} &= -\$35{,}000(0.3811) + \$10{,}500(0.3111)\\ &= -\$10{,}071.95\end{aligned}$$

Computation of Capital Recovery for Year 4

For year 4, we use P = \$35,000; MV = \$3,500 (from the data given in the problem). Additionally, we have the following compound interest factors:

(A/P, 7%, 4 years) = 0.2952
(A/F, 7%, 4 years) = 0.2252

Substituting the above values into Equation 3 we obtain the following:

$$\text{Capital recovery for year 4} = -\$35{,}000(0.2952) + \$3{,}500(0.2252) = -\$9543.8$$

Computation of Capital Recovery for Year 5

For year 5, we use P = \$35,000; MV = \$0.00 (from the data given in the problem). Additionally, we have the following compound interest factors:

(A/P, 7%, 5 years) = 0.2439
(A/F, 7%, 5 years) = 0.1739

Substituting the above values into Equation 3, we obtain the following:

$$\text{Capital Recovery for year 5} = -\$35{,}000(0.2439) + \$0(0.1739) = -\$8536.5$$

Computation of EUAC of Operating Costs for Year 1

EUAC of operating costs for year 1 = (present worth of AOC_1) (A/P, 7%, 1 year).

From the interest rate tables, we observe the following:

(P/F, 7%, 1 year) = 0.9346
(A/P, 7%, 1 year) = 1.07

Substituting these values into the equation, we get the following:

$$\text{EUAC of operating costs for year 1} = (\$8750)(0.9346)(1.07) = \$8750$$

Computation of EUAC of Operating Costs for Year 2

EUAC of operating costs for year 2 = [(present worth of AOC for year 1) + (present worth of AOC for year 2)] (A/P, 7%, 2 years).

From the interest rate tables, we observe the following:

(P/F, 7%, 1 year) = 0.9346
(P/F, 7%, 2 years) = 0.8734
(A/P, 7%, 2 years) = 0.5531

Substituting these values into the equation, we get the following:

$$\text{EUAC of operating costs for year 2} = [(\$8750)(0.9346) + (\$11{,}500)(0.8734)](0.5531) = \$10{,}078.5$$

Computation of EUAC of Operating Costs for Year 3

EUAC of operating costs for year 3 = [(present worth of AOC for year 1) + (present worth of AOC for year 2) + (present worth of AOC for year 3)] (A/P, 7%, 3 years).

From the interest rate tables, we observe the following:

(P/F, 7%, 1 year) = 0.9346
(P/F, 7%, 2 years) = 0.8734
(P/F, 7%, 3 years) = 0.8163
(A/P, 7%, 3 years) = 0.3811

Substituting these values into the equation, we get the following:

EUAC of operating costs for year 3 = [($8750) (0.9346) + ($11,500) (0.8734) + ($13,500) (0.8163)] (0.3811)
= $11,144.08

Computation of EUAC of Operating Costs for Year 4

EUAC of operating costs for year 4 = [(present worth of AOC for year 1) + (present worth of AOC for year 2) + (present worth of AOC for year 3) + (present worth of AOC for year 4)] (A/P, 7%, 4 years).

From the interest rate tables, we observe the following:

(P/F, 7%, 1 year) = 0.9346
(P/F, 7%, 2 years) = 0.8734
(P/F, 7%, 3 years) = 0.8163
(P/F, 7%, 4 years) = 0.7629
(A/P, 7%, 4 years) = 0.2952

Substituting these values into the equation, we get the following:

EUAC of operating costs for year 4 = [($8750) (0.9346)
+ ($11,500) (0.8734)
+ ($13,500) (0.8163)
+ ($16,000) (0.7629)] (0.3811)
= $12,235.54

Computation of EUAC of Operating Costs for Year 5

EUAC of operating costs for year 5 = [(present worth of AOC for year 1) + (present worth of AOC for year 2) + (present worth of AOC for year 3) + (present worth of AOC for year 4) + (present worth of AOC for year 5)] (A/P, 7%, 5 years).

From the interest rate tables, we observe the following:

(P/F, 7%, 1 year) = 0.9346
(P/F, 7%, 2 years) = 0.8734
(P/F, 7%, 3 years) = 0.8163
(P/F, 7%, 4 years) = 0.7629
(P/F, 7%, 5 years) = 0.7130
(A/P, 7%, 5 years) = 0.2439

Substituting these values into the equation, we get the following:

$$\begin{aligned}\text{EUAC of operating costs for year 5} &= [(\$8750)\ (0.9346) + (\$11{,}500)\ (0.8734) \\ &\quad + (\$13{,}500)\ (0.8163) + (\$16{,}000)\ (0.7629) \\ &\quad + (\$21{,}000)\ (0.7130)]\ (0.2439) \\ &= \$15{,}800.61\end{aligned}$$

The above results are tabulated as follows.

Year	Market Value (MV)	Annual Operating Costs	Capital Recovery	EUAC of Operating Costs	Total EUAC
1	$17,500	–$8750	–$19,950	–$8750.193	–$28,700.19
2	$14,500	–$11,500	–$12,353.55	–$10,078.50	–$22,432.05
3	$10,500	–$13,500	–$10,071.95	–$11,144.08	–$21,216.03
4	$3500	–$16,000	–$9543.80	–12,235.54	–$21, 779.34
5	$0	–$21,000	–$8536.50	–$13,761.15	–$22,297.65

The different costs can also be plotted on a graph in order to enable easy visualization. The graphical solution is depicted below. Notice the decreasing trend in capital recovery costs and corresponding increasing trend in annual operating and maintenance costs. Simultaneously, also notice the concave shape of the curve for total costs. Year 3 clearly points to the time at which total costs are minimum.

Algebraically, a cursory examination of the costs presented in the above table will reveal that the total EUAC declines until year 3 and then starts to advance. Thus, the economic service life of the truck is 3 years.

Solution

In general, the value of economic service life (ESL) is computed when the expected life of the defender and/or challenger is not specified. If the expected life is specified, no calculation of ESL is necessary. When expected life is not fixed, it is essential to know a series of market values and salvage values in order to compute the capital recovery curve. This is a pretty easy task. For instance, it is not unreasonable to assume that an asset that costs $X can lose about 20% of its value every year. Thus, end of year market value could be tabulated as follows:

Year	Market Value ($)
0	X
1	0.8X
2	0.64X
3	0.512X
4	0.41X

Estimates of market value and those of annual operating costs can be added together to produce what is referred to as the marginal cost for an asset.

Marginal costs for an asset are defined as the annual estimates of the cost involved to own and operate an asset. These can be estimated on an annual basis for any given year. The three components of marginal cost are as follows:

- Amount of money required to own the asset. This is best expressed as loss in market value.
- Past interest on market value at the beginning of each year.
- Annual operating cost for every year.

Marginal costs can be converted into equivalent uniform annual costs using the (A/P) factor. It should be clear from the preceding discussion that the EUACs of the first two components as mentioned above corresponds to capital recovery. Similarly, the sum of the EUACs of all three marginal cost components would yield the same value as the EUAC of total costs over "n" number of years.

14.5 Replacement Study

Replacement studies are performed with or without a particular study period being specified. The main objective of a replacement study is to ascertain the point in time when a challenger would replace a defender that has currently been installed. The study is deemed to be complete if and when the challenger does indeed replace the defender. If, however, the defender is chosen to be retained for the present moment in time, the study is extended over a number of years that is equal to the life of the defender. At the end of the defender's life, the challenger would replace the defender. The procedure to conduct a replacement study is detailed as follows:

14.5.1 New Replacement Study

The roles of challenger and defender are assigned based on the better EUAC of either. When the challenger has been selected, the defender should be replaced at that moment in time. If, on the other hand, it is deemed that the defender should be retained, plan to keep it for up to its entire lifespan. Next year, the following steps should be undertaken:

14.5.2 Cost Analysis One Year Later

1. Make certain that all estimates for cost are current at that time (one year later). This is especially true with respect to first cost, market value, and annual operating costs. If either of the aforementioned components is not current, adopt the procedure specified in step 2. If the current year coincides with the expiration year (corresponding to the maximum life span) of the defender, and if the challenger's

cost estimates are still current, the defender should be replaced now. If the current year, on the other hand, does not coincide with the expiration year of the defender, retain the defender for another year and perform the same study next year. This procedure can be repeated until that point in time when the current year does coincide with the expiration year of the defendant.

2. If the cost estimates for challenger and defender have changed, they should be determined and updated. Consequently, a new replacement study should be undertaken.

   It should be borne in mind that a replacement study can be undertaken every year or more frequently than one year in order to determine whether the defender needs to be replaced in case a suitable and competitive challenger was to emerge.

Consider Example 14.2 in order to better understand the application of economic service life principle to a replacement study performed over an unspecified time horizon.

EXAMPLE 14.2

A manufacturing company invested $25,000,000 in some new robotic welding equipment two years ago. The deal consisted of installing 250 units of robotic welders obtained at a unit price of $100,000 per welder. Due to evolving quality control requirements, all welders will need to be upgraded this year at an additional cost of $20,000 per unit. Needless to say, this amount of money will need to be spent in addition to the annual operation and maintenance costs. A new system based on emerging technology can be considered to replace existing robot welders this year. Interest rate is 7% and details for challenger and defender are given below. Perform the replacement study based on the following figures.

Challenger

- First Cost: $75,000.
- Future market values decrease at the rate of 20% per year.
- Annual operating cost estimated to be about $7500 in the first year, increasing at the rate of $3000 every year.
- Useful life of the asset is a maximum of 10 years.

Defender

- Current market value: $22,000.
- Future market values decrease at the rate of 20% per year.
- Annual operating cost estimated to be about $5500 in the first year, increasing at the rate of $5500 every year.
- Additional amount of $20,000 to be spent per unit towards upgrades.
- Useful life of the asset is a maximum of 3 more years.

Objective

1. Compute the values of EUAC for total cost and economic service life (ESL).
2. Perform the replacement study.
3. If the expected market value for the defender after one year is \$17,600, it is forecast to drop to abysmally low prices in view of the better technology inherent in the challenger. Next year, the market prices for the defender are expected to be as low as \$3000 and will drop to zero thereafter. The annual operating cost for the defender is expected to be \$8250 next year and \$11,000 thereafter. Based on these figures, a follow up replacement study needs to be performed.

Solution

1. In order to compute the economic service life, we perform the same analysis that we did in problem 14.1. Note that the first cost of defender in year "0" is \$100,000 and that of the challenger in year "0" is \$75,000. We compute the capital recovery and EUAC of operating costs for the challenger as follows:

Challenger Cost Computation: Capital Recovery

Year 1:

First cost = \$75,000.
Market value of challenger = 80% of \$75,000 = \$60,000.

Thus, capital recovery for 1st year is obtained as follows:

$$-\$75{,}000(A/P, 7\%, 1) + \$60{,}000(A/F, 7\%, 1).$$
$$= -\$75{,}000(1.07) + \$60{,}000(1.00) = -\$20{,}250.$$

Year 2:

First cost = \$75,000.
Market value of challenger = 80% of \$60,000 = \$48,000.

Thus, capital recovery for 1st year is obtained as follows:

$$-\$75{,}000(A/P, 7\%, 2) + \$48{,}000(A/F, 7\%, 2).$$
$$= -\$75{,}000(0.5531) + \$48{,}000(0.4831) = -\$18{,}293.7$$

Year 3:

First cost = \$75,000.
Market value of challenger = 80% of \$48,000 = \$38,400.

Thus, capital recovery for 1st year is obtained as follows:

$$-\$75{,}000(A/P, 7\%, 3) + \$38{,}400(A/F, 7\%, 3).$$
$$= -\$75{,}000(0.3811) + \$38{,}400(0.3111) = -\$16{,}636.3$$

Year 4:

First cost = \$75,000.
Market value of challenger = 80% of \$38,400 = \$30,720.

Thus, capital recovery for 1st year is obtained as follows:

$-\$75{,}000(A/P, 7\%, 4) + \$30{,}720(A/F, 7\%, 4).$
$= -\$75{,}000(0.2952) + \$30{,}720(0.2252) = -\$15{,}222.$

Year 5:

First cost = \$75,000.
Market value of challenger = 80% of \$30,720 = \$24,576.

Thus, capital recovery for 1st year is obtained as follows:

$-\$75{,}000(A/P, 7\%, 5) + \$24{,}576(A/F, 7\%, 5).$
$= -\$75{,}000(0.2439) + \$24{,}576(0.1739) = -\$14{,}019.$

Year 6:

The capital recovery for year 6 can be obtained by a recursive iteration of the process demonstrated above. The same process can be used to compute the value of capital recovery for years 7 through 10.

Challenger Cost Computation: EUAC of Annual Costs

Year 1:

The problem statement states that the annual operation and maintenance cost in the first year = \$7500.

EUAC of operating costs for year 1 = present worth of AOC_1)(A/P, 7%, 1 year).

From the interest rate tables, we observe the following:

(P/F, 7%, 1 year) = 0.9346
(A/P, 7%, 1 year) = 1.07

Substituting these values into the equation, we get the following:

EUAC of operating costs for year 1 = (\$7,500) (0.9346) (1.07) = \$7,500.

Year 2:

EUAC of operating costs for year 2 = [(present worth of AOC for year 1) + (present worth of AOC for year 2)] (A/P, 7%, 2 years).

From the interest rate tables, we observe the following:

(P/F, 7%, 1 year) = 0.9346
(P/F, 7%, 2 years) = 0.8734
(A/P, 7%, 2 years) = 0.5531

Substituting these values into the equation, we get the following:

EUAC of operating costs for year 2 = [($7500) (0.9346)
+ ($10,500) (0.8734)] (0.5531)
= $8949.28

Year 3:
EUAC of operating costs for year 3 = [(present worth of AOC for year 1) + (present worth of AOC for year 2) + (present worth of AOC for year 3)] (A/P, 7%, 3 years).

From the interest rate tables, we observe the following:

(P/F, 7%, 1 year) = 0.9346
(P/F, 7%, 2 years) = 0.8734
(P/F, 7%, 3 years) = 0.8163
(A/P, 7%, 3 years) = 0.3811

Substituting these values into the equation, we get the following:

EUAC of operating costs for year 3 = [($7500) (0.9346) + ($10,500) (0.8734)
+ ($13,500) (0.8163)] (0.3811)
= $10,364.8

Similarly, the EUAC for every subsequent year can be calculated using the above logic.

The capital recovery costs and the EUAC for annual operating costs as well as EUAC of total costs is presented in the accompanying table with figures in parenthesis denoting negative values.

Challenger: ESL Computation

Year	Value ($)	Annual Cost ($)	Capital Recovery	EUAC (Costs)	Total EUAC
1	−60,000	−7500	($20,250.00)	($7500.00)	($27,750.00)
2	−48,000	−10,500	($18,293.48)	($8949.28)	($27,242.75)
3	−38,400	−13,500	($16,634.49)	($10,364.79)	($26,999.28)
4	−30,720	−16,500	($15,223.10)	($11,746.61)	($26,969.71): ESL
5	−24,576	−19,500	($14,018.26)	($13,094.85)	($27,113.12)
6	−19,661	−22,500	($12,986.19)	($14,409.65)	($27,395.84)
7	−15,729	−25,500	($12,099.00)	($15,691.18)	($27,790.17)
8	−12,583	−28,500	($11,333.65)	($16,939.62)	($28,273.28)
9	−10,066	−31,500	($10,671.08)	($18,155.22)	($28,826.30)
10	−8053	−34,500	($10,095.45)	($19,338.21)	($29,433.67)

It will be observed from the above table that the lowest cost for challenger occurs in year 4. The curve depicting EUAC of total costs is presented

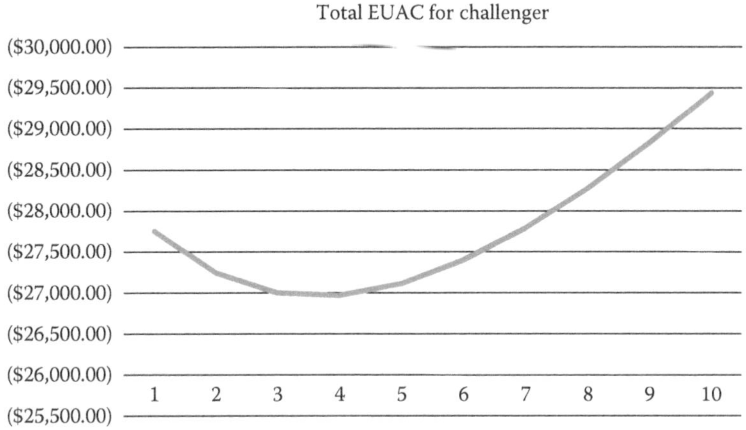

FIGURE 14.2
Figure for example 14.2.

below. The fact that the challenger ESL occurs in year 4 is once again confirmed by the ESL curve depicted in the following Figure 14.2.

A similar analysis for the defender is presented in the table below. Once again, figures in parenthesis represent negative values. Notice that the annual cost for the defender for year 1 includes the cost of upgrade. This, being a one time cost, is not present for subsequent years.

Defender: ESL Computation

Year	Value ($)	Annual Cost	Capital Recovery	EUAC (Costs)	Total EUAC
1	−17,600	−25,500	($5940.00)	($25,500.00)	($31,440.00)
2	−14,080	−11,000	($5366.09)	($18,495.17)	($23,861.26)
3	−11,264	−16,500	($4879.45)	($17,874.57)	($22,754.02): ESL

It will be observed from the above table that the lowest cost for challenger occurs in year 3. The curve depicting EUAC of total costs is shown in (Figure 14.3).

2. In order to perform the replacement study at the current moment in time (year 3, since the robotic welding equipment was purchased two years ago), a comparison between defender and challenger reveals the following:

In year 3, the total EUAC of challenger = $26,999.28 based on the challenger ESL table.

In year 3, the total EUAC of defender = $22,754.02.

Thus, EUAC (defender) < EUAC (challenger).

Due to this reason, we choose to keep the defender and conduct the study again next year.

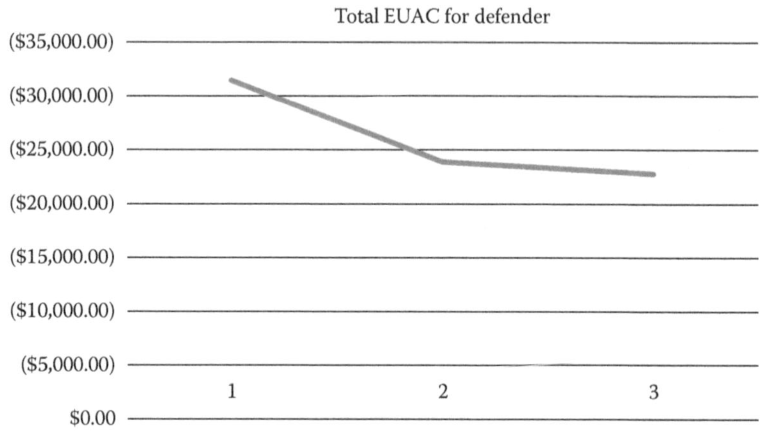

FIGURE 14.3
Figure for example 14.2: EUAC of total costs for challenger.

3. We make the following observations the following year:
   - Given the clear specter of obsolescence slowly but surely creeping up on the defender, the replacement analysis for the defender undergoes a drastic change, both in terms of market value and annual operating costs. In view of a drastic downward revaluation of market value, the capital recovery values will change, as will the values pertaining to EUAC of annual operating costs. The new analysis is presented below for a maximum of two more years of study. This is one year less than was estimated the previous year.

Defender: ESL Re-Computation One Year Later

Year	Value ($)	Annual Cost	Capital Recovery	EUAC (Costs)	Total EUAC
0	−17,600	–	($5940.00)	–	–
1	−3000	−8250	($15,832.00)	($8250.00)	($24,082.00)
2	0	−11,000	($9734.42)	($9578.50)	($19,312.92)

At this point in time, notice that the total EUAC for the challenger is −$26,969.71 whereas the new EUAC for the defender is −$19,312. Thus, the decision is to retain the defender. Perform the study again next year.

Replacement Study over a Particular Period of Time

We have considered two examples so far wherein the replacement analysis was not conducted over any particular period of time. Sometimes, the objective is to conduct the analysis over a prespecified period of time. In such instances, the EUAC for the total costs for the challenger and the remaining life of the defender are not based on ESL. Any changes in cost structure that occur after the study period has expired are not considered to be a part of the replacement study. When a replacement study is performed over a fixed time duration, it is important that the estimates that are used in the computation of EUAC of total costs are accurate and that they are included in the study. It should be borne in mind that when the remaining life of the defender is shorter than the duration of the replacement study, the cost of providing defender services from the "expiration" of the defender to the end of the study period can only be estimated at best. It is important, however, that these estimates should be as accurate as possible.

Let us consider another example to help illustrate such a case.

EXAMPLE 14.3

An engineering manager is considering replacing some equipment used for automated assembly. The current system is 7 years old. All capital recovery costs on this equipment have been exhausted. There are none remaining. It has a current market value of $105,000. It is estimated that this equipment can "live" another 3 years. Thus, its total service life is estimated to be 10 years, of which 7 years have already been exhausted. In 3 years, the equipment will have no salvage value at the end of its useful life. The annual operating costs are about $45,000. The engineering manager can choose to either replace the system now or wait 3 more years to replace it.

The manager has been able to locate a possible challenger to the existing system. It will cost $1,100,000 to procure and install the new equipment. It is expected to last for 10 years and have no salvage value at the end of its useful life. It costs $100,000 per year to operate and maintain.

The manager deems the challenger to be a worthy successor to the current defender. He plans to procure the challenger and the current defender will be used for some parallel application. He estimates that when used on a different application, the first cost of the current defendant would be $1,900,000 three years from now. Also, it would cost about $110,000 per year in operating expenses. Using an interest rate of 7%, perform a replacement study for a fixed period of 10 years.

Solution

In light of the fact that this problem deals with a fixed study period, there is no need to compute economic service life. Also, no projected market values are available for any time in the future. The engineering manager has two options available to him as stated in the problem:

- Retain the defender for three more years until its useful life is exhausted and then replace it with the challenger.
- Procure the challenger and use it for all ten years.

The cash flow diagram for option 1 can be depicted as shown in the following Figure 14.4.

The cash flow computation for this option is presented below.

The defender will be retained for the first three years. The capital recovery for the follow through period (next 7 years for the defender) will

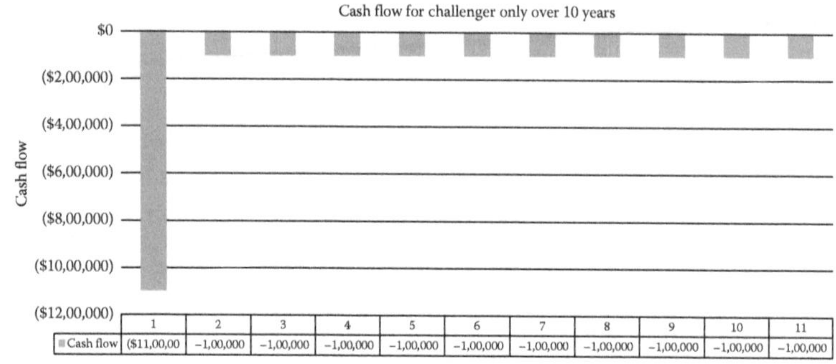

FIGURE 14.4
Cash flow diagram for option 2 for Example 14.3.

be added to this figure. The capital recovery will be determined over the entire ten year useful life. Here is the cash flow analysis for this option:

First Three years: Defender

Market value = $105,000
Annual operating costs = $45,000
Useful life (n) = 3 years
Salvage value (S) = $0
Thus,

EUAC = [–$105,000 – $45,000(P/A, 7%, 10)] (A/P, 7%, 10)
= [–$105,000 – $45,000(7.024)] (0.1424)
= [–$105,000 – $316,080] (0.1424)
= [–$421,080] (0.1424) = –$59,961.79

When the defender is assigned to another application, the cash flow for this situation is given below:

P = –$1,900,000.
Annual operating costs = –$110,000 for years 4 through 10.

The capital recovery computation will be performed over ten years. Thus, the value of "n" for this purpose is 10 years.

Salvage value (S) = $0.
Capital recovery = –$1,900,000(A/P, 7%, 10) = –$1,900,000 (0.1424)
= –$270,560.
EUAC of total expenses = (–$270,560 – $110,000) (F/A, 7%, 7) (A/F, 7%, 10).
= (–$380,560) (8.654) (0.0724)
= –$238,439.72

Thus, total EUAC for the defender in this scenario is given by the sum of the two values as follows:

Total EUAC of defender = –$59,961.79 + (–$238,439.72)
= –$298,401.51. **Subsolution 1**

Next, we consider option 2. The cash flow diagram for option 2 is depicted below.

Option 2 consists of purchasing the challenger and retaining it for all 10 years. This is a pretty straightforward computation and is presented as follows:

Purchase price, P = –$1,100,000
Annual operating expenses = –$100,000
Life of the asset (n) = 10 years
Salvage value (S) = $0

Thus,

EUAC of total expenses for challenger = –$1,100,000(A/P, 7%, 10)
–$100,000

Thus,

The EUAC value for the challenger = –$1,100,000(0.1424) – $100,000.
= –$156,640 – $100,000 = –$256,640 **Subsolution 2**

Comparing subsolutions 1 and 2, we observe that since $256,640 < $298,401 on an annual cost basis, it is advisable to purchase and install the challenger now. **Solution**

14.6 Replacement Value

Replacement value is defined as the break even value between the defender and the challenger. It is a function of the minimum market value of the defender that is necessary to make the challenger attractive. This implies that if the currently installed defender can be sold for this value (either in the form of salvage value or trade-in value), it makes the challenger looks economically attractive. The basic criterion for replacement is that if the market trade-in is greater than the break even replacement value, it deems the challenger a better alternative than the defender.

14.7 Marginal Costs

Marginal costs are defined as the annual costs of retaining an asset. Thus, by definition, the marginal cost for any asset is always based on a one year consideration. This definition is contrasted with EUAC which is derived over any number of consecutive years (based on the definition of ECUA as discussed in Chapter 1). When dealing with replacement analysis, marginal costs can include items such as capital recovery, annual operation and maintenance costs, taxes, insurance, and any expenses accrued over the course of that year. Marginal costs are based on the estimates of an asset's market value on a yearly basis as well as estimates of its annual operating expenses. The following example will help comprehend the concept of marginal cost and its computation.

EXAMPLE 14.4

New construction equipment can be purchased for $50,000. It costs $4000 to operate and maintain in the first year. This cost increases by $1000 per year for every subsequent year. The equipment is expected to last 7 years. The annual cost for risk of breakdown is $10,000 per year for three years. Beginning with the fourth year, these costs increase by $3000 per year. If the minimum attractive rate of return is 10%, compute the marginal cost of retaining the asset over its entire useful life. Additionally, estimates for market value for each year are also presented below.

Year	Market Value ($)
1	36,000
2	26,000
3	18,000
4	12,000
5	8000
6	6000
7	5000

Solution

The computation of marginal costs is presented in the following table.

Year	Loss in Market Value ($)	Interest ($)	Operation and Maintenance Costs ($)	Cost of Breakdown Risk ($)	Total Marginal Cost ($)
1	14,000	5000	4000	10,000	33,000
2	10,000	3600	5000	10,000	28,600
3	8000	2600	6000	10,000	26,600
4	6000	1800	7000	13,000	27,800
5	4000	1200	8000	16,000	29,200
6	2000	800	9000	19,000	30,800
7	1000	600	10,000	22,000	33,600

The various entries are explained as follows:

Loss in market value in year 1 = \$50,000 – \$36,000 = \$14,000

Loss in market value in year 2 = \$36,000 – \$26,000 = \$10,000 and so on for every successive year.

Interest in any given year is calculated by multiplying the market value at the beginning of that year by the MARR.

Thus,

Interest charges in year 1 = \$50,000 × 0.10 = \$5000.

Notice that the MARR is 10% = 0.10.

Interest charges in year 2 = \$36,000 × 0.10 = \$3600 and so on for every successive year.

Data pertaining to operation and maintenance costs for every year are already given in the problem statement and are substituted as such in the table.

Data pertaining to cost of breakdown risk for every year are already given in the problem statement as well and are substituted as such in the table.

Total marginal cost is computed by adding the following entities:

Total marginal cost = loss in market value + interest + operation and maintenance costs + cost of breakdown risk.

Consider another problem dealing with replacement analysis wherein marginal costs are computed for both defender and challenger.

EXAMPLE 14.5

Defender: A machine was purchased 5 years ago for \$150,000. It can be sold today for \$30,000. It costs \$20,000 to operate the machine this year. This expense will increase by \$3000 per year going forward. The market value of the machine is estimated to decrease by \$2000 per year for the next five years. The company is considering replacing this machine with a newer alternative.

Challenger: The alternative can be purchased for \$50,000. It costs \$4000 to operate and maintain in the first year. This cost increases by \$1000 per year for every subsequent year. The equipment is expected to last 7 years. The annual cost for risk of breakdown is \$10,000 per year for three years. Beginning with the fourth year, these costs increase by \$3000 per year. Using a MARR of 10%, determine when a replacement decision should be made.

Solution

If marginal costs of the defender are increasing, these costs can be compared with the minimum EUAC of the challenger. We compute the EUAC of retaining the defender for each year as follows:

Year	Challenger Marginal Cost ($)	Present Cost ($)	EUAC ($)
1	33,000	30,000.3	33,000.33
2	28,600	53,635.34	30,904.68
3	26,600	73,619.92	29,602.57
4	27,800	92,607.32	29,217.61
5	29,200	11,0737.6	29,212.58
6	30,800	12,8124.2	29,417.32
7	33,600	14,5367.7	29,858.53

The various entries are explained as follows:
Recall from Example 14.4 that the challenger in this problem is the same machine that we analyzed in Example 14.4. Thus, the challenger marginal costs are obtained directly from the solution for Example 14.4.

$$\textit{Present cost in year 1} = \text{Marginal cost (P/F, 10\%, 1)}$$

In this case,

$$(P/F, 10\%, 1) = 0.9091$$

Thus,

$$\text{Present cost} = \$33{,}000 \times 0.9091 = \$30{,}000.3$$

Similarly,

$$\textit{Present Cost in Year 2} = \text{Present cost in year 1} + \text{Marginal cost in year 2} \times (P/F, 10\%, 2).$$

Thus, in this example, we have (P/F, 10%, 2) = 0.8264
Substituting this value into the equation, we obtain the following:

$$\text{Present cost in year 2} = \$30{,}000 + \$28{,}600 \times (0.8264) = \$53{,}635$$

This procedure will be continued recursively until present costs for all years have been computed.

$$\text{EUAC in year 1} = \text{Present cost in year 1} \times (A/P, 10\%, 1).$$

The value of (A/P, 10%, 1) = 1.10
Thus,

$$\text{EUAC in year 1} = \$30{,}000.3 \times 1.10 = \$33{,}000.33$$

$$\text{EUAC in year 2} = \$53{,}635.34 \times (A/P, 10\%, 2).$$

The value of (A/P, 10%, 2) = 0.5762
Thus,

$$\text{EUAC in year 1} = \$53{,}635.34 \times 0.5762 = \$30{,}904.68$$

This method will also be continued recursively until all values have been obtained. Notice from the above table that the minimum EUAC occurs in year 5 corresponding to $29,212.58. This is the challenger's economic service life.

Next, we compare this value against the marginal costs of the defender. These costs are calculated using the principle demonstrated in Example 14.4.

Year	Loss in Market Value ($)	Interest ($)	Operation and Maintenance Costs ($)	Total Marginal Cost ($)
1	2000	3000	20,000	25,000
2	2000	2800	23,000	27,800
3	2000	2600	26,000	30,600
4	2000	2400	29,000	33,400
5	2000	2200	32,000	36,200

Comparing the EUAC value of $29,212 against the total marginal costs for the defender, we notice that for year 1, $25,000 < 29,212. Thus the conclusion is to keep the defender for one year. For year 2, $27,800 < $29,212, thus leading to the conclusion that the defender should be kept for that year. However, for year 3, $30,600 > $29,212. Thus the challenger should replace the defender at the end of two more years. **Solution**

14.8 Conclusion

This chapter dealt with the concept of equipment replacement, conducting a replacement study, and various quantitative criteria used to make replacement decisions in industry. Concepts such as capital recovery, annual operating costs, and EUAC of total costs were examined in detail. Economic service life was defined and its role in an effective replacement study was clearly delineated. Cost curves were used to graphically illustrate the changing nature of total cost per year. A clear distinction was drawn between replacement studies using ESL and their counterparts using fixed life time periods. We also defined the concept of marginal cost and applied it to solve problems related to replacement strategy.

PROBLEMS

1. A machine can be purchased for $150,000. It suffers an annual drop in market value by as much as 20%. It costs $10,000 for the first year to operate and maintain the machine. Every year, the operation and maintenance costs rise by $4000. Compute the minimum EUAC for the machine and its economic service life if the value of $i = 10\%$.
2. A truck is purchased for $50,000. Its market value declines by 20% annually. The older the truck, the better its "collector value." This so called "collector value" increases by $500 every year. If the interest rate is 7%, compute the ESL and minimum EUAC for this vehicle.
3. An automatic machine has a current salvage value of $25,000. The machine is one year old. The salvage value of the machine will decrease by $5000 every year. The maintenance costs for the next four years are as follows:

Year	Cost ($)
1	2000
2	3000
3	3750
4	4500

   Compute the marginal cost to extend service to each of the four years if the interest rate is 7%.
4. A construction firm purchases a $75,000 machine. The interest rate is 10%. It costs $7000 to install the machine. Compute the economic service life of the machine based on the market values and operation costs as presented in the following table.

Year	1	2	3	4	5
Market value ($)	65,000	56,000	47,000	38,000	28,000
Operation and maintenance costs ($)	15,000	26,000	37,000	48,000	59,000

5. A robotic welding arm has a first cost of $1,500,000 and a salvage value of $300,000 in any year. It costs $250,000 for the first year to operate and maintain the robot. Thereafter, the cost increases on an arithmetic gradient of $80,000 per year. Compute the ESL for this machine if the interest rate is 7%.
6. A construction firm purchases a $100,000 machine. The interest rate is 7%. It costs $15,000 to install the machine. Compute the economic service life of the machine based on the market values and operation costs as presented in the following table.

Year	1	2	3	4	5
Market value ($)	70,000	60,000	50,000	40,000	30,000
Operation and maintenance costs ($)	15,000	26,000	40,000	52,000	60,000

7. A hospital is purchasing a new $40,000 x-ray machine that will have no salvage value after installation. Maintenance is estimated to be $3500 per year as long as the machine is owned. After 8 years, the radioactive source will have caused sufficient damage to the point where the machine must be scrapped. The most economic life of this machine is:
   a. 1 year because there will be no salvage value after installation
   b. 8 years because the maintenance will not increase
   c. Less than 8 years, but you need more information on the economic life
8. A truck salesperson is quoted as:"Even though our list price has gone up to $78,000, I'll sell you a new truck for the old price of $70,000, an immediate savings of $8000, and give you a trade-in allowance of $20,000, so your cost is only ($70,000 – 20,000) = $50,000. The book value of your old vehicle is $10,000, so you're making an additional ($20,000 – 10,000) = $10,000 on the deal. Actually I am giving you more trade-in for your old truck than the current market value of $19,500, so you are saving an extra ($20,000 – 19,500) = $500."
   a. In a proper replacement analysis, what is the first cost of the defender?
   b. In a proper replacement analysis, what is the first cost of the challenger?

For problems 9–10, consider the following.

Newton Construction is conducting a routine review of existing equipment. They use a MARR of 21%. This includes a replacement evaluation of a machine currently in use. The machine was purchased 5 years ago for $250,000. Its current value is $170,000 and annual operating and maintenance costs are as follows:

Year (n)	Operating Cost	Maintenance Cost	Market Value
1	$20,000	$10,000	$85,000
2	$20,000	$12,000	$74,000
3	$21,000	$15,000	$63,000
4	$26,000	$22,000	$51,000
5	$31,000	$25,000	$40,000
6	$37,000	$29,000	$35,000
7	$41,000	$32,000	$28,000

9. Data for a new machine has been analyzed. Its most economic life is at 6 years with a minimum EUAC of $72,000. When should the current machine be replaced?
10. What if the economic life is at 8 years with a minimum EUAC of $54,000? When should the current machine be replaced?
11. In this problem, the following facts are known: Initial cost ($11,000), the annual maintenance (none for the first 4 years, $3000 at the end of the 5th year, $3000 at the end of the 6th year, increasing to $3500 per year after the 6th year). The salvage value in any year is zero. Assume a 9% interest rate. Calculate the life having the lowest EUAC.
12. A construction crane has an initial cost of $950,000 and a salve value of $200,000 in any year. Operating and maintenance costs are $250,000 for the first year and increase $75,000 annually. Determine the ESL for the crane if the interest rate is 8%.
13. A machine has a first cost of $45,000. Its market value declines by 15% annually. Operating and maintenance costs are $3000 annually and increase by $500 per year. The company's MARR is 8%. What is the minimum EUAC for the machine and its economic life?
14. The cash flow diagram below indicates the costs associated with a piece of equipment. The investment cost is $9000 and there is no salvage value. During the first 3 years, the equipment is under warranty so there are no maintenance costs. Then the estimated maintenance costs over 15 years follow the pattern shown in the cash flow diagram. Determine the equivalent annual cost for n = 12 if the MARR is 18%. Use gradient and uniform series factors in your solution.

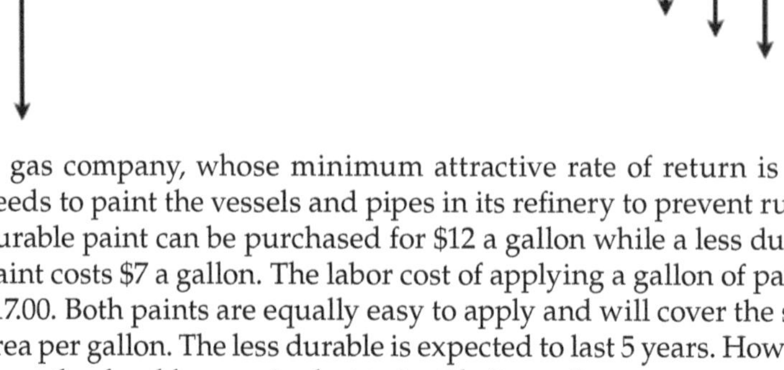

15. A gas company, whose minimum attractive rate of return is 13%, needs to paint the vessels and pipes in its refinery to prevent rust. A durable paint can be purchased for $12 a gallon while a less durable paint costs $7 a gallon. The labor cost of applying a gallon of paint is $17.00. Both paints are equally easy to apply and will cover the same area per gallon. The less durable is expected to last 5 years. How long must the durable promise last to justify its use?

16. An industrial machine has a current salvage value of $37,000. The machine is two years old. The salvage value of the machine will decrease by $2500 every year. The maintenance costs for the next four years are as follows:

Year (n)	Cost ($)
1	2000
2	3100
3	5000
4	5900

Compute the marginal cost to extend service to each of the four years if the interest rate is 5%.

17. Jason owned a car for 4 years. Being a clever young man, he wondered what his uniform annual cost for his vehicle's maintenance and repairs had been. He assembled data. Determine the EUAC assuming 9% interest.

Year	Annual Maintenance and Repair Costs
1	$820
2	$1400
3	$990
4	$2100

18. Jason reexamined his calculations and discovered that he had reversed years 2 and 3 maintenance and repair costs in his table. The correct table is below. Recalculate the EUAC.

Year	Annual Maintenance and Repair Costs
1	$820
2	$990
3	$1400
4	$2100

19. A company is considering 2 different devices to reduce costs. Both cost $1000 and have lives of six years with no salvage value. Device A is estimated to save $270 annually. Device B will save $550 the first year, but decline $75 annually, making the second year savings $475, the third year savings $400, and so on. With 9% interest, which device should the company purchase?

20. Consider a scenario where Alternative A has a 9-year life and a salvage value at the end of that period. The replacement cost at the end is estimated to be close to the original cost. If the replacement ceases to be used prior to 9 years, it will have a value that exceeds its end-of-life value. Alternative B has a 13-year life and terminal value whenever it is retired.
    a. If the analysis period lasts 11 years, set up the equations to properly calculate the EUAC for each alternative.
    b. Which alternative saves the company more money?
21. The city of Denver wishes to construct an aqueduct to expand the water supply. There are two alternatives for a portion of the aqueduct. Either a tunnel can be constructed through a mountain or a pipeline is to be laid to go around the mountain.
    a. If there is a permanent need for the aqueduct, should the tunnel or the pipeline be selected for this portion of the aqueduct? Assume a 5% interest rate.
    b. Calculate the annual cost if the tunnel is expected to have a 75-year life.

15

The Make v/s Buy Decision

15.1 The Concept: Make versus Buy Decision

Should we produce a component, part or assembly in house or purchase it from an external vendor? This is the gist of any make versus buy decision. By its very nature, it is one of the more critical decisions that have to be made early in the life of a production enterprise. This decision is continually modified as a business evolves over time. The make versus buy decision is based primarily on the priorities of the business at any given time as well as the changing nature of the marketplace. Two main activities are involved within the realm of the make versus buy decision. They are described as follows:

- *Insourcing*: Insourcing refers to a decision to manufacture an item in house instead of purchasing it from an external vendor.
- *Outsourcing*: Outsourcing refers to a decision wherein it is decided that the item be purchased from an external vendor. In essence, the outsourcing decision reverses a previous insourcing decision.

The make versus buy decision is of crucial importance to the productivity and competitiveness of any organization. Traditionally, management strategy lent itself heavily towards manufacturing items in house. This is referred to as the "make decision." The make decision requires that almost all of the value added manufacturing and assembly operations be performed in house. This leads to an organization owning several manufacturing and assembly facilities. This process is referred to as backward integration. For instance, in its early years, the Ford Motor Company owned not only the manufacturing and assembly plants to manufacture its automobiles, but the coal mines and steel mills as well. Raw material procurement is perhaps the only time that the organization buys from an external vendor. Raw materials are then subjected to various value adding processes that transform the raw material into a finished product. In the case of the make decision, almost all of the value adding processes are performed in house.

The make decision imparts the manufacturer with a significant degree of control over product quality. Since all manufacturing and assembly processes are performed in house, the parent organization can have strict control over process parameters that affect product quality. Examples of such parameters include feed rate, cutting speed, and types of material being used among others. It also imparts to the organization a greater degree of control over its scheduling operations at every step of each process. This enables the organization to meet customer demand.

One of the greatest disadvantages of the make decision is its lack of flexibility. A business that is saddled with inventory as well as manufacturing and assembly facilities cannot be nimble to respond to rapid changes in the marketplace. This lack of agility on the part of the business renders it incapable of meeting and exceeding customer requirements, leading to serious consequences on the productivity and competitiveness of the organization.

With the passage of time, managers have recognized the serious drawback of the make decision. As a result, contemporary managers are more open to purchasing an item from an outside vendor if such a decision meets certain criteria. This imparts the much needed "agility" to the organization. A large amount of flexibility can be gained without compromising essential attributes such as quality of the product. Such a decision is referred to as the "buy decision."

It is obvious that one of the principal disadvantages of the buy decision lies in the lack of control that the parent organization has over incoming product quality. Since the finished product is purchased from an external vendor, the parent organization (in this case, the organization that purchases the finished product) has very little control over the process parameters used by the vendor in manufacturing the product. This obstacle can be overcome by fostering an atmosphere of mutual trust between vendor and customer. It is the onus of the purchaser to develop and sustain world class suppliers that can be relied upon to produce high quality products. Most of the problems related to quality control can be easily addressed by adhering to a strict code of standards prevalent in industry. As a matter of fact, industry wide quality assurance is one of the main objectives of developing standards in the first place.

The recent outsourcing phenomenon has seen many US companies "farm out" a large number of their activities to external vendors. These activities used to be performed in house. Examples include activities such as computer programming, engineering, information systems, maintenance, etc. The following section presents the reasons a company chooses to make a product in house instead of buying it from an external vendor.

15.2 Circumstances Favoring a "Make" Decision

A company can choose to manufacture ("make") products in house in a variety of situations. The most noteworthy examples are listed below.

- A company may have no choice but to make the items in house. This situation can arise if the order quantity is too small. It can also arise if the product in question is a highly specialized product that is not available in the market. Similarly, if no supplier is willing to make the product, the company would have to resort to making the product in house.
- A company may have to make the product in house if the quality requirements are too stringent. This situation can also arise if the quality requirements are too specific (one of a kind) for most external vendors to supply.
- In some situations, it might be imperative for the company to safeguard its technology. In this case, the company has to resort to making the product themselves so as not to divulge their valuable technological secrets to their competitors.
- Outsourcing is generally resorted to in order to gain a cost advantage. However, with rising wages in developing countries as well as at external suppliers, the cost advantage is lost over time. In such instances where no distinct cost advantage can be gained, it may be advisable for the company to make the items themselves.
- If the organization resorts mostly to in house manufacturing, it ensures a steady supply of quality products that are used to satisfy customer demand. External vendors can be relied upon solely in instances where customer demand exceeds in house supply.
- Sometimes, a company could be dealing with a situation that involves idle plant capacity. In manufacturing, idle capacity is a major liability. Ideally, capacity utilization should be 100%. This implies that all the resources invested in building the capacity are being fully utilized. The production process is being used to produce a return on investment. The concept of return on investment (ROI) has been dealt with in various chapters throughout this book. Idle capacity implies that the return on investment is zero as far as that particular capacity is concerned. This is an example of wasted resources. Wasted capacity impacts the profitability of a company adversely and should always be avoided. Thus, in such a situation, it might make economic sense to make the product in house and thus utilize the capacity productively that would otherwise remain idle.
- On a comparative basis, it is better to make the product in house than purchasing it from an external vendor if the purchase option is just too expensive.
- If an item is purchased from an external supplier, transportation distance and associated transportation costs play an important role in the make versus buy decision. In such cases, if the closest reliable supplier is too far, it would be sensible to make the item in house.

- Sometimes, a sizeable customer could require that the item be made in house and may be willing to pay the price for in house manufacturing. In such instances, the company loses nothing by making the item in house.
- A company could choose to make a certain item in house if the demand for that item is too great and is forecast to keep rising significantly in the future. If the market potential for the item is expanding rapidly, it makes more sense for the company to adopt this option instead of purchasing the item from an external vendor.
- In a charged political atmosphere, it might make more sense for a company to make an item in house rather than purchase it from a vendor. This goes to the very core of manufacturing and its role as a wealth creator. Manufacturing creates wealth and jobs. A company that is located in a high unemployment area may be rewarded with significant tax breaks by the government. These tax breaks render it more attractive to make the item in house. Similarly, companies that are good social citizens could voluntarily decide to make an item in house in order to help the community.
- A company could choose to make an item in house if future shortages for that item have been forecast. A similar course of action could be followed if market prices for that item are forecast to rise rapidly.

This section dealt with the various reasons a company could choose to make an item in house. However, this option does not always represent the most viable or practical solution. Several reasons can exist wherein it makes more sense for a company to resort to purchasing an item instead of making it in house. These reasons are enumerated in the following section.

15.3 Circumstances Favoring a "Buy" Decision

A company can choose to purchase products from outside vendors in a variety of situations. The most noteworthy examples are listed below.

- One of the more obvious reasons for a company to choose to buy from an external vendor is due to the fact that it does not possess the requisite technical and managerial expertise required to produce the item under consideration.
- A company can only produce an item based on its production capacity. Production capacity can only be expanded so much before significant capital expenses have to be incurred in terms of buildings and equipment. Thus, a company could choose to buy a product from

an external vendor if there just is no capacity for the company to produce the item in house.

- If the manufacturing process required to produce the item represents a noncore activity for the company, it may be too challenging for the company to maintain a technological and economic viability for such an activity for an extended period of time. In such instances, it would be better to purchase the item externally rather than making it in house.
- A decision to manufacture in house is accompanied by considerable economic pressures. Resources need to be committed to the project that involve not just economic resources, but human resources as well. Overall, a "make" decision is difficult to reverse, especially in light of labor interests as well as management inertia. Thus, it might be easier to decide to purchase the item externally from the outset.
- When an item is purchased externally, it is very easy to keep records of transactions. There are no cost overruns. The cost incurred by the company in purchasing the item is exactly what it paid for it. Nothing more and nothing less. This ensures transparency and makes bookkeeping easy.
- If an item is offered for sale by several suppliers, the purchaser can pick and choose from a variety of suppliers based on price and quality. Furthermore, the deal can be negotiated based on order volume and length of the contract going forward, thus contributing even more to greater flexibility.
- A company may possess the technological wherewithal, expertise, and capacity to manufacture an item in house. However, the order size may be too small to justify making the item in house. Based on volume and cost considerations, it may be best for the company to choose to buy the item from an external vendor.
- Sometimes, a sizeable customer could require that the item be purchased from an external vendor and may be willing to pay the price for it. In such instances, the company loses nothing by purchasing the item from a vendor.
- Agile enterprises are able to swiftly respond to adverse changes in the marketplace. One of the key principles of agile manufacturing is the ability to stay lean. This involves focusing on core competencies only at the expense of noncore competencies. A company with a focus on lean manufacturing could choose to purchase an item from a vendor if that item corresponds to a noncore competency from the company's perspective.
- The decision to purchase from vendors could open new markets for companies. In this case, it is advisable to resort to this option.
- Purchasing an item from a vendor cuts back on the lead time to bring a product to market. Shorter lead times imply a significant advantage

over the competition. This can lead to better sales and higher profit margins. Thus, a company could choose to buy the item instead of making it if it reduces the lead time and enhances the company's ability to bring the product to market faster.

- The nature of an item could be such that the underlying technology has an uncertain future ahead of it. The demand for the item could be uncertain as well. In such cases, if the company deems the project to be too risky, but still desires to share in the upside (in terms of sales), it may choose to purchase the item from a vendor instead of investing in capacity to make it in house.

The preceding sections presented an overview of the myriad situations in which a make decision is favored. They also presented the many situations in which a buy decision could be favored. In reality, a make versus buy decision is specific to the particular circumstance under consideration. Several factors need to be accounted for in order to make an intelligent, well informed decision. These factors can be qualitative as well as quantitative in nature. A checklist of factors that influence this decision is presented in the following section.

15.4 The Make versus Buy Decision: A Checklist of Influential Factors

An intelligent buy versus make decision depends on a variety of factors, all of which should be taken into careful consideration before reaching a final decision. This section presents a checklist of such factors that directly affect the outcome of the buy versus make decision.

- *Bill of Materials*: The bill of materials is a vitally important component of a production operation. Every component is made out of some material. Most products are an assembly of several individual components. Thus, a comprehensive bill of materials forms the centerpiece of any production operation. The bill of materials includes all hardware, fasteners, bolts, bearings, drive mechanisms, motors, pulleys, and any other component that is deemed as a necessary component of the product under consideration. Strenuous effort should be made to locate suppliers who would be willing to supply each item listed in the bill of materials in the requisite quantity and desired quality. This will make for an accurate and objective buy versus make comparison.
- *Volume of Output*: The actual volume of output required over any time period is a vital factor influencing the make versus buy decision.

Higher volumes are easier to work with. They contribute to the viability of making the concerned products in house. As we have seen before, in house manufacturing ensures integration in addition to assuring quality. Furthermore, owing to the higher batch size, the cost difference between a part that is manufactured in house and one that is sourced externally could be minimal. When contemplating a make versus buy decision, careful attention needs to be paid to whether the manufacturing operation fits into one of the core competencies of the company. If it does not fit, the make decision could lead the company astray from its primary objectives. As such, the investment in time and labor necessary to make the item in house may not produce the intended effect and such a project could be abandoned. Market considerations and projected sales are some other factors that need to be taken into consideration. Production volume is a function of market demand. While higher production volumes make it easier to manufacture a product in house, the company should be equipped with the requisite capacity to undertake the manufacturing activity. If this is not the case, no matter how high the volume, the decision could be skewed toward purchasing the item from an external vendor.

- *Labor Requirements*: A company that intends to manufacture an item in house should be well equipped with the requisite technical expertise to undertake such a project. A manufacturing operation involves activities such as prototype design, development of individual components, and fabrication of a working prototype that is reliable, viable, and one that will function as intended. This is in addition to the actual manufacturing operation that involves machining, assembly, and testing. Time to market can be significantly small if the company is well equipped to handle the design, development, and fabrication tasks. If this is not the case, the product should be purchased externally.
- *Support Staff*: Support staff is required to perform auxiliary operations. For instance, maintenance operations can be classified as support operations. Related costs such as salaries paid to such staff are often classified as labor costs. Support operations and related costs are unpredictable. Answers need to be sought for questions such as: How long would it take to repair a failed item? What would be the consequences of a failed item?
- *Functionality of the Part*: Every item is designed to satisfy a specific function. Whether it fits into a household product or an industrial product, every item should fit seamlessly with the remained of the assembly. The parts of the assembly could be sourced from a variety of suppliers. The ability to seamlessly integrate is a vital factor that cannot be ignored when contemplating a make versus buy decision.

- *Core Expertise*: Every company needs to carefully consider the make versus buy decision based on the core expertise of its engineering and manufacturing staff. While it is tempting to retain engineering in house, a symbiotic partnership with a supplier that specializes in manufacturing a certain item can be beneficial. This is especially true when such specialized skills do not form the core competency of the purchasing company's engineering and manufacturing staff.
- *Ownership Risk*: A company that chooses to manufacture an item in house assumes the responsibility of product design, development, and manufacture. This includes investment in specialized staff, manufacturing and testing facilities, design equipment, material, and inventory. The fixed costs for such an operation are quite high. As we have seen before, it generally makes more sense to manufacture an item in house if the production volume is high. This implies that variable costs are high as well. The amount of resources (such as time, money, effort) invested into the project make it a comparatively high risk project. On the other hand, the risk is considerably minimized if the same part was purchased from a reputable supplier that specializes in the manufacturing of such parts.

Table 15.1 depicts the various factors that play an important role in deciding whether an item should be made in house or purchased from a vendor. Some factors have a positive impact on a specific decision. If, for instance, customization has a positive impact on the make decision, a plus sign (+) is put into the corresponding cell. On the other hand, if a certain factor, such as process control, has a negative impact on the buy decision, a minus sign (–) is allocated to the corresponding cell. The comprehensive list of factors of consideration is presented below.

TABLE 15.1

Factors That Affect a Buy versus Make Decision

Factor of Consideration	Make	Buy
High volume	–	+
Number of labor hours	–	+
Support staff	–	+
Time to develop	–	+
Degree of customization	+	–
Level of integration	+	–
Level of in house expertise	–	+
Degree of process control	+	–
High degree of risk exposure	–	+
Materials and inventory	–	+
Lead time from concept to production	–	+

15.5 Factors to Consider When Selecting a Vendor

Product quality is one of the main areas of concern if a company chooses to buy an item from a vendor instead of making it in house. In such cases, factors such as the reputation of the vendor, the core competencies of the vendor, their specialized skill sets, etc., should be considered very carefully before making the final decision. Remember that failure of just one critical component is all it takes for an entire product to fail. Thus, vendor choice is of crucial importance, especially when it comes to purchasing critical components. There are several probing questions a company should ask before deciding to buy from a particular vendor. A list of these questions is presented below:

Q1: What is the vendor's experience in the particular industry?

Implication: When considering the feedback to this question, the following factors need to be carefully studied:

- Quality of the vendor's product needs to be carefully considered.
- Tested performance results of the aforementioned product need to be thoroughly reviewed.
- Vendor's engineers and support staff can be interviewed and their level of experience in designing, developing, manufacturing, testing, and repairing the particular part can be assessed.
- The ability of the vendor's staff to work with customers in developing a solution can be assessed by interviewing them as well.

Q2: What is the vendor's record with respect to delivering the product on time? What are the return rates for this particular vendor?

Implication: The following factors need to be considered carefully:

- Past clients can be interviewed to study the vendor's history in fulfilling orders on time. A log recording the vendor's order receipt, order delivery, and expected delivery dates can be studied as well.
- Since quality assurance and control is a major issue in making a buy decision, the rate of product return should be analyzed carefully. A product is returned when it fails to meet specifications or is otherwise defective. The rate of product returns is directly related to the probability of defective items being manufactured. This is one factor that needs to be carefully scrutinized.

Q3: What is the vendor's commitment to customer support?

Implication: The vendor's commitment to customer support is vital. This is because when an item is purchased from a vendor, the purchaser assumes the role of a customer. Vendor support in terms of order fulfillment, support services, technical support, etc., is vital to maintain a long term vendor-client relationship. The vendor's understanding of customer needs can be gauged by studying the solutions offered to other customers. Ask the vendor to provide names of references so that the vendor's commitment to customer support can be verified.

Q4: What is the vendor's product line?

Implication: A vendor's product line should be examined carefully. This is a comparatively easy task. The product line should be studied from the perspective of a potential fit with the client's current as well as future needs. Consideration of future needs is important to build a potential long term relationship.

Q5: Does the vendor offer customized solutions?

Implication: Despite the push towards ever greater degrees of standardization, there are always instances in which a customized solution is required to solve a problem. In such cases, a vendor who is able to work with the client in order to customize their solutions to suit the client's particular needs is preferable to one who can only offer standardized solutions. The vendor's ability to offer customized solutions can be verified by checking their track record in such situations. A record of past clients is helpful in such instances. The purchaser can call each client and discuss the uniqueness of their situation as well as the role played by the vendor in offering customized solutions to solve the problem.

Q6: What is the vendor's lead time in developing a prototype that can be tested?

Implication: As a rule of thumb, shorter lead times in developing a prototype fit for testing are preferred over longer lead times. Information on development lead times can be determined by studying the vendor's engineering and development data.

Q7: What information and materials does the vendor offer prior to sales?

Implication: Information such as sizing and selection information as well as downloadable CAD files, etc., are helpful to ensure seamless integration of the purchased item. Provision of such materials can be helpful from the vendor's perspective.

Q8: What type of after sales support does the vendor offer?

Implication: After sales support are an integral part of any reputed vendor's services. Checking pertinent documentation, as well as communicating with past clients, can verify this type of support. Clients can be questioned about their experience with the vendor when availing themselves of the vendor's after sales support. The quality of the vendor's after sales support services can be verified by following this procedure.

A better understanding of the make versus buy decision can be gained by solving a few problems. Consider Example 15.1.

EXAMPLE 15.1

Auto, Inc., an automotive manufacturer, produces a component for a unit cost of $25. The fixed cost for producing the component is $50,000. Currently, the firm manufactures 15,000 units. The total cost structure for the manufacturing operation is presented in the following table.

Element of Cost	Amount ($)
Direct materials	7.50
Direct labor	10.00
Variable cost: overhead	5.00
Fixed cost: overhead	2.50

1. The company has recently learned of another manufacturer, ABC Inc, that is willing to sell the exact same component for a total cost of $23.50. The engineering manager at Auto, Inc. is trying to figure out whether the component should be manufactured in house or bought from the external manufacturer.
2. ABC, Inc is willing to reduce their price to $20.00 per unit. What should the engineering manager do in this case?
3. What role does the fixed cost play in the decision making process?

Solution

The problem is solved in three distinct parts as follows:

1. It will be observed from the data presented in the problem that the total fixed cost per unit amounts to $2.50. The remaining elements of cost constitute elements of variable cost. As we have already seen in previous chapters, the following are elements of variable cost:
    - Direct material
    - Direct labor
    - Overhead cost: variable

   The data presented to the engineering manager indicates that the total variable cost per unit amounts to $22.50.

   This figure is less than the cost of the component sold by the external manufacturer, ABC, Inc. ($23.50).

Since the variable component of the in house production cost (\$22.50) < purchase price if the component was bought from ABC, Inc. (\$23.5), the engineering manager should choose to manufacture the component in house. **Solution 1**

2. We have already computed the variable cost per unit if the component was made in house. This cost amounts to \$22.50. In the second scenario, the unit cost would be reduced if the component was purchased from ABC, Inc. at a unit price of \$20.00.

   This is because \$22.50 < \$20.00.

   In this case, the component should be purchased from ABC Inc. **Solution 2**

3. The \$50,000 invested in the form of fixed costs would have no bearing on the decision as to whether the component should be made in house or purchased from an external vendor. **Solution 3**

EXAMPLE 15.2

General Electronics, a manufacturer of electronics equipment, currently manufactures some products for household use. The cost structure of its manufacturing operation is presented in the following table.

Element of Cost	Amount (\$)
Direct materials	10.00
Direct labor	20.00
Variable cost: overhead	15.00

Additionally, an amount equal to \$1500 needs to be spent in the form of fixed costs.

The same products can be purchased from another manufacturer for \$50 per unit. The fixed cost would have to be incurred regardless of whether the products are made in house or purchased externally. If the products are purchased externally, the fixed cost per unit would be decreased by as much as \$2.50. In view of this situation, answer questions 1 and 2:

1. Should General Electronics make the products in house or buy them from the external manufacturer?
2. If the products are purchased from the external manufacturer, it results in some production capacity at General Electronics being released. What would the make versus buy decision be if the idle capacity could be rented to a subcontractor at the rate of \$7 per unit?

Solution

1. The cost situation for question 1 is presented in the following table

Element of Cost/Unit	Make ($/Unit)	Buy ($/Unit)
Direct material	10.00	
Direct labor	20.00	
Factory overhead: variable	15.00	
Total variable cost/unit	45.00	
Purchase price from market		50.00
Reduction in fixed cost		–2.50
Total cost per unit	45.00	47.50

Make Decision

It will be observed from the above table that if the components were made in house, the variable cost per unit would amount to $45.

Buy Decision

On the other hand, if the components were purchased from the market, the cost per unit would amount to $50.

Furthermore, as given in the problem statement, if the components were to be purchased directly from the market, it would result in a cost reduction in the amount of $2.50 per unit. This yields a cost per unit of $47.50 if the components were purchased from the market.

Comparing cost to make versus cost to buy, we get the following relationship:

Unit cost to make = $45
Unit cost to buy = $47.50

Since $45 < $47.50, we choose to make the product in house under the given constraints of the problem. **Solution 1**

2. The cost situation for question 2 is presented in the following table. From the calculation in part 1, the respective cost to make versus cost to buy are listed in the table.

Element of Cost/Unit	Make ($/Unit)	Buy ($/Unit)
Cost to make	$45.00	—
Cost to buy		$47.50
Income from renting idle capacity	—	–$7.00
Net cost/unit	$45.00	$40.50

The following factors will be observed from the above table:

- If the components were manufactured in house, the cost per unit = $45.00.
- There is no additional income that would reduce this unit cost if the items were manufactured in house.

- If the components were purchased from the market, cost per unit = \$47.50.
- Purchasing components from the market would result in idle capacity at General Electronics. As stated in the problem statement, this idle capacity can be rented to a subcontractor. If this is the case, income from renting idle capacity = \$7.00 per unit.
- This would result in cost reduction in the net purchase price if the component was purchased from the market.
- Net purchase price in this situation = \$47.50 − \$7.00 = \$40.50.

In this case, since \$40.50 < \$45.00, the components should be purchased from the market. **Solution 2**

EXAMPLE 15.3

Blimpy, Inc., a manufacturer of electronic parts, is trying to ascertain whether to make a particular component in house or buy it from the open market. The component can be made in house for a unit cost of \$15.00. The same component can be purchased from the open market for a unit cost of \$13.50. Both components are exactly identical in terms of product features and part quality. The cost breakdown for the in house manufacturing process is presented in the following table.

Element of Cost/Unit	Amount (\$)
Direct material	5.00
Direct labor	4.00
Other variable costs	4.00
Fixed cost	2.00
Total cost	15.00

1. Should Blimpy, Inc. make the product in house? How would the decision be affected if idle capacity can be used profitably in another project? What would be the effect if idle capacity cannot be used profitably?
2. If the component could be bought on the market for \$10.00, what would you recommend to the engineering manager at Blimpy, Inc.?

Solution

1. Upon cursory examination of the situation, it seems that since the purchase price on the open market is lower than the unit cost to manufacture the component, it would make sense to purchase the component on the open market. However, an accurate analysis of the cost structure is essential in arriving at the correct answer.

Situation 1: Capacity is Idle

From the cost structure presented in the above table, we can see the following:

$$\text{Total variable cost/unit} = \text{Direct material cost} + \text{Direct labor cost} + \text{Other variable costs.}$$

Thus,

$$\text{Total variable cost} = \$5.00 + \$4.00 + \$4.00 = \$13.00/\text{unit.}$$

Insofar as the fixed cost is concerned, these costs need to be incurred by the company regardless of whether the component is made in house or sourced externally. That capacity has already been "spoken for." Thus, the variable elements of cost materially affect the buy versus make decision.

$$\text{Total variable cost/unit} = \$13.$$

Comparing this to the purchase price on the open market, we notice that the purchase price = $13.50.

Since $13 < $13.50, it is advisable for Blimpy, Inc. to make the product in house.

Situation 2: Capacity can be Utilized Profitably Elsewhere

If the decision was to buy the component on the open market, this would result in the production capacity being rendered idle. However, this capacity can be utilized profitably elsewhere such as:

- Using the idle capacity to manufacture another product
- Renting the idle capacity to another manufacturer

If the aforementioned options were available, it would result in significant savings for Blimpy, Inc. The cost savings would take the form of fixed costs as well as variable costs. The total cost savings in this case would amount to the entire cost per unit, corresponding to $15.00/unit.

Since an identical component is available for $13.50, it would be advisable in this case to purchase the component on the open market. **Solution 1**

2. If the component could be bought on the market for $10.00, it would be advisable to buy the component instead of making it. Whether or not the idle capacity can be utilized profitably elsewhere would have no bearing this decision. This is because $10.00 < $13.00, corresponding to the variable cost per unit of the product. **Solution 2**

EXAMPLE 15.4

Due to anticipated changes in the market place for an electronic device, a manufacturing firm is planning on discontinuing the production of gadget Y. Instead, it plans to start producing gadget Z. The cost structure of gadget Y is outlined in the following table.

Element of Cost	Cost/Unit
Direct material	10.00
Direct labor	5.00
Prime cost	15.00
Overhead cost: fixed	3.333
Overhead cost: variable	1.667
Total cost	20.00

Gadget Y can be purchased on the open market for $17.50.

What is your recommendation in terms of making the gadget in house versus buying it? Make your decision in light of the following constraints.

1. If gadget Y is discontinued, capacity of the plant would remain idle.
2. If gadget Y is discontinued, resulting plant capacity can be used profitably elsewhere.

Solution

Upon cursory examination of the cost structure presented in the problem statement, it would appear as if it is cheaper to buy gadget Y on the open market. This is because of the following reasons:

- The total manufacturing cost of the gadget is $20/unit.
- The purchase price of the gadget on the open market is listed as $17.50.
- This would mean that since $\$17.50 < \$20.00$, it would be advisable to buy gadget Y on the market.

However, this would be a misleading conclusion for the following reason:

- Fixed costs are also referred to as "sunk costs." The implication is that the corresponding amount of money has already been spent or committed. It cannot be recovered. This means that the manufacturing company would continue to incur these costs regardless of whether they manufacture gadget Y. Given this situation, we have the following two options:
  - *Gadget Y discontinued and capacity of the plant would remain idle*: If the plant capacity cannot be utilized productively in some other endeavor, a comparison between the following costs needs to be made:
    - Total manufacturing cost: $20.00.
    - Purchase price + fixed overhead cost : $\$17.50 + \$3.333 = \$20.833$.

It can be observed from the above comparison that \$20.00 < \$20.833. Thus, in this situation, the company should keep manufacturing gadget Y.

- *Capacity of the plant can be utilized profitably in some other endeavor*: If the idle plant capacity can be utilized elsewhere profitably, the fixed costs invested in the capacity can be recouped. This decision has the following repercussions:
  - The decision to not manufacture gadget Y can lead to \$3.33 being saved in the form of fixed overhead costs.
  - Not manufacturing gadget Y also implies that money will not have to be spent on direct material and direct labor as well as variable overhead costs. Thus, the total amount of money being saved in this scenario would be \$20.00 per unit. Such being the case, it would be better to buy at \$17.50 on the open market.

A general rule of thumb: If capacity can be utilized profitably elsewhere, fixed costs should be treated as if they were variable costs. The total amount of money saved in this case is compared to the market price of the product to be purchased.

15.5.1 Minimum Production Volume (MPV)

Examples 15.1 through 15.4 dealt with situations characterized by excess capacity available to the manufacturer. Excess capacity implies capacity that is not being utilized productively. In the aforementioned examples, existing excess capacity could be used productively without any additional investment on the part of the manufacturer. In such cases, there is no increase in fixed costs. This is because money has already been spent on building up the existing capacity.

However, there are cases wherein unutilized capacity can be put to productive use only through incremental investment in additional pieces of equipment. How can the make versus buy decision be resolved if this was the case?

Minimum production volume is one of the critical factors that need to be considered in such cases. This is defined as the minimum volume of production that is guaranteed in the future to justify in house manufacturing instead of purchasing from a vendor. Minimum production volume (MPV) is expressed by the following formula:

$$\text{MPV} = \frac{\text{Increase in fixed costs}}{\text{Market price of product-incremental variable cost of production}}. \quad \text{(Equation 15.1)}$$

Consider Examples 15.5 and 15.6 to better understand the concept of minimum production volume (MPV).

Example 15.5

ABC, Inc. purchases 35,000 units of a part from an external source for $5.00 per unit. If the part was to be manufactured by ABC, Inc. in their factory, the company would require to purchase and install a new machine. The machine costs $75,000 and has a capacity of 52,500 units. It is expected to last five years at the end of which it would be worthless. Additionally, a foreman would also have to be hired. ABC, Inc. expects to pay the foreman a salary of $2500 per month. The following are elements of prime cost necessary to be borne by ABC, Inc. in order to manufacture the part in house:

Direct material: $0.85 per unit
Direct labor: $0.65 per unit

Variable overhead costs are calculated as 150% of direct labor costs. ABC, Inc. currently incurs fixed costs in the amount of $15,000. If the cost of capital is 7% per annum, should the firm continue to purchase the part on the open market? Should it manufacture the part in house? What do you suggest?

Solution

The following table represents the expected structure of costs if the part was manufactured in house:

Description	Amount ($)
Direct material cost: $0.85/unit × 35,000 units	29,750.00
Direct labor cost: $0.65/unit × 35,000 units	22,750.00
Variable overhead costs: 150% of direct labor	34,125.00
Foreman's salary: 2500/month × 12 months	30,000.00
Depreciation: straight line depreciation over 5 years	15,000.00
Amortization on $75,000 at 7% per annum	18,292.50
Total manufacturing cost of manufacturing 35,000 units in house	149,917.50

It will be observed from the above table that it would cost ABC, Inc. $149,917.50 to manufacture the part in house.

ABC, Inc. incurs an additional cost in the shape of existing fixed costs in the amount of $15,000. These costs would have to be incurred regardless and have no bearing on the final make versus buy decision in this context.

If the same part was bought on the open market at the rate of $5.00 per unit, it would cost the company $175,000.

Comparing the two costs, we can deduce that since $149,917.50 < $175,000, the company should manufacture the part in house. **Solution**

Example 15.6

Robotica, Inc. purchases a component on the open market for $15 per unit. In order to improve the profitability of business operations, it has been proposed to start manufacturing this component in house. In order to accomplish this objective, a new machine will have to be purchased. As a consequence, the following costs will have to be incurred:

- Purchase price of the machine: $100,000
- Foreman salary: $2000 per month
- Direct material cost: $3.50 per unit
- Direct labor cost: $1.50 per unit
- Variable overhead cost: 150% of direct labor cost
- Cost of capital is 7% per annum

The machine has a capacity of 30,000 units and is expected to last for 10 years at the end of which it will be worthless. Demand for the component is assured for the next fifteen years.

Is it advisable for the company to start manufacturing the component in house?

Solution

We will solve this example using a slightly different method than the one used to solve Example 15.5. We will use the concept of capacity utilization to solve this example.

If the component were to be manufactured in house, the following additional fixed costs would have to be incurred annually:

Description	Amount ($)
Foreman salary: $2000/month × 12 months	24,000
Amortization on $100,000 at 7% per annum over 10 years	14,240
Depreciation of machine: Straight line over 10 years	10,000
Additional fixed cost incurred per year	48,240

The amount of money saved per unit by manufacturing the component in house is presented in the following table.

Description	Cost ($/Unit)
Purchase price from open market	15.00
Direct material cost: in house manufacturing	−3.50
Direct labor cost: in house manufacturing	−1.50
Variable overhead: in house manufacturing	−2.25
Total unit cost: in house manufacturing	−7.25
Savings per unit by manufacturing in house	7.75

Robotica, Inc. can start manufacturing the component in house if the additional fixed cost incurred per year is recovered fully by the savings accrued from manufacturing in house. We calculate the annual capacity required to accomplish this objective as follows:

$$\text{Minimum annual capacity required to recover additional costs} = \frac{\$48,240}{\$7.75} = 6224.50 \text{ units.}$$

This implies that if the annual capacity over the life of the new machine is at least equal to 6225 units, Robotica, Inc. can invest in in house manufacturing. From the data given in the problem, we know that the machine has a capacity of 30,000 units. This is well over the minimum capacity of 6225 units.

Also, since there is a guaranteed demand for this component for the next 15 years, the second criterion, namely market guaranteed demand over the life of the machine, is met as well.

It would thus be advisable to invest in the new machine in order to manufacture the component in house. **Solution**

15.6 Conclusion

The very crucial concept of whether a company should manufacture a part in house or purchase it from a vendor was addressed in this chapter. The make versus buy decision is influenced by parameters that are both qualitative as well as quantitative in nature. The various qualitative parameters that affect this decision were examined in detail. Different quantitative parameters were examined in detail as well and illustrated through several examples. The concept of minimum production volume was introduced and demonstrated through actual examples. The various concepts learned over the course of this book such as fixed costs, variable costs, breakeven analysis, present value, overhead costs, etc. are summarized in this final chapter through their systematic application to the make versus buy decision; a decision that can not only have long lasting strategic consequences, but large scale socio-economic consequences as well. Thus, this is a decision that should only be taken only after careful consideration of all pertinent parameters.

PROBLEMS

1. Electra, Inc., an electronics manufacturer, produces a component for a unit cost of $37.50. The fixed cost for producing the component is $80,000. Currently, the firm manufactures 15,000 units. The total cost structure for the manufacturing operation is presented in the following table.

Element of Cost	Amount ($)
Direct materials	11.25
Direct labor	15.00
Variable cost: overhead	7.50
Fixed cost: overhead	3.75

- The company has recently learned of another manufacturer, ABC, Inc., which is willing to sell the exact same component for a total of $35. The engineering manager at Auto, Inc. is trying to figure out whether the component should be manufactured in house or bought from the external manufacturer.

- ABC, Inc. is willing to reduce their price to $30 per unit. What should the engineering manager do in this case?

2. A manufacturer of household goods currently manufactures some products for household use. The cost structure of its manufacturing operation is presented in the following table.

Element of Cost	Amount ($)
Direct materials	7.50
Direct labor	15.00
Variable cost: overhead	11.25

Additionally, an amount equal to $1100 needs to be spent in the form of fixed costs.

The same products can be purchased from another manufacturer for $37.50 per unit. The fixed cost would have to be incurred regardless of whether the products are made in house or purchased externally. If the products are purchased externally, the fixed cost per unit would be decreased by as much as $1.90. In view of this situation, answer questions 1 and 2:

- Should General Electronics make the products in house or buy them from the external manufacturer?
- If the products are purchased from the external manufacturer, it results in some production capacity at General Electronics being released. What would the make versus buy decision be if the idle capacity could be rented to a subcontractor at the rate of $7 per unit?

3. An auto manufacturer is trying to ascertain whether to make a particular component in house or buy it from the open market. The component can be made in house for a unit cost of $30.00. The same component can be purchased from the open market for a unit cost of $28.00. Both components are exactly identical in terms of product features and part quality. The cost breakdown for the in house manufacturing process is presented in the following table.

Element of Cost/Unit	Amount ($)
Direct material	10.00
Direct labor	8.00
Other variable costs	8.00
Fixed cost	4.00
Total cost	30.00

- Should the product be made in house? How would the decision be affected if idle capacity can be used profitably in another project? What would be the effect if idle capacity could not be used profitably?
- If the component could be bought on the market for $25.00, what would you suggest?

4. A manufacturing firm is planning on discontinuing the production of gadget A. Instead, it plans to start producing gadget B. The cost structure of gadget A is outlined in the following table.

Element of Cost	Cost/Unit
Direct material	25.00
Direct labor	12.50
Prime cost	37.50
Overhead cost: fixed	8.3325
Overhead cost: variable	4.1675
Total cost	50.00

Gadget Y can be purchased on the open market for $35.00.

What is your recommendation in terms of making the gadget in house versus buying it? Make your decision in light of the following constraints.

- If gadget A is discontinued, capacity of the plant would remain idle
- If gadget A is discontinued, resulting plant capacity can be used profitably elsewhere

Use this information to answer the questions below

The annual demand for an item that may be purchased or produced is 8500 units. The associated costs, along with alternatives, are listed below.

	Purchase	Produce
Item cost	$6.82	$6.45
Purchase cost per purchase	$23.00	—
Setup cost per setup	—	875.00
Holding cost per item annually	$3.75	$3.75
Production rate per year	—	42,000

5. Should the item in question be produced in house or purchased?
6. What is the economic "batch" size for the cheapest alternative?
7. Should the item be produced in house or purchased if:
   a. The item cost, whether it is purchased or produced, goes up by $0.30
   b. The purchase cost per purchase is $20.50

Use this information to answer the questions below.

Musical Tools Exclusive, Inc. produces a component at a cost of $33.95. The fixed cost for producing the component is $28,000. Currently, they manufacture 22,000 units. The total cost structure for the manufacturing operation is below.

Element of Cost	**Amount ($)**
Direct materials	19.25
Direct labor	23.00
Variable cost: overhead	15.50
Fixed cost: overhead	11.71

8. The company has recently come across another manufacturer, Toned Beat, Inc., which is willing to sell the exact same component for a total of $28. The engineering manager at Sound Siren, Inc. is trying to figure out whether the component should be manufactured in house or bought from the external manufacturer.
9. Musical Tools Exclusive, Inc. is willing to reduce their price to $31 per unit. Considering this new information, what decision should the engineer manager make?
10. The Soft Premium Plush Toy Company is planning on discontinuing the production of its Missy Blue Teddy Bear, only to replace it with its new Little Pippin Teddy Bear. The cost structure of the Missy Blue Teddy Bear is shown in the table below.

Element of Cost	**Cost Per Unit ($)**
Direct material	40.00
Direct labor	15.00
Prime cost	62.00
Overhead cost: fixed	10.5233
Overhead cost: variable	9.6671
Total cost	75.00

The Little Pippin Teddy Bear can be purchased on the market for $50. Would you recommend making the Little Pippin Teddy Bear in house? Explain your decision. Keep in mind the following when making your decision.

- If Missy Blue were discontinued, the capacity of the plant would remain idle.
- If Missy Blue is discontinued, the resulting plant capacity could be used for other means.

11. The annual demand for an item that may be purchased or produced is 45,000 units. The associated costs, along with alternatives, are listed below.

	Purchase	Produce
Item cost	$25.00	$24.38
Purchase cost per purchase	$96.00	—
Setup cost per setup	—	1541.00
Holding cost per item annually	$15.09	$15.09
Production rate per year	—	124,000

Should the item in question be produced in house or purchased? Explain your decision.

Use the following information below to answer the questions below.

Office Lodge is manufacturing office products. The cost structure of its manufacturing operation is shown below.

Element of Cost	Amount ($)
Direct materials	9.12
Direct labor	17.70
Variable cost: overhead	13.05

Additionally, any amount equal to $3000 needs to be spent in the form of fixed costs. The same products can be purchased from another manufacturer for $43.90 per unit. The fixed cost would have to be incurred regardless of whether the products are made in house or purchased. If the products were purchased, the fixed cost per unit would decrease by $2.17. Keeping this information in mind, answer the following questions.

12. Should Office Lodge make the products in house or buy them from the external manufacturer?
13. What would the decision be if the same products could be purchased from the external manufacturer for $46 per unit?
14. If the products are purchased from the external manufacturer, it results in some production capacity being released at Office Lodge. What would the make versus buy decision be if the idle capacity could be rented to a subcontractor at $6 per unit?
15. What if the idle capacity was rented to a subcontract at $17.50 per unit?

16

Vendor Selection and Vendor Cost Verification

16.1 Introduction

There was a time when manufacturing companies used to manufacture every single component that went into their products. For instance, auto manufacturers used to produce the steel, glass, and rubber that were used to manufacture their cars. They even used to own the coal mines that facilitated steel production. That practice is now a relic of a bygone era. It is not economical for manufacturers to own and maintain every single aspect of their supply chain. Modern manufacturers rely on third party vendors and suppliers to supply them with the parts and subassemblies that go into the final product. While this practice has several distinct advantages, it has its share of disadvantages as well. For instance, monitoring the quality of incoming products from the supplier is a major issue with almost every manufacturer. Ensuring and maintaining timely delivery of products from the vendor is another problem. The concept of Just in Time (JIT) manufacturing was formulated in order to address this problem. We have discussed this concept at some length in the chapter on inventory costing. However, selecting the right vendor to supply a company's parts and subassemblies remains a major issue of concern.

Vendor selection can be a complicated and emotional process that can be fraught with uncertainty and risk. Selecting the right vendor can be a great blessing for any company. On the other hand, selecting a vendor that does not fit in properly with the company's overall goals can end up being a great liability. This chapter will address that overall issue. The following concepts will be discussed in this chapter:

- Vendor selection process
- Verification of vendor cost

16.2 The Process of Selecting a Vendor

The process of selecting a vendor involves multiple steps. These steps are described as follows:

1. *Analysis of Business Requirements*: Analyzing business requirements is the first step of the vendor selection process. This is a team activity. The team consists of personnel who have a vested interest in the vendor selection process. It involves precisely defining the product, material or service for which a vendor is sought. The technical and business requirements also need to be defined in as much detail as possible. Next, the requirements as far as vendor qualifications are concerned should also be listed. Finally, a document containing the aforementioned information should be published and input of the relevant constituents should be sought. The team should analyze the comments received from the constituents and create a final document.
2. *Conducting a Comprehensive Search for a Suitable Vendor*: The number of possible vendors that can be evaluated depends on the scope of the vendor selection process. Typically, a larger scope of the vendor selection process translates to a greater number of vendors that are examined. It should be borne in mind that not all vendors will meet the minimum requirements of the search process. It is the responsibility of the team to decide which vendors need to be pursued for detailed information. Once these prospective vendors have been short listed, a request for information (RFI) is sent to them.
3. *Request for Proposal (RFP) and Request for Quotation (RFQ)*: This is the third step of the vendor selection process. During the first two steps, the business requirements are clearly defined and vendors that meet and exceed requirements are short listed. During the third step, a request for proposal or request for quotation is drafted.

   The RFP document is a critical document in order to obtain a vendor who is capable of fulfilling the business requirements of the company. Thus, it is important that the document contain clear requirements and specifications and reasonable expectations. If it contains ambiguous requirements or if some requirements are missing or if unreasonable results are expected of the vendor, it would result in a failure of the search process. The RFP process would result in incomplete bids at best and no bids at worst.

   Every RFP document is based on specific requirements that have been defined by the customer. These requirements are based on the particular needs of the customer. For instance, some RFPs may focus entirely on price while others could focus on delivery times and quality requirements of deliverables.

In terms of structure, each RFP document should contain a specific date and time on which all potential suppliers are expected to have submitted their individual bids. Any bids submitted after this deadline are discarded. Potential vendors should be given an adequate amount of time to submit a bid. This is especially important if the requirements are complex. In such situations, the vendor often needs to assemble a response team of their own in order to fully comprehend the requirements and structure an appropriate bid. Timelines that are too short for complicated projects do not offer the vendors adequate time to understand the requirements and assemble the bid. In such cases, due to partial understanding of the requirements, the response is likely to be inaccurate. This can result in a failed RFP process.

Sometimes, vendors could have questions regarding the specifications. In such cases, the RFP document should describe the process whereby such queries can be answered. Vendors should not be allowed to contact the end customer with said questions so as to encourage transparency in terms of information availability. Transparency allows for all vendors to have access to precisely the same information. The best option is to arrange a conference call that enables the representatives of each vendor to ask questions on behalf of their clients. They can also relay the information received back to their clients.

The RFP document must clearly state the format in which the vendors should respond. It is necessary that all vendors use the same format to respond. If this system is not adhered to, it would result in a situation whereby comparison of competing bids is rendered impossible. The response format should clearly indicate the information that is required from each vendor in each section. This includes information such as resumes of potential team members, list of references, a format for pricing, and a document outlining the vendor's explicit understanding of the requirements.

This document contains the following details:

- Details of the submission.
- Introduction and executive summary. This section introduces a prospective vendor to the client organization, range of products and services offered, nature of the proposal, etc.
- Overview of business activities and background. This section serves to provide an overview of the different business activities conducted by the organization. The scope of these activities in the context of the specific RFQ document can also be emphasized.
- Detailed specification of products and services that the vendor would be required to provide.
- A clear listing of all assumptions made regarding the requirements as well as constraints under which the vendor would have to operate.

- A clear listing of terms and conditions under which business will be conducted. These conditions are applicable to the contract between vendor and client.
- An objective listing of selection criteria is an important element of this document. It clearly delineates the criteria on which prospective vendors will be selected. These criteria can pertain to not only specific products and services, but to support services such as servicing, customer support, etc. Cost can also be one of the important criteria that are explicitly mentioned in this listing.

4. *Evaluation of the Proposal and Selection of the Vendor*: This is the fourth step of the vendor selection process. The main objective of this step is to select a vendor by reaching an objective decision based on quotes submitted by vendors. The final decision should serve the best interests on the company only and should be devoid of any emotional interference and political positioning. The investigation that leads to the final decision should be as thorough as possible. It should be based on input collected from all concerned stakeholders. The team leader can rely on the following methodology to reach a consensus in terms of the final vendor selection decision:
   - After vendor quotes are received, a preliminary review of all vendor proposals should be conducted. All proposals are reviewed for clarity and completeness. If any omissions and/or ambiguities are detected, they should be clarified with the vendor who submitted the proposal. This practice will ensure the thoroughness and efficiency of the evaluation and selection process.
   - Business requirements and vendor requirements should be recorded. This process consists of listing the business requirements as well as the vendor requirements compiled during the first step. A detailed and comprehensive listing of all requirements is necessary to ensure a fair decision.
   - Each requirement should be assigned an "importance value." This is an important detail that should not be overlooked. Every RFQ is based on a set of requirements that are ranked in order of importance. This order of importance is based on the specific needs of the company at that given moment in time. Thus, assignment of an "importance value" to each requirement serves to align the vendor quotation with the best interests of the company as expressed through the RFQ document. The importance value can range from a value of extremely unimportant (designated by a score of "1") to extremely important (designated by a score of "10").
   - Sometimes, the vendor selection team may not be able to agree upon a specific importance value. In such cases, individual scores of team members should be collected and the value should be

averaged out across the entire team. Sometimes, a particular team member may feel as if they are not qualified enough to provide an opinion on a particular requirement. In such cases, they may refrain from submitting a score for that requirement. In such cases, only the average score of values that have been submitted should be used in order to reach a decision.

- A requirement can sometimes be binary. This means that the vendor either meets the requirement or does not. In such cases, that requirement should be marked: "Pass/Fail." For instance, if all external contractors are required to be fully bonded and insured, any vendor who does not meet this requirement will automatically be excluded from any further consideration.
- Next, a performance value should be assigned to each requirement. The assignment of a performance value to each requirement is perhaps the most involved part of the entire process. As such, it is the most time consuming step in the entire process. During this step, the team assigns a value to each vendor. This value is based on their perception as to how they believe each vendor performs on each requirement. If the team is unable to agree upon a particular performance value, then all performance values obtained from the team should be aggregated and averaged out across the team. If any team member chooses to abstain from assigning a value, the average of all submitted scores should be used.
- Total performance score is calculated using a combination of the performance value and importance value as described in steps 3 and 4. The total performance score is obtained by multiplying the individual importance value by the vendor's performance value. The total performance score is calculated by adding all the performance scores.
- The winning vendor is selected based on the performance score. It is clear that the aforementioned process uses objective criteria to reach a final decision. One of the greatest advantages of an objective decision lies in the fact that it is devoid of emotion and as such constitutes the best decision that can be made at that time based only on the facts available to the selection team. It should be borne in mind, however, that the total performance score is not an absolute value that is used to determine the fate of the vendor's proposal. It is more of a guide that can be used to emphasize differences between vendors. This can further serve as a means to initiate discussions between team members. These discussions can be used to eventually select the appropriate vendor.

5. *Strategies to Negotiate a Contract*: A contract negotiation strategy is the final step in the vendor selection process. The objective of a good negotiation strategy is to establish a mutually beneficial partnership

with the prospective vendor so that both entities are able to meet their respective corporate goals and objectives. A successful contract negotiation strategy consists of looking for positive attributes that benefit both parties through a deal that is fair in nature. The following strategies can be used to plan contract negotiations:

- *Ranking of Priorities along with Alternatives*: The contract negotiation strategy is an ongoing, continuously evolving process. Given its complex nature, it is often not possible to negotiate all the different components of a contract at the same time. Thus, it is important to keep one's priorities straight. It is important to know at all times the most important element in the contract. Make sure that this element is discussed and agreed upon. Only then should the team move on to less important items. Prioritization is very important in a contract negotiation process. Knowing the least important items also helps in obtaining what is most important. The least important items in a contract negotiation are often easy to sacrifice in order to obtain the most important item.
- *Knowing the Difference between Wants and Needs*: A company's priorities are subject to frequent change. Thus, it is important to continuously keep track of these priorities throughout the contract negotiation process until the process is finalized. Do not be afraid to ask uncomfortable questions such as: "Is something really a priority for a company? Must we really have this or would it just be nice to have it?" When prioritizing between wants and needs, an objective measure needs to be used in order to separate the wheat from the chaff.
- *Knowledge of the "Bottom Line" Enables You to Know When to Walk Away from the Negotiating Table*: Once again, the bottom line consists of the one or two top priorities that are an absolute must have during any contract negotiation. These must be categorically listed along with the underlying rationale. These top priorities are typically non-negotiable. If the vendor does not agree to them, then no matter what other concessions are offered, it is wise to just walk away from the negotiating table thereby saving time and effort.
- *Time Constraints and Benchmarks Should Be Clearly Defined*: Every substantial project involves performance measurement standards that are expected from vendors. These are essential to the success of the business If this is the case, it is imperative that the vendor meet them each and every time. If the vendor is unable to meet these standards, a fair penalty for noncompliance should be negotiated. Examples of such constraints include project completion date, delivery date for the first shipment of parts, lead times, etc.
- *Assessing Potential Liabilities and Risks*: Every project is littered with potential pitfalls. It is important to plan for such contingencies.

The following questions should be addressed: "What is the potential for something to go wrong?" "What would happen if unforeseen costs are incurred?" "If regulations are violated, who will be responsible for the transgression?" It is important to be prepared should the worst actually happen.

- *Ensuring Confidentiality, Noncompete, Dispute Resolution and Handling Changes in Requirements*: Issues pertaining to confidentiality, noncompete agreements, dispute resolution, and changes in requirements are just some more deal breakers in any contract negotiation. For instance, it is essential to insert a confidentiality clause into the contract should the vendor be exposed to any confidential information. In such cases, the liability is assumed by the vendor.
- *Repeat the Process from the Point of View of the Vendor*: After conclusion of the contract negotiation process for the company, it should be repeated from the point of view of the vendor. Issues pertaining to topics that could be important to the vendor should be addressed. Try to ascertain the risks and liabilities the vendors might want the client (you) to assume. This practice will help the company look at the contract from the vendor's point of view. This is the basis for building good long term partnerships.

The following items need to be reviewed prior to the beginning of contract negotiations:

- *Determination of the Need for Legal Counsel*: Whether or not legal counsel is required depends on the type of project for which the contract is being negotiated. Typically, large and complex contracts are likely to require legal counsel. In such cases, the services of a lawyer specializing in contract negotiations should be retained.
- *Determination of the Venue for the Negotiations to Take Place*: The venue at which the contract will be negotiated should be agreed upon beforehand. If it is too expensive to travel to the vendor's site, a teleconference can be setup to facilitate the negotiation. Video conferences are preferable in such cases since a lot can be surmised by studying the body language of a person.
- *Determine That the Person Negotiating on Behalf of the Vendor is Authorized to Do So*: This is a crucial yet often overlooked issue. The person negotiating on behalf of the vendor should have complete authority to negotiate. If this is not the case, traveling to the vendor's site for the negotiating session would end up being a huge waste of time.

An example of questions typically asked of a vendor on a raw material supplier questionnaire are depicted in Table 16.1.

TABLE 16.1

Example of Typical Questions Asked of a Raw Materials Supplier During the Vendor Qualification Process

Information on the Organization	
1. What is the total number of employees and the organization structure within the company including the names of key management personnel?	Answer:
2. How is the quality control unit organized and how many employees does it contain?	Answer:
3. How is the quality assurance unit organized and how many employees does it contain?	Answer:
4. How is the research and development unit organized and how many employees does it contain?	Answer:
5. How many employees work in the production process?	Answer:
6. How many employees work in the logistics department?	Answer:
Internal Audit Program	
1. Does an internal audit program exist for the different departments?	Yes No
2. Is a written report produced after each audit?	Yes No
3. If a written report is produced after each report, is the report written using a particular report format that is produced by the quality assurance department?	Yes No
4. Are internal audits conducted regularly by authorized personnel from within the company? If not, are they conducted by third party auditors who are sufficiently well qualified to assess compliance with quality requirements?	Yes No
Raw Materials	
1. If you are a distributor, do you have a standard procedure for the qualification of your supplier of raw materials distributed?	Yes No
2. If yes, please specify the method	Audit Questionnaire Other:____________
3. If you are a manufacturer, do you have a standard procedure for the qualification of your suppliers?	Yes No
4. If yes, please specify the method	Audit Questionnaire Other:____________
Quality Control	
1. Are all the raw material specifications freely available to us?	Yes No

(*Continued*)

TABLE 16.1 (*Continued*)

Example of Typical Questions Asked of a Raw Materials Supplier During the Vendor Qualification Process

Information on the Organization	
2. For every batch that you deliver, will you provide us with a full certificate of analysis in advance or one along with the product/raw material? If yes, please attach an example of a certificate of analysis	Yes No
3. Do you have a qualified person who is in charge of approving materials and releasing products? If yes, please specify qualifications and position within your company	Yes No
4. Is there a standard procedure in place to qualify each machinery, installation, and manufacturing equipment?	Yes No
5. Is there a standard procedure in place to validate manufacturing processes?	Yes No
6. Is there a system for in-process control in place?	Yes No
7. Do you keep a batch record for every batch, undersigned and dated, that reports the production rate, registration of production operations, registration of control, and release by the qualified person?	Yes No
8. If yes, are batch records kept under controlled access?	Yes No
9. If yes, how long are batch records retained?	______________.
10. If yes, how long are reference samples retained?	______________.
11. Is there a complete certification of packaging material?	Yes No
12. Do you have an internal quality lab? If yes, please list and specify which analyses are performed internally: ______________	Yes No
13. Do you subcontract your analytical control to an independent quality lab? If yes, please specify the name and certification of the lab: ________	Yes No

The next section of this chapter will address the process of verifying a vendor's cost through an analytical process.

16.3 Vendor Cost Verification

As we have discussed in the previous section, the request for quotation document contains the following details:

- Engineering bill of materials, specifications, and designs
- Final date for the completion of the estimate
- Quantity to be produced, production rate, and production schedule
- Details of any special test to be performed, inspection, and quality control requirements
- Any special packaging and shipping instructions
- Marketing details and related information

The total cost is obtained by first estimating the labor and material cost. This number is further extended by adding the overhead costs. The number thus obtained is referred to as total cost. It is then multiplied by the markup rate in order to compute the selling price. We have already discussed this principle in detail in Chapter 10. The entire process is initiated with the bill of materials (BOM).

The bill of materials is an engineering document that lists all the parts required for a product. It is a basic design document. Bill of material lists are usually listed on the corner of a drawing or could be separately listed on other documents. This list contains information on the number of the drawing, the name of the part or assembly, the material to be used for the part, and the quantity of each unit for the final product assembly.

The cost estimate is the principal summary document for setting the price. Cost estimates are typically organized into a bill of materials when a product contains many parts and assemblies. A bill of materials can be depicted in the

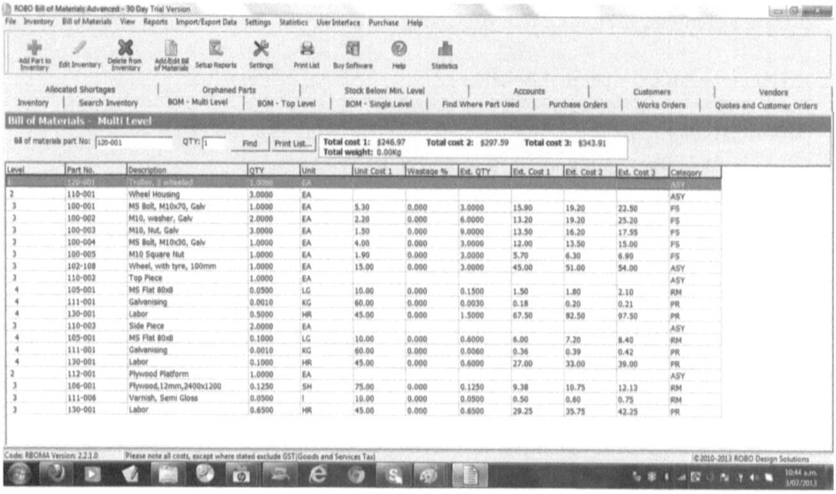

FIGURE 16.1
An example of a costed bill of materials. (From www.billofmaterials.net.)

form of a tree or a table. The tree is a graphical illustration of the bill of materials. Typically though, the bill of materials is depicted in a tabular format. An example of a bill of materials expressed in a tabular format is shown in Figure 16.1.

In the tabular format, level 1 corresponds to the final products. Level 2 corresponds to the next lower subassembly. Level 3 corresponds to either raw materials or parts purchased for level 2. Costs roll up from the bottom level. In this case, costs roll up from level 3 and proceed upwards. The cost of items listed at level 1 is the sum of costs of items listed at level 2. The costs include material costs as well as labor costs. Thus, the total cost of items listed at level 1 is the sum of material cost and labor cost of items listed at level 2. The cost of each item at level 2 is the sum of material cost and labor cost of items listed at level 3.

Other business documents rely upon the information presented in the bill of materials. The bill of materials constitutes their primary source of information. Purchase orders are identified using this document if the item is purchased from a vendor or to initiate the production process. It should be borne in mind that factors such as scrap, waste, and material shrinkage are other estimated values that contribute to the cost of material used. We have discussed this concept in Chapter 3.

As far as new parts are concerned, new operation estimates are prepared. In such cases, the BOM constitutes a comprehensive list of parts. This includes standard commercial materials, parts, and raw materials as well as subcontracts. The bill of materials assures the production manager that nothing is overlooked in terms of materials, whether they are purchased or fabricated.

Design engineering assigns the control and assignment numbers for various types of engineering data. This includes drawings, lists of bill of materials, and engineering instructions. The design numbering system is capable of serving the diverse departments of engineering, manufacturing, and sales. Yet it can be uniquely identified.

The bill of materials provides a comprehensive amount of information that leads to other documents, products, and cost estimates. New product designs often incorporate parts and subassemblies of other manufactured products. In such cases, the engineer can use the bill of materials to locate previously prepared estimates for corresponding parts and subassemblies. Old estimates can be updated using indexes, new productive hour cost rates, etc. The concept of cost indexes was presented in Chapter 2 of this book. Thus, this practice does away with the requirement to estimate parts of a product from scratch.

By following this process, it is possible for a company to verify the cost estimates submitted by its vendor for authenticity. If any discrepancies are found to be present between the actual cost and the quote submitted by the vendor, it can be brought up as a legitimate cause for concern during the negotiation process.

16.4 Conclusion

In this chapter, we studied the various nuances of the vendor selection process. The importance of operating as a team during this process was emphasized. An example of typical questions asked of prospective vendors was presented. Furthermore, the concept of vendor cost verification was introduced and the pivotal role played by the bill of materials in this process was studied. Given the importance of external vendors in the modern manufacturing environment, the importance of vendor selection and cost verification cannot possibly be overlooked.

17

Influence of Assembly Automation and Market Factors on Production Economics

17.1 Introduction

In the face of increasing global competition, manufacturers are continually forced to look for ways to improve productivity. This has now become a dire necessity instead of just a passing fancy. It is evident that factory automation is revolutionizing global manufacturing. The increasing role played by automation in manufacturing operations is clearly evident. Automation is now all pervasive. It is not confined merely to manufacturing operations. An ever increasing number of industries have automated their material handling, assembly, and subassembly operations as well. The rise of artificial intelligence (AI) technologies has revolutionized entire industries. One of the major difficulties confronting engineers lies in the basic design of their products: products are either not designed for automated assembly or are poorly designed.

Assembly automation is considered quite important to the economic well-being of a manufacturing entity. This has prompted an increasing number of designers to adopt design for assembly (DFA) procedures. DFA procedures consist of certain design guidelines that have been developed to simplify the design and assembly of products. For example, snap fits constitute a common feature in most products that have been designed for automated assembly. Besides the design simplicity and assembly convenience, DFA products tend to lower material and inventory costs by virtue of requiring fewer components and material handling costs.

The practice of DFA is a relatively recent development. However, many companies have been practicing DFA for a long time. General Electric (GE) published an internal manufacturing producibility handbook in the 1960's. The principal objective of this was to serve as a set of guidelines and manufacturing data for designers. These guidelines included many of the now known principles of DFA. In general, there are three types of assembly processes. They are described as follows.

- *Manual Assembly*: Manual assembly is an assembly process in which operations are performed manually with or without the aid of simple tools such as screw drivers and pliers. One of the distinguishing characteristics of this process is that the unit cost is constant and the process requires little initial investment. Hand tools are generally used to facilitate easy assembly. There is usually an upper limit to the production volume, and labor costs (including benefits, cases of workers compensation due to fatigue and injury, overhead for maintaining a clean, healthy environment, etc.) are higher.
- *Automatic Assembly*: Automatic assembly is also referred to as fixed automation or Detroit type assembly. It uses either synchronous indexing machines and part feeders or nonsynchronous machines. The parts are handled by a free transfer device. A system using automatic assembly is generally built for a single product and the cost/unit decreases with increasing volume of production. This process involves a custom built machine that assembles one and only one specific product. As production volume increases, the fraction of the capital investment compared to the total manufacturing cost decreases. Indexing tables, parts feeders, and automatic controls typify this inherently rigid assembly method.
- *Robotic Assembly*: Robotic assembly or soft automation is the most flexible assembly method in existence. It can achieve volumes closer to those achieved by automatic assembly methods. It uses robotic assembly systems to achieve its objective. This can be in the form of a single robot or a multi-station robotic assembly cell. All activities are simultaneously controlled and coordinated by a programmable logic controllers (PLC) or computer. This process offers high flexibility, which tends to offset high capital requirements.

Several factors influence the cost of manual assembly and flexible automated assembly. They are as follows:

- Performance time
- Production quantity
- Wage rate
- Interest rate
- Reliability of robots

Additionally, local values of some of the aforementioned variables can lead to different decisions in different countries. Each of these variables will be discussed at length in the following pages.

17.2 Effect of Performance Time on Cost of Different Assembly Methods

In order to evaluate the impact of performance time on assembly cost, two DFA products (a standard 13 amp plug and a coffee maker) were considered and their assembly time and annual assembly costs were determined. For the flexible automated assembly method, the assembly cycle times were estimated using the robot time measurement system (RTM). The methods time measurement system (MTM) was used to estimate assembly cycle times for the manual assembly method.

17.2.1 Standard 13 Amp Plug

The assembly sequence for the standard 13 amp plug was as follows:

1. Insert the cover screw in the fixture.
2. Insert the base over the cover screw.
3. Insert, in any sequence, the ground pin, the live pin, the neutral pin, and the redesigned cord grip.
4. Insert the fuse.
5. Place the cover.
6. Drive the cover screw.

A special fixture was required for the last step for flexible automated assembly. It was estimated that 1,000,000 units of the electrical plug could easily be sold every year. For the assembly, a Seiko robot could be used. The workstation layout for the assembly is depicted in Figure 17.1.

Next, the cycle time and assembly cost were determined for both flexible automated and manual methods. Table 17.1 shows the assembly sequence, RTM and MTM elements, and associated times, expressed in time measurement units (TMUs).

Based on these cycle times and production volume, annual costs for the two assembly methods were estimated as follows:

Flexible automated assembly:

Production volume: 1,000,000 units/year

Assembly cycle time: 0.4623 minutes

Capacity/workstation: 2.3458 pieces/minute = 281,496 pieces/year

(This is assuming 2000 hours/year and one shift per day).

If only one shift is operated daily throughout the year, the number of workstations required to meet the demand would be 3.5524 or 4. The least

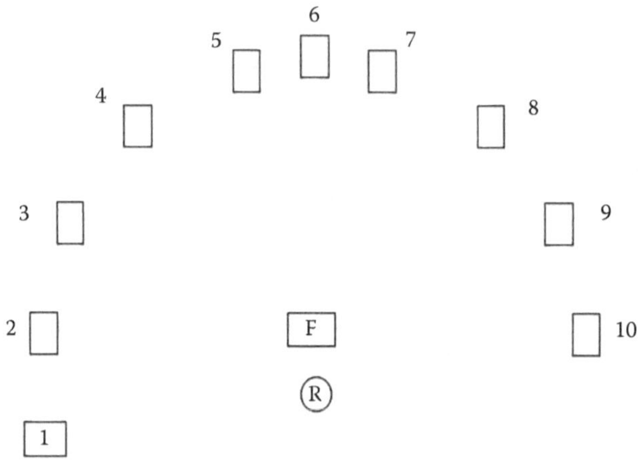

FIGURE 17.1
Workstation schematic for assembling the standard 13-amp plug. 1 – Assembled plugs bin. 2 – Screw feeder. 3 – Plug base feeder. 4 – Cord grip feeder. 5 – Earth pin feeder. 6 – Neutral pin feeder. 7 – Live pin feeder. 8 – Fuse clip feeder. 9 – Fuse feeder. 10 – Plug cover feeder. F – Fixture. R – Robot.

investment alternative to meet the demand would be to have two workstations and operate two shifts every day.

- Fixed cost/workstation:

  Cost of robot: $100,000

  Feeders (10): $500,000. (This is computed as $50,000/feeder. 10 feeders are required to feed components)

  Gravity chutes: $1500 ($150/chute)

  Bins: $1500 ($150/bin)

  Work bench: $1500

  Fixtures: $6000

  Total: $610,500
- Variable cost/workstation:

  Maintenance: $2500

  Energy: $6000 (@$0.12/KWH)

  Total: $8500

Thus,

$$\text{Total variable cost} = \$8500 \times 2 = \$17{,}000.$$

Thus, total investment = $1,221,000. This is because we assumed two workstations for the study.

TABLE 17.1

Estimated Cycle Times for Robotic and Manual Assembly of 13 amp Plug

Assembly Sequence	RTM Element	Time (TMU)	MTM Element	Time (TMU)
Place cover screw in fixture using right hand	R2 76	26	R12 C	14.2
	ST 0.5	3	G1 A	2
	GR 1.5	7	M12 C	15.2
	M2 30	23	P2 SS	25.6
	SF 0.5	3		
	RE 3.0	10		
Place plug base using right hand	R2 30	23	R12 C	14.2
	ST 0.5	3	G1 A	2
	GR 3.0	10	M12 C	15.2
	M2 30	23	P2 NS	21
	SF 0.5	3		
	RE 0.5	4		
Place cord grip (right hand) and ground pin (left hand)	R2 30	23	R12 C (L&R)	14.2
	ST 0.5	3	G1 A (L&R)	2
	GR 1.5	7	M12 C (L&R)	15.2
	M2 30	23	P2 SS	25.6
	SF 0.5	3	M5 A	7.3
	RE 0.5	4	P2 SS	25.6
	R2 30	23		
	ST 0.5	3		
	GR 3.0	10		
	M2 30	23		
	SF 0.5	3		
	RE 3.0	10		
Place neutral pin (right hand) and live pin (left hand)	R2 30	23	R12 C (L&R)	14.2
	ST 0.5	3	G1 A (L&R)	2
	GR 3.0	10	M12 C (L&R)	15.2
	M2 30	23	P2 SS	25.6
	SF 0.5	3	M5 A	7.3
	RE 3.0	10	P2 SS	25.6
	R2 30	23		
	ST 0.5	3		
	GR 3.0	10		
	M2 30	23		
	SF 0.5	3		
	RE 3.0	10		

(Continued)

TABLE 17.1 (*Continued*)

Estimated Cycle Times for Robotic and Manual Assembly of 13 amp Plug

Assembly Sequence	RTM Element	Time (TMU)	MTM Element	Time (TMU)
Place fuse clip (right hand) and fuse (left hand)	R2 30	23	R12 C (L&R)	14.2
	ST 0.5	3	G1 A (L&R)	2
	GR 3.0	10	M12 C (L&R)	15.2
	M2 30	23	P2 SS	25.6
	SF 0.5	3	M5 A	7.3
	RE 3.0	10	P2 SS	25.6
	R2 30	23		
	ST 0.5	3		
	GR 1.5	7		
	M2 30	23		
	SF 0.5	3		
	RE 0.5	3		
Place cover	R2 30	23		58.7
	ST 0.5	3		
	GR 3.0	10		
	M2 30	23		
	SF 0.5	3		
	RE 3.0	10		
Drive (tighten) cover screw	Index work head	10		34.7
	T: Rotate assembly	20		
	Index work head	10		
Drop (release) assembly	R2 1	10.5	G1 A	2
	GR 0.5	4	R30 B	24.3
	M2 76	26	RL 1	2
	RE 1.5	7		
Total cycle time (TMU)	710.5		500.8	
Total cycle time (minutes)	0.4263		0.300	

Assuming an annual interest rate of 5%, a 10 year life of investment, and robot salvage value of $14,000/robot, the annual cost of flexible automated assembly is computed as follows:

$$\text{Annual cost} = \$17{,}000 + \$1{,}221{,}000(A/P, 5\%, 10) - \$28{,}000(A/F, 5\%, 10).$$

Thus,

$$\text{Annual cost} = \$17{,}000 + \$1{,}221{,}000(0.1295) - \$28{,}000(0.0795).$$

Thus,

Annual cost = \$17,000 + \$158,119.5 − \$2,226 = \$172,893.50.

Next, we consider the manual assembly operation.

Manual assembly:
- Assembly cycle time: 0.300 minutes.
- Capacity/workstation: 3.3333 pieces/minute = 400,000 pieces/year assuming 2000 hours/year and 1 shift.

The least cost alternative to achieve the required production volume would be to have three workstations. This would imply one worker per workstation and assuming one shift per day. The unit workstation cost is computed as follows:

Cost of fixture: \$1500

Workbench: \$1500

Bins (10): \$1500

Total: \$4500

Thus,

Total cost for three workstations = \$13,500.
Annual labor cost = \$120,000. (\$40,000/worker).

Given an annual interest rate of 5%, a 10 year useful life of the investment, zero salvage value, and \$18.00/hour wage rate, the assembly cost is computed as follows:

Annual cost = \$120,000 + \$13,500 (A/P, 5%, 10) = \$121,748.25.

The cost advantage of the manual assembly method over the flexible assembly automated method in this case is \$51,145.25 per year.

17.2.2 Norelco Coffee Maker

Next we consider the case of a redesigned Norelco Coffee Maker. The coffee maker was redesigned for assembly by a Rensselaer Polytechnic team. The new design has 26% fewer parts and 70% fewer fasteners. The final assembly consists of four modules.

The cycle times for flexible automation and manual assembly methods were estimated for the coffee maker as well. Table 17.2 depicts these times.

TABLE 17.2

Estimation of Robotic and Manual Assembly Cycle Times for Redesigned Coffee Maker

Assembly Sequence	RTM Element	Time (TMU)	MTM Element	Time (TMU)
Place module 1 in fixture	R2 100	26	R30 B	25.8
	ST 0.5	3	G1 A	2
	GR 1.5	7	M30 C	22.7
	M2 100	26	P1 NS	10.4
	SF 0.5	3	RL 1	2
	RE 1.5	7		
Place module 2	R2 100	26	R30 B	25.8
	ST 0.5	3	G1 A	2
	GR 1.5	7	M30 C	22.7
	M2 100	52	P2 NS	21
	SF 0.5	3	RL 1	2
	RE 1.5	7		
Place module 3	R2 100	26	R30 B	25.8
	ST 0.5	3	G1 A	2
	GR 1.5	7	M 30 C	22.7
	M2 100	26	P2 NS	21
	SF 0.5	3	RL 1	2
Place module 4	R2 100	26	R 30 B	25.8
	ST 0.5	3	G1 A	2
	GR 1.5	7	M 30 C	22.7
	M2 100	26	P2 NS	21
	SF 0.5	3	RL 1	2
Total cycle time (TMU):	314		283.4	
Total cycle time (minutes):	0.1884		0.170	

It was estimated that 500,000 coffee makers could easily be sold each year. Using this quantity as well as estimated assembly cycle times, the cost of flexible automation and manual assembly methods are computed as follows:

Flexible automated assembly:

Production volume: 500,000

Cycle time: 0.1884 minutes

Capacity/workstation: 5.3079 units/minute = 636,948 units/year based on a 2000 hour year and 1 operating shift

- Fixed cost/workstation:

Cost of robot: $100,000

Feeders (5): $250,000. (This is computed as $50,000/feeder. 5 feeders are required to feed components)

Work bench: $1500

Fixtures: $3000

Total: $354,500

- Variable cost/workstation:

  Maintenance: $2500

  Energy: $6000 (@$0.12/KWH)

  Total: $8500

Assuming an annual interest rate of 5%, a 10 year life of investment, and robot salvage value of $14,000/robot, the annual cost of flexible automated assembly is computed as follows:

$$\text{Annual cost} = \$8500 + \$354{,}500(A/P, 5\%, 10) - \$14{,}000(A/F, 5\%, 10).$$

Thus,

$$\text{Annual cost} = \$8500 + \$354{,}500(0.1295) - \$14{,}000(0.0795).$$

Thus,

$$\text{Annual cost} = \$8500 + \$45{,}907.75 - \$1113 = \$53{,}294.75.$$

Manual assembly:

Cycle time: 0.17 minutes

Capacity/workstation: 5.8824 units/machine = 705,888 units/year

- Fixed cost:

  Cost of fixture: $1500

  Workbench: $1500

  Bins (5): $750

  Total: $3750

- Variable cost:

  Labor cost = $40,000/worker/year

Assuming an annual interest rate of 5% per year, a workstation life of 10 years, and zero salvage value, the annual cost is calculated as follows:

$$\text{Annual cost} = \$40{,}000 + \$3750(A/P, 5\%, 10) = \$40{,}000 + \$3750(0.1295) = \$40{,}485.63.$$

In this case, the manual assembly method would have a cost advantage of as much as $12,809.125 per year.

In real world situations, while the cycle time for a robot does not change, its reliability certainly does. The actual production capacity would be less than that represented by the reciprocal of the cycle time. This would tend to favor manual assembly.

Based on the estimates of cycle times and fixed and variable costs, it was observed that the manual assembly of DFA products is more economical than their assembly using flexible automated methods. However, before such a general conclusion can be drawn, the effect of factors such as production volume and labor costs, which heavily influence fixed and variable costs on assembly costs, must be determined. For instance, in the case of the electrical plug, if the production volume increased to 1,688,976 units (total capacity of robotic workstations in three shifts), the economic advantage would favor flexible automated assembly. Reduced labor costs, on the other hand, would turn the economic advantage in favor of manual assembly. If the production volume were to increase to a point that new investment in robots and workstations becomes necessary, the economic advantage would again favor manual assembly.

Similar observations can also be made in the case of the coffee maker. For instance, if the production volume were to double, the economic advantage of manual assembly would be lost and flexible automated assembly would be more desirable. Clearly, it is not possible to make a sweeping statement that flexible automated assembly is more economical or manual assembly is always more economical. Before deciding on the particular assembly method, careful economic analysis must be performed and the effects of changing market conditions must be determined. The impact of production and market factors is considered in subsequent sections.

17.3 Effect of Variation in Production Volume on Cost of Different Assembly Methods

Market demand has the most profound effect on production volume, and therefore, on investment needs and the choice of assembly method. As production volume changes, the workstation needs as well as equipment and manpower needs also change. For increased production, more resources will be required. If production needs decline, resource requirements will diminish accordingly. In order to determine the impact of production volume on assembly costs, variations in the production volume of a DFA product were considered. The case of a compressor cylinder head valve assembly is presented in this section. Other factors such as interest rates and wage rates are assumed to remain unchanged.

The DFA product was chosen for this case study. It has twelve elements (exhaust vale leaf, leaf lock, two rivets, valve plate, valve body, gasket, suction valve plate, and four screws). It is assumed that 300 assembled units are required per hour. The assembly sequence is as follows:

1. Pick valve plate and place on rivets already fed and placed in the riveting machine.
2. Pick bent exhaust valve leaf and place it on valve plate.
3. Pick leaf lock and place it on valve plate and leaf.
4. Perform riveting.
5. Pick up all the riveted components.
6. Rotate the subassembly 90 degrees.
7. Place it upon valve body and gasket located at another station.
8. Pick and place gasket on valve body.
9. Pick up suction valve plate and place it upon the riveted subassembly.
10. Pick up the entire group of parts and place it upon four preplaced screws in a pneumatic screwing machine.
11. Complete screwing operation.
12. Pick up and place the assembled group on a belt.

Note that special feeders and chutes are required for feeding parts.

Distances moved by the robot and the human worker were determined. These distances and the types of movement were used to estimate assembly cycle time. For flexible automated assembly, the PUMA assembly robot was considered. The estimates of assembly time were based on MTM for manual assembly and RTM for flexible automated assembly. These estimates are presented in Table 17.3. From these cycle times, the corresponding costs for each method of assembly were computed. Other factors such as wage rate, interest rate, etc., were kept constant. The wage rate was assumed to be $20/hour and the interest rate was 5% per annum.

The cost calculations for a production volume of 300 units per hour are shown below:

Flexible automated assembly:

Assembly cycle time: 0.2589 minutes

Production capacity/workstation: 1/cycle time

= 3.8625 units/minute

= 231.75 units/hour

Production volume: 300 units/hour

TABLE 17.3

Estimates of Assembly Time Using RTM and MTM Elements for Assembling Compressor Cylinder Head Valve Assembly

Assembly Sequence	RTM Element	Time (TMU)	MTM Element	Time (TMU)
Pick valve plate and place on rivets	R2 100	26	R30 C	26.7
	ST 0.5	3	G1 A	2
	GR 0.5	4	M30 C	30.7
	M2 75	27	P1 S	5.6
	ST 0.5	3	RL 1	2
	RE 0.5	4		
Pick bent exhaust valve leaf and place it on valve plate (robot arm 1)	R2 75	26	R30 C	26.7
	ST 0.5	3	G1 A	2
	GR 0.5	4	M 30 C	30.7
	M2 75	27	P1 S	5.6
	ST 0.5	3	RL 1	2
	RE 0.5	4		
Perform riveting		30		30
Pick up riveted components, rotate 90 degrees, and place it upon valve body and gasket located elsewhere	R2 30	23	R10 A	8.7
	ST 0.5	3	G1 A	2
	GR 0.5	4	T 90	5.4
	Rotate assy	20	M30 C	30.7
	M2 100	29	P1 S	5.6
	ST 0.5	3	RL 1	2
	RE 0.5	4		
Place gasket on valve body (robot arm 2)	R2 100	26	R30 C	26.7
	ST 0.5	3	G1 A	2
	GR 0.5	4	M30 C	30.7
	M2 75	26	P1 S	5.6
	ST 0.5	3	PL 1	2
	RE 0.5	4		
Pick up suction valve plate and place it upon riveted subassembly	R2 100	26	R30 C	26.7
	ST 0.5	3	G1 A	2
	GR 0.5	4	M30 C	30.7
	M2 75	27	P1 S	5.6
	ST 0.5	3		
	RE 0.5	4		
Pick up all grouped parts and place them upon four screws (robot arm 2)	R1 1	10.5	G1 A	2
	ST 0.5	3	M30 C	30.7
	GR 0.5	4	P1 S	5.6
	M2 100	29	RL 1	2
	ST 0.5	3		
	RE 0.5	4		

(Continued)

TABLE 17.3 (*Continued*)

Estimates of Assembly Time Using RTM and MTM Elements for Assembling Compressor Cylinder Head Valve Assembly

Assembly Sequence	RTM Element	Time (TMU)	MTM Element	Time (TMU)
Complete screwing operation		40		40
Pick up and place subassembly on	R2 100	26	R10 C	12.9
belt	ST 0.5	3	G1 A	2
	GR 0.5	4	M30 C	30.7
	M2 100	29	RL 1	2
	ST 0.5	3		
	RE 0.5	4		
Total cycle time (TMU)		431.5		408.33
Total cycle time (minutes)		0.2589		0.2450

This implies that if only one work shift was operated per day, two workstations will be required. The least investment alternative would be to have two shifts per day. Thus, we have the following:

- Fixed cost for the workstation:
  Robot (PUMA assembly robot) = \$100,000
  Feeders (5) = \$250,000. The cost/feeder = \$50,000
  Riveting and screwing fixtures = \$6000
  Bins and work bench = \$3500
  Miscellaneous (foundation, etc.) = \$30,000
  Total investment \$389,500
- Variable cost for workstation:
  Maintenance: \$6500
  Energy: \$7000. (@\$0.12/KWH)
  Total: \$13,500

$$\text{Annual cost} = \$13{,}500 + \$389{,}500\ (A/P, 5\%, 10)$$

Thus,

$$\text{Annual cost} = \$13{,}500 + \$50{,}440.25 = \$63{,}940.25$$

Manual assembly method:

$$\text{Assembly cycle time} = 0.245 \text{ minutes}$$
$$\text{Production capacity/workstation} = 1/\text{cycle time}$$
$$= 4.0816 \text{ units/minute} = 244.896 \text{ units/hour}$$
$$\text{Production volume} = 300 \text{ units/hour}$$

This implies that two workstations will be required if only one shift is run per day. With one workstation, operators will need to work 10 hours per day or the equivalent of 1.25 shifts. The least investment alternative would be to have one workstation and run the operation 10 hours per day.

Thus, the fixed cost for the workstation would be as follows:

Bins (9 bins):	\$1350
Work bench:	\$1500
Riveting and screwing fixtures:	\$6000
Total	\$8850

Variable cost is computed as follows:

Labor cost:	\$40,000/year
Utilities:	\$1750/year
Total:	\$41,750/year

$$\text{Annual cost} = \$41{,}750 + \$8850\ (A/P, 5\%, 10) = \$42{,}896$$

In the above manner, costs of manual and flexible automated methods can be computed for production volumes ranging from 100 units per hour to 800 units per hour. It will be noticed that production volume has a cyclical effect. It alternately favors manual and flexible assembly methods. Thus, fluctuations in market demand would have a profound effect on the economic desirability of a particular assembly method. Once a method has been implemented, it is not possible to change it without incurring additional costs. If a particular method has been chosen based on production volume, the following question should asked: "Can we sell the entire production volume?" and "Would this mean not meeting the market demand?"

17.4 Effect of Variation in Interest Rates on Cost of Different Assembly Methods

Investments are invariably made through the use of borrowed money. The cost of money is, therefore, a major consideration in choosing an alternative as we have seen throughout this book. In general, lower interest rates would favor large investments in equipment and vice versa. When interest rates are high, assembly methods that have low capita costs tend to be more attractive. In the previous example, production volume and wage rates were kept unchanged and interest rates are varied, thus the influence of interest rates on economic decision making can be determined.

In general, variations in interest rates from as low as 4% to as high as 18% would cover a very broad range. The costs of flexible automated assembly and manual assembly are determined as follows:

17.4.1 Flexible Automated Assembly

Consider the example of the compressor cylinder head valve assembly that was presented in the preceding section. We know the following metrics:

Assembly cycle time: 0.2589 minutes
Production capacity/workstation: 1/cycle time = 231.75/hour

This implies that more than one shift would be necessary. The least investment alternative would be to have one workstation and operate it for eleven hours every day.

Thus,

Fixed cost/workstation = $389,500

Variable cost/workstation = $13,500 including maintenance and energy costs.

Thus,

Total annual cost = $13,500 + $389,500 (A/P, i, 10)

By substituting different values of "i," the cost of assembly can be ascertained.

17.4.2 Manual Assembly

For the manual assembly method, the production capacity for the workstation would be equal to 1/0.245 = 244.896 units per hour. The least cost alternative would be to run one 10 hour shift. The fixed cost would total $8850 per year and the variable cost would be $41,750 per year.

Thus,

Total annual cost = $41,750 + $8850 (A/P, i, 10)

Once again, we can substitute different values of "i" in order to compute the cost of manual assembly. Since the flexible automated assembly method requires substantial investment in robots, feeders, etc., it continues to become economically less desirable as interest rates rise. The manual assembly method, which requires little investment, is economically more attractive.

17.5 Effect of Variation in Wage Rates on Cost of Different Assembly Methods

In the previous analysis, the wage rate was assumed to be $20 per hour. According to the bureau of labor statistics, this is about the typical wage rate of a manufacturing worker in the United States. It is important that we understand how this factor influences the cost of manual assembly. It is obvious that the cost of flexible automated assembly will not be affected by changes in the wage rate.

In order to determine the influence of wages, the wage rate was varied from $6 per hour to $16 per hour and the costs of assembling the compressor cylinder head valve manually were computed. Table 17.4 summarizes

TABLE 17.4

Influence of Wage Rates on Choice of Flexible Automation vs Manual Method for Assembly Operations

Wage Rate $/Hour	Flexible Assembly Automation (FAA) Cost $	Manual Assembly (MA) Cost $	Choice
8	63,940.25	22,896.08	MA
10	63,940.25	27,896.08	MA
12	63,940.25	32,896.08	MA
14	63,940.25	37,896.08	MA
16	63,940.25	42,896.08	MA
18	63,940.25	47,896.08	MA
20	63,940.25	52,896.08	MA
22	63,940.25	57,896.08	MA
24	63,940.25	62,896.08	MA
26	63,940.25	67,896.08	FAA
28	63,940.25	72,896.08	FAA
30	63,940.25	77,896.08	FAA
32	63,940.25	82,896.08	FAA
34	63,940.25	87,896.08	FAA
36	63,940.25	92,896.08	FAA

these costs. It is obvious that higher labor costs encourage automation and vice versa. Also, as we have noted before, flexible automated assembly is insensitive to wage rate fluctuations.

17.6 Interactive Effects of Production and Market Factors on Cost of Different Assembly Methods

So far in this chapter, we have investigated the effect of variation of one factor while all other factors remain unchanged. In practice, however, several factors can and often do change simultaneously. Thus, the influence of all production factors and market factors on assembly costs must be considered. To this end, the production volume as well as the labor rate can be varied simultaneously and its effect on total assembly cost can be ascertained.

17.7 It Pays When Products are Designed for Assembly

Regardless of the assembly method used, it is definitely beneficial to design products for ease of assembly. Table 17.5 compares the assembly costs of the 13 amp plug and the coffee maker (discussed at length in preceding sections of this chapter) for both flexible automated as well as manual methods before and after redesigning them for assembly. It is clear from Table 17.5 that the assembly of redesigned products takes less time and consequentially is more economical. The economic advantage of redesigning the two products also extends to their assembly manually.

TABLE 17.5

Estimated Assembly Time Reduction Using DFA Methods for a 13 amp Plug and a Coffee Maker

Product	Design	Assembly Method	Assembly Time (minutes)
13 Amp plug	Original	Flexible automated	0.6315
		Manual	0.4868
	DFA	Flexible automated	0.4263
		Manual	0.3000
Coffee maker	Original	Flexible automated	0.7848
		Manual	0.5901
	DFA	Flexible automated	0.1884
		Manual	0.1700

17.8 Equipment Reliability and Economic Advantage

In all cost calculations, it has been assumed that the production capacity of a robotic workstation remains unchanged. In reality, the robot reliability is not 100%. As a matter of fact, after taking into consideration issues such as drifting, dropped products, downtime, etc., reliability may actually be as low as 50%. This implies that for flexible automated assembly, we must have larger built in production capacity than is indicated by the reciprocal of cycle time. Additional investment in this capacity would have quite a profound impact on the assembly cost of the flexible automated assembly method.

For instance, consider the case of the 13 amp plug. If the robot reliability was indeed 50%, the workstation capacity would actually be reduced to 140,748 pieces per year. In order to meet the demand, eight workstations would be required. The minimum investment alternative in this case would be to have three workstations operating three shifts per day. The annual cost is represented as follows:

$$\text{Annual cost} = \$25{,}500 + \$1{,}831{,}500\ (A/P, 5\%, 10) - \$42{,}000\ (A/F, 5\%, 10).$$

Thus,

$$\text{Annual cost} = \$25{,}500 + \$237{,}179.25 - \$3339$$

This implies that the annual cost = \$259,340.25.

This would imply that the advantage of manual assembly over flexible automated assembly would increase to \$137,592, a substantial improvement. A similar case can be made for the coffee maker and the cylinder valve assembly. Clearly, the reliability issue is a critical one that cannot possibly be overlooked.

17.9 Effect of Local Values of Production and Market Factors on the Choice of Assembly Methods

The preceding discussion clearly demonstrates that the choice of assembly method for DFA products is a function of production and market factors. Since these factors vary from one geographical location to another, they are expected to influence assembly costs significantly. Thus, an assembly method while economically desirable in one location may not be as desirable in another location. To demonstrate this point, assembly times of the cylinder

TABLE 17.6

Estimated Cycle Times for Flexible Automated and Manual Assembly of Compressor Cylinder Head Valve Assembly in Four Different Countries

DFA Product	Performance Time in Minutes	
Compressor Head Valve Subassembly	**Robot**	**Human**
Germany	0.2589	0.1960
Japan	0.2589	0.1715
UK	0.2589	0.2450
USA	0.2589	0.2450

valve assembly in four different countries is shown in Table 17.6. The countries of choice include Japan, United Kingdom, USA, and Germany. Both manual as well as flexible automated assembly methods were considered.

RTM was used to estimate the assembly cycle time for flexible automated assembly. For manual assembly, MTM was used for estimating cycle times in USA and UK. Work factor was used to estimate performance times of German workers. Performance times obtained by work factor were reduced by 10% to obtain performance times of Japanese workers. A PUMA assembly robot was used to assemble the product. In view of the size of various components and the equipment, a workstation layout for assembly was configured. From this, distance moved by the robot and human arms was determined. These distances and type of movements were used to estimate cycle times.

From the assembly times and production volumes, annual costs for the two methods can be estimated. The hourly earnings of workers were estimated using International Labor Organization (ILO's) Year Book of Labor Statistics. Table 17.7 presents a summary of worker hourly earnings and interest rates in each country as of the writing of this book. Note that while the interest rates are currently very low, they are nevertheless on an upward swing and are expected to rise and be substantially higher in the future. Interest rates were obtained using an online resource (tradingeconomics.com). For the purposes of exchange rate conversion, one euro was the equivalent of 1.19 USD. One British pound was equal to 1.36 USD. One Japanese yen was equal to 0.0090 USD.

TABLE 17.7

A Comparison of Interest Rates and Hourly Wages (Expressed in US Dollars) in Different Countries

Country	Interest Rate (%)	Hourly Wage Rate (US$)
Germany	0%	26.60
Japan	−0.10%	30.00
UK	0.25%	16.42
USA	1.25%	20.00

Also, the hourly wage rate for each country was approximately calculated using a 2000 hour work year. The hourly wage rate was calculated using the data presented on tradingeconomics.com.

The number of workdays was considered to be the same in all countries. It will be noted that the manual option is more economically attractive in UK, USA, and Germany, while the flexible automation option is cheaper in Japan. Note the negative interest rate currently prevalent in Japan (from Table 17.7). The difference in costs between the manual and flexible automated assembly options was relatively small in Germany. When the initial investment rises, the flexible automation option is still cheaper in Japan.

As shown in Table 17.7, note also that interest rates and hourly wage rates differ substantially across all four countries. Low interest rates in Japan lead to lower investment costs and encourage more capital expenditure. Higher hourly rates also favor automation. Relatively smaller hourly wages and higher interest rates in western countries tend to favor manual assembly. Thus, it can be concluded that the impact of local values of production and market factors could be very profound and as such, should be considered before a particular method of assembly is chosen.

17.10 Conclusion

The purpose of this chapter was to demonstrate that production and market factors have a significant impact on the specific choice of assembly method. Specifically, factors such as interest rate, wage rate, production volume, equipment reliability, and performance times are essential considerations in ascertaining assembly costs.

While there is no doubt that designing products for ease of assembly is cost effective, the cost analyses conducted in this chapter do cast some doubt on the belief that flexible automation will always lead to increased productivity. As a matter of fact, it was demonstrated in this chapter that manual assembly of DFA products can be economically more attractive under certain conditions. The disadvantages of flexible automated assembly are primarily created by the slow performance of robots, their low reliability, high interest rates, and declining wages. In general, low interest rates and high labor costs tend to favor automation and vice versa. It is clear, then, that sound economic analysis is a prerequisite to choosing an assembly method. There is no absolute guarantee that advanced technologies such as the ones presented in this chapter will always enhance productivity. The choice of method is seriously influenced by the local value of production and market factors. As such, solutions must be tailor made for each specific situation, geographical region, and country.

Appendix: Interest Rate Tables

Interest Rate 1/4 %	Single Payment		Uniform Payment Series				Arithmetic Gradient	
	Compound Amount	Present Worth	Sinking Fund	Capital Recovery	Compound Amount	Present Worth	Gradient Uniform Series	Gradient Present Worth
n	(F/P)	(P/F)	(A/F)	(A/P)	(F/A)	(P/A)	(A/G)	(P/G)
1	1.00250	0.99751	1.00000	1.00250	1.00000	0.99751	0.00000	0.00000
2	1.00501	0.99502	0.49938	0.50188	2.00250	1.99252	0.49938	0.99502
3	1.00752	0.99254	0.33250	0.33500	3.00751	2.98506	0.99834	2.98009
4	1.01004	0.99006	0.24906	0.25156	4.01503	3.97512	1.49688	5.95028
5	1.01256	0.98759	0.19900	0.20150	5.02506	4.96272	1.99501	9.90065
6	1.01509	0.98513	0.16563	0.16813	6.03763	5.94785	2.49272	14.82630
7	1.01763	0.98267	0.14179	0.14429	7.05272	6.93052	2.99001	20.72235
8	1.02018	0.98022	0.12391	0.12641	8.07035	7.91074	3.48689	27.58391
9	1.02273	0.97778	0.11000	0.11250	9.09053	8.88852	3.98335	35.40614
10	1.02528	0.97534	0.09888	0.10138	10.11325	9.86386	4.47940	44.18420
11	1.02785	0.97291	0.08978	0.09228	11.13854	10.83677	4.97503	53.91328
12	1.03042	0.97048	0.08219	0.08469	12.16638	11.80725	5.47025	64.58858
13	1.03299	0.96806	0.07578	0.07828	13.19680	12.77532	5.96504	76.20532
14	1.03557	0.96565	0.07028	0.07278	14.22979	13.74096	6.45943	88.75874
15	1.03816	0.96324	0.06551	0.06801	15.26537	14.70420	6.95339	102.24409
16	1.04076	0.96084	0.06134	0.06384	16.30353	15.66504	7.44694	116.65666
17	1.04336	0.95844	0.05766	0.06016	17.34429	16.62348	7.94008	131.99172
18	1.04597	0.95605	0.05438	0.05688	18.38765	17.57953	8.43279	148.24459
19	1.04858	0.95367	0.05146	0.05396	19.43362	18.53320	8.92510	165.41059
20	1.05121	0.95129	0.04882	0.05132	20.48220	19.48449	9.41698	183.48508
21	1.05383	0.94892	0.04644	0.04894	21.53341	20.43340	9.90845	202.46341
22	1.05647	0.94655	0.04427	0.04677	22.58724	21.37995	10.39951	222.34096
23	1.05911	0.94419	0.04229	0.04479	23.64371	22.32414	10.89014	243.11313
24	1.06176	0.94184	0.04048	0.04298	24.70282	23.26598	11.38036	264.77534
25	1.06441	0.93949	0.03881	0.04131	25.76457	24.20547	11.87017	287.32301
26	1.06707	0.93714	0.03727	0.03977	26.82899	25.14261	12.35956	310.75160
27	1.06974	0.93481	0.03585	0.03835	27.89606	26.07742	12.84853	335.05657
28	1.07241	0.93248	0.03452	0.03702	28.96580	27.00989	13.33709	360.23340
29	1.07510	0.93015	0.03329	0.03579	30.03821	27.94004	13.82523	386.27760
30	1.07778	0.92783	0.03214	0.03464	31.11331	28.86787	14.31296	413.18468
31	1.08048	0.92552	0.03106	0.03356	32.19109	29.79339	14.80027	440.95017
32	1.08318	0.92321	0.03006	0.03256	33.27157	30.71660	15.28716	469.56963
33	1.08589	0.92091	0.02911	0.03161	34.35475	31.63750	15.77364	499.03863
34	1.08860	0.91861	0.02822	0.03072	35.44064	32.55611	16.25970	529.35276
35	1.09132	0.91632	0.02738	0.02988	36.52924	33.47243	16.74535	560.50760
40	1.10503	0.90495	0.02380	0.02630	42.01320	38.01986	19.16735	728.73988
45	1.11892	0.89372	0.02102	0.02352	47.56606	42.51088	21.57895	917.34000
50	1.13297	0.88263	0.01880	0.02130	53.18868	46.94617	23.98016	1125.77667
55	1.14720	0.87168	0.01698	0.01948	58.88194	51.32644	26.37098	1353.52863
60	1.16162	0.86087	0.01547	0.01797	64.64671	55.65236	28.75142	1600.08454
65	1.17621	0.85019	0.01419	0.01669	70.48391	59.92461	31.12148	1864.94272
70	1.19099	0.83964	0.01309	0.01559	76.39444	64.14385	33.48117	2147.61109
75	1.20595	0.82922	0.01214	0.01464	82.37922	68.31075	35.83048	2447.60694
80	1.22110	0.81894	0.01131	0.01381	88.43918	72.42595	38.16942	2764.45681
85	1.23644	0.80877	0.01057	0.01307	94.57528	76.49010	40.49801	3097.69631
90	1.25197	0.79874	0.00992	0.01242	100.78845	80.50382	42.81623	3446.86997
95	1.26770	0.78883	0.00934	0.01184	107.07969	84.46774	45.12410	3811.53113
100	1.28362	0.77904	0.00881	0.01131	113.44996	88.38248	47.42163	4191.24173

Interest Rate	Single Payment		Uniform Payment Series				Arithmetic Gradient	
1/2 %	Compound Amount	Present Worth	Sinking Fund	Capital Recovery	Compound Amount	Present Worth	Gradient Uniform Series	Gradient Present Worth
n	(F/P)	(P/F)	(A/F)	(A/P)	(F/A)	(P/A)	(A/G)	(P/G)
1	1.00500	0.99502	1.00000	1.00500	1.00000	0.99502	0.00000	0.00000
2	1.01003	0.99007	0.49875	0.50375	2.00500	1.98510	0.49875	0.99007
3	1.01508	0.98515	0.33167	0.33667	3.01502	2.97025	0.99667	2.96037
4	1.02015	0.98025	0.24813	0.25313	4.03010	3.95050	1.49377	5.90111
5	1.02525	0.97537	0.19801	0.20301	5.05025	4.92587	1.99003	9.80260
6	1.03038	0.97052	0.16460	0.16960	6.07550	5.89638	2.48545	14.65519
7	1.03553	0.96569	0.14073	0.14573	7.10588	6.86207	2.98005	20.44933
8	1.04071	0.96089	0.12283	0.12783	8.14141	7.82296	3.47382	27.17552
9	1.04591	0.95610	0.10891	0.11391	9.18212	8.77906	3.96675	34.82436
10	1.05114	0.95135	0.09777	0.10277	10.22803	9.73041	4.45885	43.38649
11	1.05640	0.94661	0.08866	0.09366	11.27917	10.67703	4.95013	52.85264
12	1.06168	0.94191	0.08107	0.08607	12.33556	11.61893	5.44057	63.21360
13	1.06699	0.93722	0.07464	0.07964	13.39724	12.55615	5.93018	74.46023
14	1.07232	0.93256	0.06914	0.07414	14.46423	13.48871	6.41896	86.58346
15	1.07768	0.92792	0.06436	0.06936	15.53655	14.41662	6.90691	99.57430
16	1.08307	0.92330	0.06019	0.06519	16.61423	15.33993	7.39403	113.42380
17	1.08849	0.91871	0.05651	0.06151	17.69730	16.25863	7.88031	128.12311
18	1.09393	0.91414	0.05323	0.05823	18.78579	17.17277	8.36577	143.66343
19	1.09940	0.90959	0.05030	0.05530	19.87972	18.08236	8.85040	160.03602
20	1.10490	0.90506	0.04767	0.05267	20.97912	18.98742	9.33419	177.23221
21	1.11042	0.90056	0.04528	0.05028	22.08401	19.88798	9.81716	195.24341
22	1.11597	0.89608	0.04311	0.04811	23.19443	20.78406	10.29929	214.06109
23	1.12155	0.89162	0.04113	0.04613	24.31040	21.67568	10.78060	233.67676
24	1.12716	0.88719	0.03932	0.04432	25.43196	22.56287	11.26107	254.08203
25	1.13280	0.88277	0.03765	0.04265	26.55912	23.44564	11.74072	275.26856
26	1.13846	0.87838	0.03611	0.04111	27.69191	24.32402	12.21953	297.22805
27	1.14415	0.87401	0.03469	0.03969	28.83037	25.19803	12.69751	319.95231
28	1.14987	0.86966	0.03336	0.03836	29.97452	26.06769	13.17467	343.43317
29	1.15562	0.86533	0.03213	0.03713	31.12439	26.93302	13.65099	367.66255
30	1.16140	0.86103	0.03098	0.03598	32.28002	27.79405	14.12649	392.63241
31	1.16721	0.85675	0.02990	0.03490	33.44142	28.65080	14.60116	418.33479
32	1.17304	0.85248	0.02889	0.03389	34.60862	29.50328	15.07499	444.76178
33	1.17891	0.84824	0.02795	0.03295	35.78167	30.35153	15.54800	471.90554
34	1.18480	0.84402	0.02706	0.03206	36.96058	31.19555	16.02018	499.75827
35	1.19073	0.83982	0.02622	0.03122	38.14538	32.03537	16.49153	528.31226
40	1.22079	0.81914	0.02265	0.02765	44.15885	36.17223	18.83585	681.33469
45	1.25162	0.79896	0.01987	0.02487	50.32416	40.20720	21.15947	850.76312
50	1.28323	0.77929	0.01765	0.02265	56.64516	44.14279	23.46242	1035.69659
55	1.31563	0.76009	0.01584	0.02084	63.12577	47.98145	25.74471	1235.26857
60	1.34885	0.74137	0.01433	0.01933	69.77003	51.72556	28.00638	1448.64580
65	1.38291	0.72311	0.01306	0.01806	76.58206	55.37746	30.24745	1675.02719
70	1.41783	0.70530	0.01197	0.01697	83.56611	58.93942	32.46796	1913.64274
75	1.45363	0.68793	0.01102	0.01602	90.72650	62.41365	34.66794	2163.75249
80	1.49034	0.67099	0.01020	0.01520	98.06771	65.80231	36.84742	2424.64551
85	1.52797	0.65446	0.00947	0.01447	105.59430	69.10750	39.00646	2695.63890
90	1.56655	0.63834	0.00883	0.01383	113.31094	72.33130	41.14508	2976.07688
95	1.60611	0.62262	0.00825	0.01325	121.22243	75.47569	43.26333	3265.32983
100	1.64667	0.60729	0.00387	0.01273	258.66740	78.54264	45.36126	3562.79343

Interest Rate 3/4 %	Single Payment		Uniform Payment Series				Arithmetic Gradient	
	Compound Amount	Present Worth	Sinking Fund	Capital Recovery	Compound Amount	Present Worth	Gradient Uniform Series	Gradient Present Worth
n	(F/P)	(P/F)	(A/F)	(A/P)	(F/A)	(P/A)	(A/G)	(P/G)
1	1.00750	0.99256	1.00000	1.00750	1.00000	0.99256	0.00000	0.00000
2	1.01506	0.98517	0.49813	0.50563	2.00750	1.97772	0.49813	0.98517
3	1.02267	0.97783	0.33085	0.33835	3.02256	2.95556	0.99502	2.94083
4	1.03034	0.97055	0.24721	0.25471	4.04523	3.92611	1.49066	5.85250
5	1.03807	0.96333	0.19702	0.20452	5.07556	4.88944	1.98506	9.70581
6	1.04585	0.95616	0.16357	0.17107	6.11363	5.84560	2.47821	14.48660
7	1.05370	0.94904	0.13967	0.14717	7.15948	6.79464	2.97011	20.18084
8	1.06160	0.94198	0.12176	0.12926	8.21318	7.73661	3.46077	26.77467
9	1.06956	0.93496	0.10782	0.11532	9.27478	8.67158	3.95019	34.25438
10	1.07758	0.92800	0.09667	0.10417	10.34434	9.59958	4.43836	42.60641
11	1.08566	0.92109	0.08755	0.09505	11.42192	10.52067	4.92529	51.81736
12	1.09381	0.91424	0.07995	0.08745	12.50759	11.43491	5.41097	61.87398
13	1.10201	0.90743	0.07352	0.08102	13.60139	12.34235	5.89541	72.76316
14	1.11028	0.90068	0.06801	0.07551	14.70340	13.24302	6.37860	84.47197
15	1.11860	0.89397	0.06324	0.07074	15.81368	14.13699	6.86055	96.98758
16	1.12699	0.88732	0.05906	0.06656	16.93228	15.02431	7.34126	110.29735
17	1.13544	0.88071	0.05537	0.06287	18.05927	15.90502	7.82072	124.38875
18	1.14396	0.87416	0.05210	0.05960	19.19472	16.77918	8.29894	139.24940
19	1.15254	0.86765	0.04917	0.05667	20.33868	17.64683	8.77592	154.86708
20	1.16118	0.86119	0.04653	0.05403	21.49122	18.50802	9.25165	171.22969
21	1.16989	0.85478	0.04415	0.05165	22.65240	19.36280	9.72614	188.32527
22	1.17867	0.84842	0.04198	0.04948	23.82230	20.21121	10.19939	206.14200
23	1.18751	0.84210	0.04000	0.04750	25.00096	21.05331	10.67139	224.66820
24	1.19641	0.83583	0.03818	0.04568	26.18847	21.88915	11.14216	243.89233
25	1.20539	0.82961	0.03652	0.04402	27.38488	22.71876	11.61168	263.80295
26	1.21443	0.82343	0.03498	0.04248	28.59027	23.54219	12.07996	284.38879
27	1.22354	0.81730	0.03355	0.04105	29.80470	24.35949	12.54701	305.63869
28	1.23271	0.81122	0.03223	0.03973	31.02823	25.17071	13.01281	327.54162
29	1.24196	0.80518	0.03100	0.03850	32.26094	25.97589	13.47737	350.08668
30	1.25127	0.79919	0.02985	0.03735	33.50290	26.77508	13.94069	373.26310
31	1.26066	0.79324	0.02877	0.03627	34.75417	27.56832	14.40277	397.06023
32	1.27011	0.78733	0.02777	0.03527	36.01483	28.35565	14.86362	421.46754
33	1.27964	0.78147	0.02682	0.03432	37.28494	29.13712	15.32322	446.47463
34	1.28923	0.77565	0.02593	0.03343	38.56458	29.91278	15.78159	472.07122
35	1.29890	0.76988	0.02509	0.03259	39.85381	30.68266	16.23872	498.24714
40	1.34835	0.74165	0.02153	0.02903	46.44648	34.44694	18.50583	637.46933
45	1.39968	0.71445	0.01877	0.02627	53.29011	38.07318	20.74209	789.71734
50	1.45296	0.68825	0.01656	0.02406	60.39426	41.56645	22.94756	953.84863
55	1.50827	0.66301	0.01476	0.02226	67.76883	44.93161	25.12233	1128.78691
60	1.56568	0.63870	0.01326	0.02076	75.42414	48.17337	27.26649	1313.51888
65	1.62528	0.61528	0.01199	0.01949	83.37085	51.29626	29.38014	1507.09100
70	1.68715	0.59272	0.01091	0.01841	91.62007	54.30462	31.46337	1708.60649
75	1.75137	0.57098	0.00998	0.01748	100.18331	57.20267	33.51631	1917.22249
80	1.81804	0.55004	0.00917	0.01667	109.07253	59.99444	35.53908	2132.14723
85	1.88725	0.52987	0.00845	0.01595	118.30013	62.68384	37.53180	2352.63748
90	1.95909	0.51044	0.00782	0.01532	127.87899	65.27461	39.49462	2577.99605
95	2.03367	0.49172	0.00726	0.01476	137.82250	67.77038	41.42768	2807.56938
100	2.11108	0.47369	0.00675	0.01425	148.14451	70.17462	43.33112	3040.74530

Interest Rate 1%	Single Payment		Uniform Payment Series				Arithmetic Gradient	
	Compound Amount	Present Worth	Sinking Fund	Capital Recovery	Compound Amount	Present Worth	Gradient Uniform Series	Gradient Present Worth
n	(F/P)	(P/F)	(A/F)	(A/P)	(F/A)	(P/A)	(A/G)	(P/G)
1	1.01000	0.99010	1.00000	1.01000	1.00000	0.99010	0.00000	0.00000
2	1.02010	0.98030	0.49751	0.50751	2.01000	1.97040	0.49751	0.98030
3	1.03030	0.97059	0.33002	0.34002	3.03010	2.94099	0.99337	2.92148
4	1.04060	0.96098	0.24628	0.25628	4.06040	3.90197	1.48756	5.80442
5	1.05101	0.95147	0.19604	0.20604	5.10101	4.85343	1.98010	9.61028
6	1.06152	0.94205	0.16255	0.17255	6.15202	5.79548	2.47098	14.32051
7	1.07214	0.93272	0.13863	0.14863	7.21354	6.72819	2.96020	19.91681
8	1.08286	0.92348	0.12069	0.13069	8.28567	7.65168	3.44777	26.38120
9	1.09369	0.91434	0.10674	0.11674	9.36853	8.56602	3.93367	33.69592
10	1.10462	0.90529	0.09558	0.10558	10.46221	9.47130	4.41792	41.84350
11	1.11567	0.89632	0.08645	0.09645	11.56683	10.36763	4.90052	50.80674
12	1.12683	0.88745	0.07885	0.08885	12.68250	11.25508	5.38145	60.56868
13	1.13809	0.87866	0.07241	0.08241	13.80933	12.13374	5.86073	71.11263
14	1.14947	0.86996	0.06690	0.07690	14.94742	13.00370	6.33836	82.42215
15	1.16097	0.86135	0.06212	0.07212	16.09690	13.86505	6.81433	94.48104
16	1.17258	0.85282	0.05794	0.06794	17.25786	14.71787	7.28865	107.27336
17	1.18430	0.84438	0.05426	0.06426	18.43044	15.56225	7.76131	120.78340
18	1.19615	0.83602	0.05098	0.06098	19.61475	16.39827	8.23231	134.99569
19	1.20811	0.82774	0.04805	0.05805	20.81090	17.22601	8.70167	149.89501
20	1.22019	0.81954	0.04542	0.05542	22.01900	18.04555	9.16937	165.46636
21	1.23239	0.81143	0.04303	0.05303	23.23919	18.85698	9.63542	181.69496
22	1.24472	0.80340	0.04086	0.05086	24.47159	19.66038	10.09982	198.56628
23	1.25716	0.79544	0.03889	0.04889	25.71630	20.45582	10.56257	216.06600
24	1.26973	0.78757	0.03707	0.04707	26.97346	21.24339	11.02367	234.18002
25	1.28243	0.77977	0.03541	0.04541	28.24320	22.02316	11.48312	252.89446
26	1.29526	0.77205	0.03387	0.04387	29.52563	22.79520	11.94092	272.19566
27	1.30821	0.76440	0.03245	0.04245	30.82089	23.55961	12.39707	292.07016
28	1.32129	0.75684	0.03112	0.04112	32.12910	24.31644	12.85158	312.50472
29	1.33450	0.74934	0.02990	0.03990	33.45039	25.06579	13.30444	333.48630
30	1.34785	0.74192	0.02875	0.03875	34.78489	25.80771	13.75566	355.00207
31	1.36133	0.73458	0.02768	0.03768	36.13274	26.54229	14.20523	377.03938
32	1.37494	0.72730	0.02667	0.03667	37.49407	27.26959	14.65317	399.58581
33	1.38869	0.72010	0.02573	0.03573	38.86901	27.98969	15.09946	422.62911
34	1.40258	0.71297	0.02484	0.03484	40.25770	28.70267	15.54410	446.15723
35	1.41660	0.70591	0.02400	0.03400	41.66028	29.40858	15.98711	470.15831
40	1.48886	0.67165	0.02046	0.03046	48.88637	32.83469	18.17761	596.85606
45	1.56481	0.63905	0.01771	0.02771	56.48107	36.09451	20.32730	733.70372
50	1.64463	0.60804	0.01551	0.02551	64.46318	39.19612	22.43635	879.41763
55	1.72852	0.57853	0.01373	0.02373	72.85246	42.14719	24.50495	1032.81478
60	1.81670	0.55045	0.01224	0.02224	81.66967	44.95504	26.53331	1192.80614
65	1.90937	0.52373	0.01100	0.02100	90.93665	47.62661	28.52167	1358.39028
70	2.00676	0.49831	0.00993	0.01993	100.67634	50.16851	30.47026	1528.64744
75	2.10913	0.47413	0.00902	0.01902	110.91285	52.58705	32.37934	1702.73397
80	2.21672	0.45112	0.00822	0.01822	121.67152	54.88821	34.24920	1879.87710
85	2.32979	0.42922	0.00752	0.01752	132.97900	57.07768	36.08013	2059.37006
90	2.44863	0.40839	0.00690	0.01690	144.86327	59.16088	37.87245	2240.56748
95	2.57354	0.38857	0.00636	0.01636	157.35376	61.14298	39.62648	2422.88110
100	2.70481	0.36971	0.00587	0.01587	170.48138	63.02888	41.34257	2605.77575

Interest Rate 1.25%	Single Payment		Uniform Payment Series				Arithmetic Gradient	
	Compound Amount	Present Worth	Sinking Fund	Capital Recovery	Compound Amount	Present Worth	Gradient Uniform Series	Gradient Present Worth
n	(F/P)	(P/F)	(A/F)	(A/P)	(F/A)	(P/A)	(A/G)	(P/G)
1	1.01250	0.98765	1.00000	1.01250	1.00000	0.98765	0.00000	0.00000
2	1.02516	0.97546	0.49689	0.50939	2.01250	1.96312	0.49689	0.97546
3	1.03797	0.96342	0.32920	0.34170	3.03766	2.92653	0.99172	2.90230
4	1.05095	0.95152	0.24536	0.25786	4.07563	3.87806	1.48447	5.75687
5	1.06408	0.93978	0.19506	0.20756	5.12657	4.81784	1.97516	9.51598
6	1.07738	0.92817	0.16153	0.17403	6.19065	5.74601	2.46377	14.15685
7	1.09085	0.91672	0.13759	0.15009	7.26804	6.66273	2.95032	19.65715
8	1.10449	0.90540	0.11963	0.13213	8.35889	7.56812	3.43479	25.99494
9	1.11829	0.89422	0.10567	0.11817	9.46337	8.46234	3.91720	33.14870
10	1.13227	0.88318	0.09450	0.10700	10.58167	9.34553	4.39754	41.09733
11	1.14642	0.87228	0.08537	0.09787	11.71394	10.21780	4.87581	49.82011
12	1.16075	0.86151	0.07776	0.09026	12.86036	11.07931	5.35202	59.29670
13	1.17526	0.85087	0.07132	0.08382	14.02112	11.93018	5.82616	69.50717
14	1.18995	0.84037	0.06581	0.07831	15.19638	12.77055	6.29824	80.43196
15	1.20483	0.82999	0.06103	0.07353	16.38633	13.60055	6.76825	92.05186
16	1.21989	0.81975	0.05685	0.06935	17.59116	14.42029	7.23620	104.34806
17	1.23514	0.80963	0.05316	0.06566	18.81105	15.22992	7.70208	117.30207
18	1.25058	0.79963	0.04988	0.06238	20.04619	16.02955	8.16591	130.89580
19	1.26621	0.78976	0.04696	0.05946	21.29677	16.81931	8.62767	145.11145
20	1.28204	0.78001	0.04432	0.05682	22.56298	17.59932	9.08738	159.93161
21	1.29806	0.77038	0.04194	0.05444	23.84502	18.36969	9.54502	175.33919
22	1.31429	0.76087	0.03977	0.05227	25.14308	19.13056	10.00062	191.31742
23	1.33072	0.75147	0.03780	0.05030	26.45737	19.88204	10.45415	207.84986
24	1.34735	0.74220	0.03599	0.04849	27.78808	20.62423	10.90564	224.92039
25	1.36419	0.73303	0.03432	0.04682	29.13544	21.35727	11.35507	242.51321
26	1.38125	0.72398	0.03279	0.04529	30.49963	22.08125	11.80245	260.61282
27	1.39851	0.71505	0.03137	0.04387	31.88087	22.79630	12.24778	279.20402
28	1.41599	0.70622	0.03005	0.04255	33.27938	23.50252	12.69106	298.27192
29	1.43369	0.69750	0.02882	0.04132	34.69538	24.20002	13.13230	317.80191
30	1.45161	0.68889	0.02768	0.04018	36.12907	24.88891	13.57150	337.77969
31	1.46976	0.68038	0.02661	0.03911	37.58068	25.56929	14.00865	358.19120
32	1.48813	0.67198	0.02561	0.03811	39.05044	26.24127	14.44376	379.02271
33	1.50673	0.66369	0.02467	0.03717	40.53857	26.90496	14.87684	400.26072
34	1.52557	0.65549	0.02378	0.03628	42.04530	27.56046	15.30788	421.89203
35	1.54464	0.64740	0.02295	0.03545	43.57087	28.20786	15.73688	443.90369
40	1.64362	0.60841	0.01942	0.03192	51.48956	31.32693	17.85148	559.23198
45	1.74895	0.57177	0.01669	0.02919	59.91569	34.25817	19.91557	682.27103
50	1.86102	0.53734	0.01452	0.02702	68.88179	37.01288	21.92950	811.67385
55	1.98028	0.50498	0.01275	0.02525	78.42246	39.60169	23.89362	946.22770
60	2.10718	0.47457	0.01129	0.02379	88.57451	42.03459	25.80834	1084.84285
65	2.24221	0.44599	0.01006	0.02256	99.37713	44.32098	27.67408	1226.54213
70	2.38590	0.41913	0.00902	0.02152	110.87200	46.46968	29.49130	1370.45134
75	2.53879	0.39389	0.00812	0.02062	123.10349	48.48897	31.26052	1515.79039
80	2.70148	0.37017	0.00735	0.01985	136.11880	50.38666	32.98225	1661.86513
85	2.87460	0.34787	0.00667	0.01917	149.96815	52.17006	34.65704	1808.05983
90	3.05881	0.32692	0.00607	0.01857	164.70501	53.84606	36.28548	1953.83026
95	3.25483	0.30724	0.00554	0.01804	180.38623	55.42113	37.86818	2098.69730
100	3.46340	0.28873	0.00507	0.01757	197.07234	56.90134	39.40577	2242.24109

Interest Rate **1.50%**	Single Payment		Uniform Payment Series				Arithmetic Gradient	
	Compound Amount	Present Worth	Sinking Fund	Capital Recovery	Compound Amount	Present Worth	Gradient Uniform Series	Gradient Present Worth
n	(F/P)	(P/F)	(A/F)	(A/P)	(F/A)	(P/A)	(A/G)	(P/G)
1	1.01500	0.98522	1.00000	1.01500	1.00000	0.98522	0.00000	0.00000
2	1.03023	0.97066	0.49628	0.51128	2.01500	1.95588	0.49628	0.97066
3	1.04568	0.95632	0.32838	0.34338	3.04522	2.91220	0.99007	2.88330
4	1.06136	0.94218	0.24444	0.25944	4.09090	3.85438	1.48139	5.70985
5	1.07728	0.92826	0.19409	0.20909	5.15227	4.78264	1.97023	9.42289
6	1.09344	0.91454	0.16053	0.17553	6.22955	5.69719	2.45658	13.99560
7	1.10984	0.90103	0.13656	0.15156	7.32299	6.59821	2.94046	19.40176
8	1.12649	0.88771	0.11858	0.13358	8.43284	7.48593	3.42185	25.61574
9	1.14339	0.87459	0.10461	0.11961	9.55933	8.36052	3.90077	32.61248
10	1.16054	0.86167	0.09343	0.10843	10.70272	9.22218	4.37721	40.36748
11	1.17795	0.84893	0.08429	0.09929	11.86326	10.07112	4.85118	48.85681
12	1.19562	0.83639	0.07668	0.09168	13.04121	10.90751	5.32267	58.05708
13	1.21355	0.82403	0.07024	0.08524	14.23683	11.73153	5.79169	67.94540
14	1.23176	0.81185	0.06472	0.07972	15.45038	12.54338	6.25824	78.49944
15	1.25023	0.79985	0.05994	0.07494	16.68214	13.34323	6.72231	89.69736
16	1.26899	0.78803	0.05577	0.07077	17.93237	14.13126	7.18392	101.51783
17	1.28802	0.77639	0.05208	0.06708	19.20136	14.90765	7.64306	113.93999
18	1.30734	0.76491	0.04881	0.06381	20.48938	15.67256	8.09973	126.94349
19	1.32695	0.75361	0.04588	0.06088	21.79672	16.42617	8.55394	140.50842
20	1.34686	0.74247	0.04325	0.05825	23.12367	17.16864	9.00569	154.61536
21	1.36706	0.73150	0.04087	0.05587	24.47052	17.90014	9.45497	169.24532
22	1.38756	0.72069	0.03870	0.05370	25.83758	18.62082	9.90180	184.37976
23	1.40838	0.71004	0.03673	0.05173	27.22514	19.33086	10.34618	200.00058
24	1.42950	0.69954	0.03492	0.04992	28.63352	20.03041	10.78810	216.09009
25	1.45095	0.68921	0.03326	0.04826	30.06302	20.71961	11.22758	232.63103
26	1.47271	0.67902	0.03173	0.04673	31.51397	21.39863	11.66460	249.60654
27	1.49480	0.66899	0.03032	0.04532	32.98668	22.06762	12.09918	267.00017
28	1.51722	0.65910	0.02900	0.04400	34.48148	22.72672	12.53132	284.79585
29	1.53998	0.64936	0.02778	0.04278	35.99870	23.37608	12.96102	302.97790
30	1.56308	0.63976	0.02664	0.04164	37.53868	24.01584	13.38829	321.53101
31	1.58653	0.63031	0.02557	0.04057	39.10176	24.64615	13.81312	340.44024
32	1.61032	0.62099	0.02458	0.03958	40.68829	25.26714	14.23553	359.69102
33	1.63448	0.61182	0.02364	0.03864	42.29861	25.87895	14.65550	379.26912
34	1.65900	0.60277	0.02276	0.03776	43.93309	26.48173	15.07306	399.16067
35	1.68388	0.59387	0.02193	0.03693	45.59209	27.07559	15.48820	419.35212
40	1.81402	0.55126	0.01843	0.03343	54.26789	29.91585	17.52773	524.35682
45	1.95421	0.51171	0.01572	0.03072	63.61420	32.55234	19.50739	635.01098
50	2.10524	0.47500	0.01357	0.02857	73.68283	34.99969	21.42772	749.96361
55	2.26794	0.44093	0.01183	0.02683	84.52960	37.27147	23.28936	868.02846
60	2.44322	0.40930	0.01039	0.02539	96.21465	39.38027	25.09296	988.16739
65	2.63204	0.37993	0.00919	0.02419	108.80277	41.33779	26.83925	1109.47518
70	2.83546	0.35268	0.00817	0.02317	122.36375	43.15487	28.52901	1231.16582
75	3.05459	0.32738	0.00730	0.02230	136.97278	44.84160	30.16306	1352.56005
80	3.29066	0.30389	0.00655	0.02155	152.71085	46.40732	31.74228	1473.07411
85	3.54498	0.28209	0.00589	0.02089	169.66523	47.86072	33.26756	1592.20953
90	3.81895	0.26185	0.00532	0.02032	187.92990	49.20985	34.73987	1709.54387
95	4.11409	0.24307	0.00482	0.01982	207.60614	50.46220	36.16018	1824.72242
100	4.43205	0.22563	0.00437	0.01937	228.80304	51.62470	37.52953	1937.45061

Interest Rate **1.75%**	**Single Payment**		**Uniform Payment Series**				**Arithmetic Gradient**	
	Compound Amount	**Present Worth**	**Sinking Fund**	**Capital Recovery**	**Compound Amount**	**Present Worth**	**Gradient Uniform Series**	**Gradient Present Worth**
n	(F/P)	(P/F)	(A/F)	(A/P)	(F/A)	(P/A)	(A/G)	(P/G)
1	1.01750	0.98280	1.00000	1.01750	1.00000	0.98280	0.00000	0.00000
2	1.03531	0.96590	0.49566	0.51316	2.01750	1.94870	0.49566	0.96590
3	1.05342	0.94929	0.32757	0.34507	3.05281	2.89798	0.98843	2.86447
4	1.07186	0.93296	0.24353	0.26103	4.10623	3.83094	1.47832	5.66334
5	1.09062	0.91691	0.19312	0.21062	5.17809	4.74786	1.96531	9.33099
6	1.10970	0.90114	0.15952	0.17702	6.26871	5.64900	2.44941	13.83671
7	1.12912	0.88564	0.13553	0.15303	7.37841	6.53464	2.93062	19.15057
8	1.14888	0.87041	0.11754	0.13504	8.50753	7.40505	3.40895	25.24345
9	1.16899	0.85544	0.10356	0.12106	9.65641	8.26049	3.88439	32.08698
10	1.18944	0.84073	0.09238	0.10988	10.82540	9.10122	4.35695	39.65354
11	1.21026	0.82627	0.08323	0.10073	12.01484	9.92749	4.82662	47.91623
12	1.23144	0.81206	0.07561	0.09311	13.22510	10.73955	5.29341	56.84886
13	1.25299	0.79809	0.06917	0.08667	14.45654	11.53764	5.75733	66.42596
14	1.27492	0.78436	0.06366	0.08116	15.70953	12.32201	6.21836	76.62270
15	1.29723	0.77087	0.05888	0.07638	16.98445	13.09288	6.67653	87.41495
16	1.31993	0.75762	0.05470	0.07220	18.28168	13.85050	7.13182	98.77919
17	1.34303	0.74459	0.05102	0.06852	19.60161	14.59508	7.58424	110.69257
18	1.36653	0.73178	0.04774	0.06524	20.94463	15.32686	8.03379	123.13283
19	1.39045	0.71919	0.04482	0.06232	22.31117	16.04606	8.48048	136.07832
20	1.41478	0.70682	0.04219	0.05969	23.70161	16.75288	8.92431	149.50799
21	1.43954	0.69467	0.03981	0.05731	25.11639	17.44755	9.36529	163.40134
22	1.46473	0.68272	0.03766	0.05516	26.55593	18.13027	9.80341	177.73847
23	1.49036	0.67098	0.03569	0.05319	28.02065	18.80125	10.23868	192.49999
24	1.51644	0.65944	0.03389	0.05139	29.51102	19.46069	10.67111	207.66706
25	1.54298	0.64810	0.03223	0.04973	31.02746	20.10878	11.10069	223.22138
26	1.56998	0.63695	0.03070	0.04820	32.57044	20.74573	11.52744	239.14512
27	1.59746	0.62599	0.02929	0.04679	34.14042	21.37173	11.95135	255.42098
28	1.62541	0.61523	0.02798	0.04548	35.73788	21.98695	12.37243	272.03215
29	1.65386	0.60465	0.02676	0.04426	37.36329	22.59160	12.79069	288.96226
30	1.68280	0.59425	0.02563	0.04313	39.01715	23.18585	13.20613	306.19544
31	1.71225	0.58403	0.02457	0.04207	40.69995	23.76988	13.61876	323.71626
32	1.74221	0.57398	0.02358	0.04108	42.41220	24.34386	14.02858	341.50971
33	1.77270	0.56411	0.02265	0.04015	44.15441	24.90797	14.43559	359.56125
34	1.80372	0.55441	0.02177	0.03927	45.92712	25.46238	14.83981	377.85673
35	1.83529	0.54487	0.02095	0.03845	47.73084	26.00725	15.24123	396.38241
40	2.00160	0.49960	0.01747	0.03497	57.23413	28.59423	17.20665	492.01087
45	2.18298	0.45809	0.01479	0.03229	67.59858	30.96626	19.10318	591.55397
50	2.38079	0.42003	0.01267	0.03017	78.90222	33.14121	20.93167	693.70101
55	2.59653	0.38513	0.01096	0.02846	91.23016	35.13545	22.69310	797.33210
60	2.83182	0.35313	0.00955	0.02705	104.67522	36.96399	24.38848	901.49545
65	3.08843	0.32379	0.00838	0.02588	119.33861	38.64060	26.01894	1005.38718
70	3.36829	0.29689	0.00739	0.02489	135.33076	40.17790	27.58564	1108.33334
75	3.67351	0.27222	0.00655	0.02405	152.77206	41.58748	29.08986	1209.77384
80	4.00639	0.24960	0.00582	0.02332	171.79382	42.87993	30.53289	1309.24819
85	4.36944	0.22886	0.00519	0.02269	192.53928	44.06500	31.91609	1406.38282
90	4.76538	0.20985	0.00465	0.02215	215.16462	45.15161	33.24089	1500.87981
95	5.19720	0.19241	0.00417	0.02167	239.84018	46.14793	34.50874	1592.50690
100	5.66816	0.17642	0.00375	0.02125	266.75177	47.06147	35.72112	1681.08862

Interest rate **2%**	**Single Payment**		**Uniform Payment Series**				**Arithmetic Gradient**	
	Compound Amount	**Present Worth**	**Sinking Fund**	**Capital Recovery**	**Compound Amount**	**Present Worth**	**Gradient Uniform Series**	**Gradient Present Worth**
n	**(F/P)**	**(P/F)**	**(A/F)**	**(A/P)**	**(F/A)**	**(P/A)**	**(A/G)**	**(P/G)**
1	1.02000	0.98039	1.00000	1.02000	1.00000	0.98039	0.00000	0.00000
2	1.04040	0.96117	0.49505	0.51505	2.02000	1.94156	0.49505	0.96117
3	1.06121	0.94232	0.32675	0.34675	3.06040	2.88388	0.98680	2.84581
4	1.08243	0.92385	0.24262	0.26262	4.12161	3.80773	1.47525	5.61735
5	1.10408	0.90573	0.19216	0.21216	5.20404	4.71346	1.96040	9.24027
6	1.12616	0.88797	0.15853	0.17853	6.30812	5.60143	2.44226	13.68013
7	1.14869	0.87056	0.13451	0.15451	7.43428	6.47199	2.92082	18.90349
8	1.17166	0.85349	0.11651	0.13651	8.58297	7.32548	3.39608	24.87792
9	1.19509	0.83676	0.10252	0.12252	9.75463	8.16224	3.86805	31.57197
10	1.21899	0.82035	0.09133	0.11133	10.94972	8.98259	4.33674	38.95510
11	1.24337	0.80426	0.08218	0.10218	12.16872	9.78685	4.80213	46.99773
12	1.26824	0.78849	0.07456	0.09456	13.41209	10.57534	5.26424	55.67116
13	1.29361	0.77303	0.06812	0.08812	14.68033	11.34837	5.72307	64.94755
14	1.31948	0.75788	0.06260	0.08260	15.97394	12.10625	6.17862	74.79992
15	1.34587	0.74301	0.05783	0.07783	17.29342	12.84926	6.63090	85.20213
16	1.37279	0.72845	0.05365	0.07365	18.63929	13.57771	7.07990	96.12881
17	1.40024	0.71416	0.04997	0.06997	20.01207	14.29187	7.52564	107.55542
18	1.42825	0.70016	0.04670	0.06670	21.41231	14.99203	7.96811	119.45813
19	1.45681	0.68643	0.04378	0.06378	22.84056	15.67846	8.40732	131.81388
20	1.48595	0.67297	0.04116	0.06116	24.29737	16.35143	8.84328	144.60033
21	1.51567	0.65978	0.03878	0.05878	25.78332	17.01121	9.27599	157.79585
22	1.54598	0.64684	0.03663	0.05663	27.29898	17.65805	9.70546	171.37947
23	1.57690	0.63416	0.03467	0.05467	28.84496	18.29220	10.13169	185.33090
24	1.60844	0.62172	0.03287	0.05287	30.42186	18.91393	10.55468	199.63049
25	1.64061	0.60953	0.03122	0.05122	32.03030	19.52346	10.97445	214.25924
26	1.67342	0.59758	0.02970	0.04970	33.67091	20.12104	11.39100	229.19872
27	1.70689	0.58586	0.02829	0.04829	35.34432	20.70690	11.80433	244.43113
28	1.74102	0.57437	0.02699	0.04699	37.05121	21.28127	12.21446	259.93924
29	1.77584	0.56311	0.02578	0.04578	38.79223	21.84438	12.62138	275.70639
30	1.81136	0.55207	0.02465	0.04465	40.56808	22.39646	13.02512	291.71644
31	1.84759	0.54125	0.02360	0.04360	42.37944	22.93770	13.42566	307.95382
32	1.88454	0.53063	0.02261	0.04261	44.22703	23.46833	13.82303	324.40346
33	1.92223	0.52023	0.02169	0.04169	46.11157	23.98856	14.21722	341.05077
34	1.96068	0.51003	0.02082	0.04082	48.03380	24.49859	14.60826	357.88170
35	1.99989	0.50003	0.02000	0.04000	49.99448	24.99862	14.99613	374.88264
40	2.20804	0.45289	0.01656	0.03656	60.40198	27.35548	16.88850	461.99313
45	2.43785	0.41020	0.01391	0.03391	71.89271	29.49016	18.70336	551.56519
50	2.69159	0.37153	0.01182	0.03182	84.57940	31.42361	20.44198	642.36059
55	2.97173	0.33650	0.01014	0.03014	98.58653	33.17479	22.10572	733.35269
60	3.28103	0.30478	0.00877	0.02877	114.05154	34.76089	23.69610	823.69753
65	3.62252	0.27605	0.00763	0.02763	131.12616	36.19747	25.21471	912.70854
70	3.99956	0.25003	0.00667	0.02667	149.97791	37.49862	26.66323	999.83432
75	4.41584	0.22646	0.00586	0.02586	170.79177	38.67711	28.04344	1084.63929
80	4.87544	0.20511	0.00516	0.02516	193.77196	39.74451	29.35718	1166.78677
85	5.38288	0.18577	0.00456	0.02456	219.14394	40.71129	30.60635	1246.02415
90	5.94313	0.16826	0.00405	0.02405	247.15666	41.58693	31.79292	1322.17008
95	6.56170	0.15240	0.00360	0.02360	278.08496	42.38002	32.91889	1395.10327
100	7.24465	0.13803	0.00320	0.02320	312.23231	43.09835	33.98628	1464.75275

Interest Rate 2.50%	Single Payment		Uniform Payment Series				Arithmetic Gradient	
	Compound Amount	Present Worth	Sinking Fund	Capital Recovery	Compound Amount	Present Worth	Gradient Uniform Series	Gradient Present Worth
n	(F/P)	(P/F)	(A/F)	(A/P)	(F/A)	(P/A)	(A/G)	(P/G)
1	1.02500	0.97561	1.00000	1.02500	1.00000	0.97561	0.00000	0.00000
2	1.05063	0.95181	0.49383	0.51883	2.02500	1.92742	0.49383	0.95181
3	1.07689	0.92860	0.32514	0.35014	3.07563	2.85602	0.98354	2.80901
4	1.10381	0.90595	0.24082	0.26582	4.15252	3.76197	1.46914	5.52687
5	1.13141	0.88385	0.19025	0.21525	5.25633	4.64583	1.95063	9.06228
6	1.15969	0.86230	0.15655	0.18155	6.38774	5.50813	2.42801	13.37377
7	1.18869	0.84127	0.13250	0.15750	7.54743	6.34939	2.90128	18.42136
8	1.21840	0.82075	0.11447	0.13947	8.73612	7.17014	3.37045	24.16658
9	1.24886	0.80073	0.10046	0.12546	9.95452	7.97087	3.83552	30.57241
10	1.28008	0.78120	0.08926	0.11426	11.20338	8.75206	4.29649	37.60320
11	1.31209	0.76214	0.08011	0.10511	12.48347	9.51421	4.75338	45.22464
12	1.34489	0.74356	0.07249	0.09749	13.79555	10.25776	5.20618	53.40376
13	1.37851	0.72542	0.06605	0.09105	15.14044	10.98318	5.65490	62.10880
14	1.41297	0.70773	0.06054	0.08554	16.51895	11.69091	6.09955	71.30926
15	1.44830	0.69047	0.05577	0.08077	17.93193	12.38138	6.54013	80.97577
16	1.48451	0.67362	0.05160	0.07660	19.38022	13.05500	6.97665	91.08015
17	1.52162	0.65720	0.04793	0.07293	20.86473	13.71220	7.40912	101.59527
18	1.55966	0.64117	0.04467	0.06967	22.38635	14.35336	7.83754	112.49509
19	1.59865	0.62553	0.04176	0.06676	23.94601	14.97889	8.26193	123.75459
20	1.63862	0.61027	0.03915	0.06415	25.54466	15.58916	8.68230	135.34974
21	1.67958	0.59539	0.03679	0.06179	27.18327	16.18455	9.09865	147.25746
22	1.72157	0.58086	0.03465	0.05965	28.86286	16.76541	9.51099	159.45562
23	1.76461	0.56670	0.03270	0.05770	30.58443	17.33211	9.91933	171.92296
24	1.80873	0.55288	0.03091	0.05591	32.34904	17.88499	10.32369	184.63909
25	1.85394	0.53939	0.02928	0.05428	34.15776	18.42438	10.72408	197.58447
26	1.90029	0.52623	0.02777	0.05277	36.01171	18.95061	11.12050	210.74034
27	1.94780	0.51340	0.02638	0.05138	37.91200	19.46401	11.51298	224.08873
28	1.99650	0.50088	0.02509	0.05009	39.85980	19.96489	11.90152	237.61243
29	2.04641	0.48866	0.02389	0.04889	41.85630	20.45355	12.28613	251.29494
30	2.09757	0.47674	0.02278	0.04778	43.90270	20.93029	12.66683	265.12048
31	2.15001	0.46511	0.02174	0.04674	46.00027	21.39541	13.04364	279.07393
32	2.20376	0.45377	0.02077	0.04577	48.15028	21.84918	13.41656	293.14081
33	2.25885	0.44270	0.01986	0.04486	50.35403	22.29188	13.78562	307.30731
34	2.31532	0.43191	0.01901	0.04401	52.61289	22.72379	14.15082	321.56018
35	2.37321	0.42137	0.01821	0.04321	54.92821	23.14516	14.51218	335.88680
40	2.68506	0.37243	0.01484	0.03984	67.40255	25.10278	16.26203	408.22200
45	3.03790	0.32917	0.01227	0.03727	81.51613	26.83302	17.91848	480.80703
50	3.43711	0.29094	0.01026	0.03526	97.48435	28.36231	19.48389	552.60805
55	3.88877	0.25715	0.00865	0.03365	115.55092	29.71398	20.96077	622.82803
60	4.39979	0.22728	0.00735	0.03235	135.99159	30.90866	22.35185	690.86565
65	4.97796	0.20089	0.00628	0.03128	159.11833	31.96458	23.65996	756.28059
70	5.63210	0.17755	0.00540	0.03040	185.28411	32.89786	24.88807	818.76427
75	6.37221	0.15693	0.00465	0.02965	214.88830	33.72274	26.03926	878.11515
80	7.20957	0.13870	0.00403	0.02903	248.38271	34.45182	27.11666	934.21807
85	8.15696	0.12259	0.00349	0.02849	286.27857	35.09621	28.12346	987.02686
90	9.22886	0.10836	0.00304	0.02804	329.15425	35.66577	29.06288	1036.54990
95	10.44160	0.09577	0.00265	0.02765	377.66415	36.16917	29.93815	1082.83807
100	11.81372	0.08465	0.00231	0.02731	432.54865	36.61411	30.75249	1125.97474

Interest Rate **3%**	**Single Payment**		**Uniform Payment Series**				**Arithmetic Gradient**	
	Compound Amount	**Present Worth**	**Sinking Fund**	**Capital Recovery**	**Compound Amount**	**Present Worth**	**Gradient Uniform Series**	**Gradient Present Worth**
n	(F/P)	(P/F)	(A/F)	(A/P)	(F/A)	(P/A)	(A/G)	(P/G)
1	1.03000	0.97087	1.00000	1.03000	1.00000	0.97087	0.00000	0.00000
2	1.06090	0.94260	0.49261	0.52261	2.03000	1.91347	0.49261	0.94260
3	1.09273	0.91514	0.32353	0.35353	3.09090	2.82861	0.98030	2.77288
4	1.12551	0.88849	0.23903	0.26903	4.18363	3.71710	1.46306	5.43834
5	1.15927	0.86261	0.18835	0.21835	5.30914	4.57971	1.94090	8.88878
6	1.19405	0.83748	0.15460	0.18460	6.46841	5.41719	2.41383	13.07620
7	1.22987	0.81309	0.13051	0.16051	7.66246	6.23028	2.88185	17.95475
8	1.26677	0.78941	0.11246	0.14246	8.89234	7.01969	3.34496	23.48061
9	1.30477	0.76642	0.09843	0.12843	10.15911	7.78611	3.80318	29.61194
10	1.34392	0.74409	0.08723	0.11723	11.46388	8.53020	4.25650	36.30879
11	1.38423	0.72242	0.07808	0.10808	12.80780	9.25262	4.70494	43.53300
12	1.42576	0.70138	0.07046	0.10046	14.19203	9.95400	5.14850	51.24818
13	1.46853	0.68095	0.06403	0.09403	15.61779	10.63496	5.58720	59.41960
14	1.51259	0.66112	0.05853	0.08853	17.08632	11.29607	6.02104	68.01413
15	1.55797	0.64186	0.05377	0.08377	18.59891	11.93794	6.45004	77.00020
16	1.60471	0.62317	0.04961	0.07961	20.15688	12.56110	6.87421	86.34770
17	1.65285	0.60502	0.04595	0.07595	21.76159	13.16612	7.29357	96.02796
18	1.70243	0.58739	0.04271	0.07271	23.41444	13.75351	7.70812	106.01367
19	1.75351	0.57029	0.03981	0.06981	25.11687	14.32380	8.11788	116.27882
20	1.80611	0.55368	0.03722	0.06722	26.87037	14.87747	8.52286	126.79866
21	1.86029	0.53755	0.03487	0.06487	28.67649	15.41502	8.92309	137.54964
22	1.91610	0.52189	0.03275	0.06275	30.53678	15.93692	9.31858	148.50939
23	1.97359	0.50669	0.03081	0.06081	32.45288	16.44361	9.70934	159.65661
24	2.03279	0.49193	0.02905	0.05905	34.42647	16.93554	10.09540	170.97108
25	2.09378	0.47761	0.02743	0.05743	36.45926	17.41315	10.47677	182.43362
26	2.15659	0.46369	0.02594	0.05594	38.55304	17.87684	10.85348	194.02598
27	2.22129	0.45019	0.02456	0.05456	40.70963	18.32703	11.22554	205.73090
28	2.28793	0.43708	0.02329	0.05329	42.93092	18.76411	11.59298	217.53197
29	2.35657	0.42435	0.02211	0.05211	45.21885	19.18845	11.95582	229.41367
30	2.42726	0.41199	0.02102	0.05102	47.57542	19.60044	12.31407	241.36129
31	2.50008	0.39999	0.02000	0.05000	50.00268	20.00043	12.66777	253.36090
32	2.57508	0.38834	0.01905	0.04905	52.50276	20.38877	13.01694	265.39935
33	2.65234	0.37703	0.01816	0.04816	55.07784	20.76579	13.36160	277.46419
34	2.73191	0.36604	0.01732	0.04732	57.73018	21.13184	13.70177	289.54367
35	2.81386	0.35538	0.01654	0.04654	60.46208	21.48722	14.03749	301.62670
40	3.26204	0.30656	0.01326	0.04326	75.40126	23.11477	15.65016	361.74994
45	3.78160	0.26444	0.01079	0.04079	92.71986	24.51871	17.15557	420.63248
50	4.38391	0.22811	0.00887	0.03887	112.79687	25.72976	18.55751	477.48033
55	5.08215	0.19677	0.00735	0.03735	136.07162	26.77443	19.86004	531.74111
60	5.89160	0.16973	0.00613	0.03613	163.05344	27.67556	21.06742	583.05261
65	6.82998	0.14641	0.00515	0.03515	194.33276	28.45289	22.18407	631.20100
70	7.91782	0.12630	0.00434	0.03434	230.59406	29.12342	23.21454	676.08687
75	9.17893	0.10895	0.00367	0.03367	272.63086	29.70183	24.16342	717.69785
80	10.64089	0.09398	0.00311	0.03311	321.36302	30.20076	25.03534	756.08652
85	12.33571	0.08107	0.00265	0.03265	377.85695	30.63115	25.83490	791.35287
90	14.30047	0.06993	0.00226	0.03226	443.34890	31.00241	26.56665	823.63021
95	16.57816	0.06032	0.00193	0.03193	519.27203	31.32266	27.23505	853.07418
100	19.21863	0.05203	0.00165	0.03165	607.28773	31.59891	27.84445	879.85405

Interest Rate **3.50%**	**Single Payment**		**Uniform Payment Series**				**Arithmetic Gradient**	
	Compound Amount	**Present Worth**	**Sinking Fund**	**Capital Recovery**	**Compound Amount**	**Present Worth**	**Gradient Uniform Series**	**Gradient Present Worth**
n	**(F/P)**	**(P/F)**	**(A/F)**	**(A/P)**	**(F/A)**	**(P/A)**	**(A/G)**	**(P/G)**
1	1.03500	0.96618	1.00000	1.03500	1.00000	0.96618	0.00000	0.00000
2	1.07123	0.93351	0.49140	0.52640	2.03500	1.89969	0.49140	0.93351
3	1.10872	0.90194	0.32193	0.35693	3.10622	2.80164	0.97707	2.73740
4	1.14752	0.87144	0.23725	0.27225	4.21494	3.67308	1.45701	5.35172
5	1.18769	0.84197	0.18648	0.22148	5.36247	4.51505	1.93123	8.71962
6	1.22926	0.81350	0.15267	0.18767	6.55015	5.32855	2.39974	12.78712
7	1.27228	0.78599	0.12854	0.16354	7.77941	6.11454	2.86253	17.50306
8	1.31681	0.75941	0.11048	0.14548	9.05169	6.87396	3.31962	22.81895
9	1.36290	0.73373	0.09645	0.13145	10.36850	7.60769	3.77103	28.68879
10	1.41060	0.70892	0.08524	0.12024	11.73139	8.31661	4.21675	35.06906
11	1.45997	0.68495	0.07609	0.11109	13.14199	9.00155	4.65681	41.91852
12	1.51107	0.66178	0.06848	0.10348	14.60196	9.66333	5.09122	49.19814
13	1.56396	0.63940	0.06206	0.09706	16.11303	10.30274	5.51999	56.87099
14	1.61869	0.61778	0.05657	0.09157	17.67699	10.92052	5.94314	64.90215
15	1.67535	0.59689	0.05183	0.08683	19.29568	11.51741	6.36068	73.25862
16	1.73399	0.57671	0.04768	0.08268	20.97103	12.09412	6.77265	81.90921
17	1.79468	0.55720	0.04404	0.07904	22.70502	12.65132	7.17905	90.82447
18	1.85749	0.53836	0.04082	0.07582	24.49969	13.18968	7.57991	99.97661
19	1.92250	0.52016	0.03794	0.07294	26.35718	13.70984	7.97525	109.33941
20	1.98979	0.50257	0.03536	0.07036	28.27968	14.21240	8.36510	118.88816
21	2.05943	0.48557	0.03304	0.06804	30.26947	14.69797	8.74948	128.59958
22	2.13151	0.46915	0.03093	0.06593	32.32890	15.16712	9.12841	138.45174
23	2.20611	0.45329	0.02902	0.06402	34.46041	15.62041	9.50193	148.42403
24	2.28333	0.43796	0.02727	0.06227	36.66653	16.05837	9.87006	158.49704
25	2.36324	0.42315	0.02567	0.06067	38.94986	16.48151	10.23283	168.65257
26	2.44596	0.40884	0.02421	0.05921	41.31310	16.89035	10.59028	178.87351
27	2.53157	0.39501	0.02285	0.05785	43.75906	17.28536	10.94243	189.14383
28	2.62017	0.38165	0.02160	0.05660	46.29063	17.66702	11.28931	199.44849
29	2.71188	0.36875	0.02045	0.05545	48.91080	18.03577	11.63097	209.77344
30	2.80679	0.35628	0.01937	0.05437	51.62268	18.39205	11.96743	220.10552
31	2.90503	0.34423	0.01837	0.05337	54.42947	18.73628	12.29873	230.43243
32	3.00671	0.33259	0.01744	0.05244	57.33450	19.06887	12.62491	240.74271
33	3.11194	0.32134	0.01657	0.05157	60.34121	19.39021	12.94600	251.02568
34	3.22086	0.31048	0.01576	0.05076	63.45315	19.70068	13.26205	261.27138
35	3.33359	0.29998	0.01500	0.05000	66.67401	20.00066	13.57308	271.47060
40	3.95926	0.25257	0.01183	0.04683	84.55028	21.35507	15.05453	321.49067
45	4.70236	0.21266	0.00945	0.04445	105.78167	22.49545	16.41701	369.30813
50	5.58493	0.17905	0.00763	0.04263	130.99791	23.45562	17.66613	414.36998
55	6.63314	0.15076	0.00621	0.04121	160.94689	24.26405	18.80778	456.35302
60	7.87809	0.12693	0.00509	0.04009	196.51688	24.94473	19.84808	495.10502
65	9.35670	0.10688	0.00419	0.03919	238.76288	25.51785	20.79324	530.59874
70	11.11283	0.08999	0.00346	0.03846	288.93786	26.00040	21.64953	562.89624
75	13.19855	0.07577	0.00287	0.03787	348.53001	26.40669	22.42316	592.12133
80	15.67574	0.06379	0.00238	0.03738	419.30679	26.74878	23.12025	618.43850
85	18.61786	0.05371	0.00199	0.03699	503.36739	27.03680	23.74678	642.03700
90	22.11218	0.04522	0.00166	0.03666	603.20503	27.27932	24.30849	663.11885
95	26.26233	0.03808	0.00139	0.03639	721.78082	27.48350	24.81089	681.89016
100	31.19141	0.03206	0.00116	0.03616	862.61166	27.65543	25.25923	698.55469

Interest Rate 4%	Single Payment		Uniform Payment Series				Arithmetic Gradient	
	Compound Amount	Present Worth	Sinking Fund	Capital Recovery	Compound Amount	Present Worth	Gradient Uniform Series	Gradient Present Worth
n	(F/P)	(P/F)	(A/F)	(A/P)	(F/A)	(P/A)	(A/G)	(P/G)
1	1.04000	0.96154	1.00000	1.04000	1.00000	0.96154	0.00000	0.00000
2	1.08160	0.92456	0.49020	0.53020	2.04000	1.88609	0.49020	0.92456
3	1.12486	0.88900	0.32035	0.36035	3.12160	2.77509	0.97386	2.70255
4	1.16986	0.85480	0.23549	0.27549	4.24646	3.62990	1.45100	5.26696
5	1.21665	0.82193	0.18463	0.22463	5.41632	4.45182	1.92161	8.55467
6	1.26532	0.79031	0.15076	0.19076	6.63298	5.24214	2.38571	12.50624
7	1.31593	0.75992	0.12661	0.16661	7.89829	6.00205	2.84332	17.06575
8	1.36857	0.73069	0.10853	0.14853	9.21423	6.73274	3.29443	22.18058
9	1.42331	0.70259	0.09449	0.13449	10.58280	7.43533	3.73908	27.80127
10	1.48024	0.67556	0.08329	0.12329	12.00611	8.11090	4.17726	33.88135
11	1.53945	0.64958	0.07415	0.11415	13.48635	8.76048	4.60901	40.37716
12	1.60103	0.62460	0.06655	0.10655	15.02581	9.38507	5.03435	47.24773
13	1.66507	0.60057	0.06014	0.10014	16.62684	9.98565	5.45329	54.45462
14	1.73168	0.57748	0.05467	0.09467	18.29191	10.56312	5.86586	61.96179
15	1.80094	0.55526	0.04994	0.08994	20.02359	11.11839	6.27209	69.73550
16	1.87298	0.53391	0.04582	0.08582	21.82453	11.65230	6.67200	77.74412
17	1.94790	0.51337	0.04220	0.08220	23.69751	12.16567	7.06563	85.95809
18	2.02582	0.49363	0.03899	0.07899	25.64541	12.65930	7.45300	94.34977
19	2.10685	0.47464	0.03614	0.07614	27.67123	13.13394	7.83416	102.89333
20	2.19112	0.45639	0.03358	0.07358	29.77808	13.59033	8.20912	111.56469
21	2.27877	0.43883	0.03128	0.07128	31.96920	14.02916	8.57794	120.34136
22	2.36992	0.42196	0.02920	0.06920	34.24797	14.45112	8.94065	129.20242
23	2.46472	0.40573	0.02731	0.06731	36.61789	14.85684	9.29729	138.12840
24	2.56330	0.39012	0.02559	0.06559	39.08260	15.24696	9.64790	147.10119
25	2.66584	0.37512	0.02401	0.06401	41.64591	15.62208	9.99252	156.10400
26	2.77247	0.36069	0.02257	0.06257	44.31174	15.98277	10.33120	165.12123
27	2.88337	0.34682	0.02124	0.06124	47.08421	16.32959	10.66399	174.13846
28	2.99870	0.33348	0.02001	0.06001	49.96758	16.66306	10.99092	183.14235
29	3.11865	0.32065	0.01888	0.05888	52.96629	16.98371	11.31205	192.12059
30	3.24340	0.30832	0.01783	0.05783	56.08494	17.29203	11.62743	201.06183
31	3.37313	0.29646	0.01686	0.05686	59.32834	17.58849	11.93710	209.95564
32	3.50806	0.28506	0.01595	0.05595	62.70147	17.87355	12.24113	218.79244
33	3.64838	0.27409	0.01510	0.05510	66.20953	18.14765	12.53956	227.56345
34	3.79432	0.26355	0.01431	0.05431	69.85791	18.41120	12.83244	236.26067
35	3.94609	0.25342	0.01358	0.05358	73.65222	18.66461	13.11984	244.87679
40	4.80102	0.20829	0.01052	0.05052	95.02552	19.79277	14.47651	286.53030
45	5.84118	0.17120	0.00826	0.04826	121.02939	20.72004	15.70474	325.40278
50	7.10668	0.14071	0.00655	0.04655	152.66708	21.48218	16.81225	361.16385
55	8.64637	0.11566	0.00523	0.04523	191.15917	22.10861	17.80704	393.68897
60	10.51963	0.09506	0.00420	0.04420	237.99069	22.62349	18.69723	422.99665
65	12.79874	0.07813	0.00339	0.04339	294.96838	23.04668	19.49093	449.20138
70	15.57162	0.06422	0.00275	0.04275	364.29046	23.39451	20.19614	472.47892
75	18.94525	0.05278	0.00223	0.04223	448.63137	23.68041	20.82062	493.04083
80	23.04980	0.04338	0.00181	0.04181	551.24498	23.91539	21.37185	511.11614
85	28.04360	0.03566	0.00148	0.04148	676.09012	24.10853	21.85693	526.93843
90	34.11933	0.02931	0.00121	0.04121	827.98333	24.26728	22.28255	540.73692
95	41.51139	0.02409	0.00099	0.04099	1012.78465	24.39776	22.65498	552.73067
100	50.50495	0.01980	0.00081	0.04081	1237.62370	24.50500	22.98000	563.12487

Interest Rate **4.50%**	**Single Payment**		**Uniform Payment Series**				**Arithmetic Gradient**	
	Compound Amount	**Present Worth**	**Sinking Fund**	**Capital Recovery**	**Compound Amount**	**Present Worth**	**Gradient Uniform Series**	**Gradient Present Worth**
n	(F/P)	(P/F)	(A/F)	(A/P)	(F/A)	(P/A)	(A/G)	(P/G)
1	1.04500	0.95694	1.00000	1.04500	1.00000	0.95694	0.00000	0.00000
2	1.09203	0.91573	0.48900	0.53400	2.04500	1.87267	0.48900	0.91573
3	1.14117	0.87630	0.31877	0.36377	3.13703	2.74896	0.97066	2.66832
4	1.19252	0.83856	0.23374	0.27874	4.27819	3.58753	1.44501	5.18401
5	1.24618	0.80245	0.18279	0.22779	5.47071	4.38998	1.91204	8.39381
6	1.30226	0.76790	0.14888	0.19388	6.71689	5.15787	2.37177	12.23329
7	1.36086	0.73483	0.12470	0.16970	8.01915	5.89270	2.82422	16.64226
8	1.42210	0.70319	0.10661	0.15161	9.38001	6.59589	3.26939	21.56456
9	1.48610	0.67290	0.09257	0.13757	10.80211	7.26879	3.70733	26.94779
10	1.55297	0.64393	0.08138	0.12638	12.28821	7.91272	4.13804	32.74314
11	1.62285	0.61620	0.07225	0.11725	13.84118	8.52892	4.56156	38.90513
12	1.69588	0.58966	0.06467	0.10967	15.46403	9.11858	4.97791	45.39143
13	1.77220	0.56427	0.05828	0.10328	17.15991	9.68285	5.38712	52.16269
14	1.85194	0.53997	0.05282	0.09782	18.93211	10.22283	5.78924	59.18234
15	1.93528	0.51672	0.04811	0.09311	20.78405	10.73955	6.18429	66.41642
16	2.02237	0.49447	0.04402	0.08902	22.71934	11.23402	6.57231	73.83346
17	2.11338	0.47318	0.04042	0.08542	24.74171	11.70719	6.95336	81.40429
18	2.20848	0.45280	0.03724	0.08224	26.85508	12.15999	7.32746	89.10189
19	2.30786	0.43330	0.03441	0.07941	29.06356	12.59329	7.69468	96.90132
20	2.41171	0.41464	0.03188	0.07688	31.37142	13.00794	8.05505	104.77954
21	2.52024	0.39679	0.02960	0.07460	33.78314	13.40472	8.40862	112.71529
22	2.63365	0.37970	0.02755	0.07255	36.30338	13.78442	8.75546	120.68901
23	2.75217	0.36335	0.02568	0.07068	38.93703	14.14777	9.09561	128.68271
24	2.87601	0.34770	0.02399	0.06899	41.68920	14.49548	9.42914	136.67989
25	3.00543	0.33273	0.02244	0.06744	44.56521	14.82821	9.75610	144.66542
26	3.14068	0.31840	0.02102	0.06602	47.57064	15.14661	10.07654	152.62549
27	3.28201	0.30469	0.01972	0.06472	50.71132	15.45130	10.39055	160.54746
28	3.42970	0.29157	0.01852	0.06352	53.99333	15.74287	10.69817	168.41987
29	3.58404	0.27902	0.01741	0.06241	57.42303	16.02189	10.99947	176.23229
30	3.74532	0.26700	0.01639	0.06139	61.00707	16.28889	11.29453	183.97529
31	3.91386	0.25550	0.01544	0.06044	64.75239	16.54439	11.58340	191.64036
32	4.08998	0.24450	0.01456	0.05956	68.66625	16.78889	11.86617	199.21986
33	4.27403	0.23397	0.01374	0.05874	72.75623	17.02286	12.14290	206.70694
34	4.46636	0.22390	0.01298	0.05798	77.03026	17.24676	12.41367	214.09550
35	4.66735	0.21425	0.01227	0.05727	81.49662	17.46101	12.67854	221.38015
40	5.81636	0.17193	0.00934	0.05434	107.03032	18.40158	13.91720	256.09859
45	7.24825	0.13796	0.00720	0.05220	138.84997	19.15635	15.02020	287.73224
50	9.03264	0.11071	0.00560	0.05060	178.50303	19.76201	15.99762	316.14501
55	11.25631	0.08884	0.00439	0.04939	227.91796	20.24802	16.85967	341.37492
60	14.02741	0.07129	0.00345	0.04845	289.49795	20.63802	17.61655	363.57070
65	17.48070	0.05721	0.00273	0.04773	366.23783	20.95098	18.27822	382.94651
70	21.78414	0.04590	0.00217	0.04717	461.86968	21.20211	18.85427	399.75032
75	27.14700	0.03684	0.00172	0.04672	581.04436	21.40363	19.35382	414.24216
80	33.83010	0.02956	0.00137	0.04637	729.55770	21.56534	19.78543	426.67970
85	42.15846	0.02372	0.00109	0.04609	914.63234	21.69511	20.15703	437.30905
90	52.53711	0.01903	0.00087	0.04587	1145.26901	21.79924	20.47591	446.35924
95	65.47079	0.01527	0.00070	0.04570	1432.68426	21.88280	20.74869	454.03937
100	81.58852	0.01226	0.00056	0.04556	1790.85596	21.94985	20.98135	460.53756

Interest Rate 5%	Single Payment		Uniform Payment Series				Arithmetic Gradient	
	Compound Amount	Present Worth	Sinking Fund	Capital Recovery	Compound Amount	Present Worth	Gradient Uniform Series	Gradient Present Worth
n	(F/P)	(P/F)	(A/F)	(A/P)	(F/A)	(P/A)	(A/G)	(P/G)
1	1.05000	0.95238	1.00000	1.05000	1.00000	0.95238	0.00000	0.00000
2	1.10250	0.90703	0.48780	0.53780	2.05000	1.85941	0.48780	0.90703
3	1.15763	0.86384	0.31721	0.36721	3.15250	2.72325	0.96749	2.63470
4	1.21551	0.82270	0.23201	0.28201	4.31013	3.54595	1.43905	5.10281
5	1.27628	0.78353	0.18097	0.23097	5.52563	4.32948	1.90252	8.23692
6	1.34010	0.74622	0.14702	0.19702	6.80191	5.07569	2.35790	11.96799
7	1.40710	0.71068	0.12282	0.17282	8.14201	5.78637	2.80523	16.23208
8	1.47746	0.67684	0.10472	0.15472	9.54911	6.46321	3.24451	20.96996
9	1.55133	0.64461	0.09069	0.14069	11.02656	7.10782	3.67579	26.12683
10	1.62889	0.61391	0.07950	0.12950	12.57789	7.72173	4.09909	31.65205
11	1.71034	0.58468	0.07039	0.12039	14.20679	8.30641	4.51444	37.49884
12	1.79586	0.55684	0.06283	0.11283	15.91713	8.86325	4.92190	43.62405
13	1.88565	0.53032	0.05646	0.10646	17.71298	9.39357	5.32150	49.98791
14	1.97993	0.50507	0.05102	0.10102	19.59863	9.89864	5.71329	56.55379
15	2.07893	0.48102	0.04634	0.09634	21.57856	10.37966	6.09731	63.28803
16	2.18287	0.45811	0.04227	0.09227	23.65749	10.83777	6.47363	70.15970
17	2.29202	0.43630	0.03870	0.08870	25.84037	11.27407	6.84229	77.14045
18	2.40662	0.41552	0.03555	0.08555	28.13238	11.68959	7.20336	84.20430
19	2.52695	0.39573	0.03275	0.08275	30.53900	12.08532	7.55690	91.32751
20	2.65330	0.37689	0.03024	0.08024	33.06595	12.46221	7.90297	98.48841
21	2.78596	0.35894	0.02800	0.07800	35.71925	12.82115	8.24164	105.66726
22	2.92526	0.34185	0.02597	0.07597	38.50521	13.16300	8.57298	112.84611
23	3.07152	0.32557	0.02414	0.07414	41.43048	13.48857	8.89706	120.00868
24	3.22510	0.31007	0.02247	0.07247	44.50200	13.79864	9.21397	127.14024
25	3.38635	0.29530	0.02095	0.07095	47.72710	14.09394	9.52377	134.22751
26	3.55567	0.28124	0.01956	0.06956	51.11345	14.37519	9.82655	141.25852
27	3.73346	0.26785	0.01829	0.06829	54.66913	14.64303	10.12240	148.22258
28	3.92013	0.25509	0.01712	0.06712	58.40258	14.89813	10.41138	155.11011
29	4.11614	0.24295	0.01605	0.06605	62.32271	15.14107	10.69360	161.91261
30	4.32194	0.23138	0.01505	0.06505	66.43885	15.37245	10.96914	168.62255
31	4.53804	0.22036	0.01413	0.06413	70.76079	15.59281	11.23809	175.23334
32	4.76494	0.20987	0.01328	0.06328	75.29883	15.80268	11.50053	181.73919
33	5.00319	0.19987	0.01249	0.06249	80.06377	16.00255	11.75657	188.13511
34	5.25335	0.19035	0.01176	0.06176	85.06696	16.19290	12.00630	194.41682
35	5.51602	0.18129	0.01107	0.06107	90.32031	16.37419	12.24980	200.58069
40	7.03999	0.14205	0.00828	0.05828	120.79977	17.15909	13.37747	229.54518
45	8.98501	0.11130	0.00626	0.05626	159.70016	17.77407	14.36444	255.31454
50	11.46740	0.08720	0.00478	0.05478	209.34800	18.25593	15.22326	277.91478
55	14.63563	0.06833	0.00367	0.05367	272.71262	18.63347	15.96645	297.51040
60	18.67919	0.05354	0.00283	0.05283	353.58372	18.92929	16.60618	314.34316
65	23.83990	0.04195	0.00219	0.05219	456.79801	19.16107	17.15410	328.69098
70	30.42643	0.03287	0.00170	0.05170	588.52851	19.34268	17.62119	340.84090
75	38.83269	0.02575	0.00132	0.05132	756.65372	19.48497	18.01759	351.07215
80	49.56144	0.02018	0.00103	0.05103	971.22882	19.59646	18.35260	359.64605
85	63.25435	0.01581	0.00080	0.05080	1245.08707	19.68382	18.63463	366.80070
90	80.73037	0.01239	0.00063	0.05063	1594.60730	19.75226	18.87120	372.74879
95	103.03468	0.00971	0.00049	0.05049	2040.69353	19.80589	19.06894	377.67742
100	131.50126	0.00760	0.00038	0.05038	2610.02516	19.84791	19.23372	381.74922

Interest rate 6%	Single Payment		Uniform Payment Series				Arithmetic Gradient	
	Compound Amount	Present Worth	Sinking Fund	Capital Recovery	Compound Amount	Present Worth	Gradient Uniform Series	Gradient Present Worth
n	(F/P)	(P/F)	(A/F)	(A/P)	(F/A)	(P/A)	(A/G)	(P/G)
1	1.06000	0.94340	1.00000	1.06000	1.00000	0.94340	0.00000	0.00000
2	1.12360	0.89000	0.48544	0.54544	2.06000	1.83339	0.48544	0.89000
3	1.19102	0.83962	0.31411	0.37411	3.18360	2.67301	0.96118	2.56924
4	1.26248	0.79209	0.22859	0.28859	4.37462	3.46511	1.42723	4.94552
5	1.33823	0.74726	0.17740	0.23740	5.63709	4.21236	1.88363	7.93455
6	1.41852	0.70496	0.14336	0.20336	6.97532	4.91732	2.33040	11.45935
7	1.50363	0.66506	0.11914	0.17914	8.39384	5.58238	2.76758	15.44969
8	1.59385	0.62741	0.10104	0.16104	9.89747	6.20979	3.19521	19.84158
9	1.68948	0.59190	0.08702	0.14702	11.49132	6.80169	3.61333	24.57677
10	1.79085	0.55839	0.07587	0.13587	13.18079	7.36009	4.02201	29.60232
11	1.89830	0.52679	0.06679	0.12679	14.97164	7.88687	4.42129	34.87020
12	2.01220	0.49697	0.05928	0.11928	16.86994	8.38384	4.81126	40.33686
13	2.13293	0.46884	0.05296	0.11296	18.88214	8.85268	5.19198	45.96293
14	2.26090	0.44230	0.04758	0.10758	21.01507	9.29498	5.56352	51.71284
15	2.39656	0.41727	0.04296	0.10296	23.27597	9.71225	5.92598	57.55455
16	2.54035	0.39365	0.03895	0.09895	25.67253	10.10590	6.27943	63.45925
17	2.69277	0.37136	0.03544	0.09544	28.21288	10.47726	6.62397	69.40108
18	2.85434	0.35034	0.03236	0.09236	30.90565	10.82760	6.95970	75.35692
19	3.02560	0.33051	0.02962	0.08962	33.75999	11.15812	7.28673	81.30615
20	3.20714	0.31180	0.02718	0.08718	36.78559	11.46992	7.60515	87.23044
21	3.39956	0.29416	0.02500	0.08500	39.99273	11.76408	7.91508	93.11355
22	3.60354	0.27751	0.02305	0.08305	43.39229	12.04158	8.21662	98.94116
23	3.81975	0.26180	0.02128	0.08128	46.99583	12.30338	8.50991	104.70070
24	4.04893	0.24698	0.01968	0.07968	50.81558	12.55036	8.79506	110.38121
25	4.29187	0.23300	0.01823	0.07823	54.86451	12.78336	9.07220	115.97317
26	4.54938	0.21981	0.01690	0.07690	59.15638	13.00317	9.34145	121.46842
27	4.82235	0.20737	0.01570	0.07570	63.70577	13.21053	9.60294	126.85999
28	5.11169	0.19563	0.01459	0.07459	68.52811	13.40616	9.85681	132.14200
29	5.41839	0.18456	0.01358	0.07358	73.63980	13.59072	10.10319	137.30959
30	5.74349	0.17411	0.01265	0.07265	79.05819	13.76483	10.34221	142.35879
31	6.08810	0.16425	0.01179	0.07179	84.80168	13.92909	10.57402	147.28643
32	6.45339	0.15496	0.01100	0.07100	90.88978	14.08404	10.79875	152.09011
33	6.84059	0.14619	0.01027	0.07027	97.34316	14.23023	11.01655	156.76807
34	7.25103	0.13791	0.00960	0.06960	104.18375	14.36814	11.22756	161.31915
35	7.68609	0.13011	0.00897	0.06897	111.43478	14.49825	11.43192	165.74273
40	10.28572	0.09722	0.00646	0.06646	154.76197	15.04630	12.35898	185.95682
45	13.76461	0.07265	0.00470	0.06470	212.74351	15.45583	13.14129	203.10965
50	18.42015	0.05429	0.00344	0.06344	290.33590	15.76186	13.79643	217.45738
55	24.65032	0.04057	0.00254	0.06254	394.17203	15.99054	14.34112	229.32225
60	32.98769	0.03031	0.00188	0.06188	533.12818	16.16143	14.79095	239.04279
65	44.14497	0.02265	0.00139	0.06139	719.08286	16.28912	15.16012	246.94502
70	59.07593	0.01693	0.00103	0.06103	967.93217	16.38454	15.46135	253.32714
75	79.05692	0.01265	0.00077	0.06077	1300.94868	16.45585	15.70583	258.45274
80	105.79599	0.00945	0.00057	0.06057	1746.59989	16.50913	15.90328	262.54931
85	141.57890	0.00706	0.00043	0.06043	2342.98174	16.54895	16.06202	265.80958
90	189.46451	0.00528	0.00032	0.06032	3141.07519	16.57870	16.18912	268.39461
95	253.54625	0.00394	0.00024	0.06024	4209.10425	16.60093	16.29050	270.43746
100	339.30208	0.00295	0.00018	0.06018	5638.36806	16.61755	16.37107	272.04706

Interest rate **7%**	**Single Payment**		**Uniform Payment Series**				**Arithmetic Gradient**	
	Compound Amount	**Present Worth**	**Sinking Fund**	**Capital Recovery**	**Compound Amount**	**Present Worth**	**Gradient Uniform Series**	**Gradient Present Worth**
n	(F/P)	(P/F)	(A/F)	(A/P)	(F/A)	(P/A)	(A/G)	(P/G)
1	1.07000	0.93458	1.00000	1.07000	1.00000	0.93458	0.00000	0.00000
2	1.14490	0.87344	0.48309	0.55309	2.07000	1.80802	0.48309	0.87344
3	1.22504	0.81630	0.31105	0.38105	3.21490	2.62432	0.95493	2.50603
4	1.31080	0.76290	0.22523	0.29523	4.43994	3.38721	1.41554	4.79472
5	1.40255	0.71299	0.17389	0.24389	5.75074	4.10020	1.86495	7.64666
6	1.50073	0.66634	0.13980	0.20980	7.15329	4.76654	2.30322	10.97838
7	1.60578	0.62275	0.11555	0.18555	8.65402	5.38929	2.73039	14.71487
8	1.71819	0.58201	0.09747	0.16747	10.25980	5.97130	3.14654	18.78894
9	1.83846	0.54393	0.08349	0.15349	11.97799	6.51523	3.55174	23.14041
10	1.96715	0.50835	0.07238	0.14238	13.81645	7.02358	3.94607	27.71555
11	2.10485	0.47509	0.06336	0.13336	15.78360	7.49867	4.32963	32.46648
12	2.25219	0.44401	0.05590	0.12590	17.88845	7.94269	4.70252	37.35061
13	2.40985	0.41496	0.04965	0.11965	20.14064	8.35765	5.06484	42.33018
14	2.57853	0.38782	0.04434	0.11434	22.55049	8.74547	5.41673	47.37181
15	2.75903	0.36245	0.03979	0.10979	25.12902	9.10791	5.75829	52.44605
16	2.95216	0.33873	0.03586	0.10586	27.88805	9.44665	6.08968	57.52707
17	3.15882	0.31657	0.03243	0.10243	30.84022	9.76322	6.41102	62.59226
18	3.37993	0.29586	0.02941	0.09941	33.99903	10.05909	6.72247	67.62195
19	3.61653	0.27651	0.02675	0.09675	37.37896	10.33560	7.02418	72.59910
20	3.86968	0.25842	0.02439	0.09439	40.99549	10.59401	7.31631	77.50906
21	4.14056	0.24151	0.02229	0.09229	44.86518	10.83553	7.59901	82.33932
22	4.43040	0.22571	0.02041	0.09041	49.00574	11.06124	7.87247	87.07930
23	4.74053	0.21095	0.01871	0.08871	53.43614	11.27219	8.13685	91.72013
24	5.07237	0.19715	0.01719	0.08719	58.17667	11.46933	8.39234	96.25450
25	5.42743	0.18425	0.01581	0.08581	63.24904	11.65358	8.63910	100.67648
26	5.80735	0.17220	0.01456	0.08456	68.67647	11.82578	8.87733	104.98137
27	6.21387	0.16093	0.01343	0.08343	74.48382	11.98671	9.10722	109.16556
28	6.64884	0.15040	0.01239	0.08239	80.69769	12.13711	9.32894	113.22642
29	7.11426	0.14056	0.01145	0.08145	87.34653	12.27767	9.54270	117.16218
30	7.61226	0.13137	0.01059	0.08059	94.46079	12.40904	9.74868	120.97182
31	8.14511	0.12277	0.00980	0.07980	102.07304	12.53181	9.94708	124.65501
32	8.71527	0.11474	0.00907	0.07907	110.21815	12.64656	10.13810	128.21199
33	9.32534	0.10723	0.00841	0.07841	118.93343	12.75379	10.32191	131.64350
34	9.97811	0.10022	0.00780	0.07780	128.25876	12.85401	10.49873	134.95074
35	10.67658	0.09366	0.00723	0.07723	138.23688	12.94767	10.66873	138.13528
40	14.97446	0.06678	0.00501	0.07501	199.63511	13.33171	11.42335	152.29277
45	21.00245	0.04761	0.00350	0.07350	285.74931	13.60552	12.03599	163.75592
50	29.45703	0.03395	0.00246	0.07246	406.52893	13.80075	12.52868	172.90512
55	41.31500	0.02420	0.00174	0.07174	575.92859	13.93994	12.92146	180.12433
60	57.94643	0.01726	0.00123	0.07123	813.52038	14.03918	13.23209	185.76774
65	81.27286	0.01230	0.00087	0.07087	1146.75516	14.10994	13.47598	190.14521
70	113.98939	0.00877	0.00062	0.07062	1614.13417	14.16039	13.66619	193.51853
75	159.87602	0.00625	0.00044	0.07044	2269.65742	14.19636	13.81365	196.10351
80	224.23439	0.00446	0.00031	0.07031	3189.06268	14.22201	13.92735	198.07480
85	314.50033	0.00318	0.00022	0.07022	4478.57612	14.24029	14.01458	199.57173
90	441.10298	0.00227	0.00016	0.07016	6287.18543	14.25333	14.08122	200.70420
95	618.66975	0.00162	0.00011	0.07011	8823.85354	14.26262	14.13191	201.55811
100	867.71633	0.00115	0.00008	0.07008	12381.66179	14.26925	14.17034	202.20008

Interest rate 8%	Single Payment		Uniform Payment Series				Arithmetic Gradient	
	Compound Amount	Present Worth	Sinking Fund	Capital Recovery	Compound Amount	Present Worth	Gradient Uniform Series	Gradient Present Worth
n	(F/P)	(P/F)	(A/F)	(A/P)	(F/A)	(P/A)	(A/G)	(P/G)
1	1.08000	0.92593	1.00000	1.08000	1.00000	0.92593	0.00000	0.00000
2	1.16640	0.85734	0.48077	0.56077	2.08000	1.78326	0.48077	0.85734
3	1.25971	0.79383	0.30803	0.38803	3.24640	2.57710	0.94874	2.44500
4	1.36049	0.73503	0.22192	0.30192	4.50611	3.31213	1.40396	4.65009
5	1.46933	0.68058	0.17046	0.25046	5.86660	3.99271	1.84647	7.37243
6	1.58687	0.63017	0.13632	0.21632	7.33593	4.62288	2.27635	10.52327
7	1.71382	0.58349	0.11207	0.19207	8.92280	5.20637	2.69366	14.02422
8	1.85093	0.54027	0.09401	0.17401	10.63663	5.74664	3.09852	17.80610
9	1.99900	0.50025	0.08008	0.16008	12.48756	6.24689	3.49103	21.80809
10	2.15892	0.46319	0.06903	0.14903	14.48656	6.71008	3.87131	25.97683
11	2.33164	0.42888	0.06008	0.14008	16.64549	7.13896	4.23950	30.26566
12	2.51817	0.39711	0.05270	0.13270	18.97713	7.53608	4.59575	34.63391
13	2.71962	0.36770	0.04652	0.12652	21.49530	7.90378	4.94021	39.04629
14	2.93719	0.34046	0.04130	0.12130	24.21492	8.24424	5.27305	43.47228
15	3.17217	0.31524	0.03683	0.11683	27.15211	8.55948	5.59446	47.88566
16	3.42594	0.29189	0.03298	0.11298	30.32428	8.85137	5.90463	52.26402
17	3.70002	0.27027	0.02963	0.10963	33.75023	9.12164	6.20375	56.58832
18	3.99602	0.25025	0.02670	0.10670	37.45024	9.37189	6.49203	60.84256
19	4.31570	0.23171	0.02413	0.10413	41.44626	9.60360	6.76969	65.01337
20	4.66096	0.21455	0.02185	0.10185	45.76196	9.81815	7.03695	69.08979
21	5.03383	0.19866	0.01983	0.09983	50.42292	10.01680	7.29403	73.06291
22	5.43654	0.18394	0.01803	0.09803	55.45676	10.20074	7.54118	76.92566
23	5.87146	0.17032	0.01642	0.09642	60.89330	10.37106	7.77863	80.67259
24	6.34118	0.15770	0.01498	0.09498	66.76476	10.52876	8.00661	84.29968
25	6.84848	0.14602	0.01368	0.09368	73.10594	10.67478	8.22538	87.80411
26	7.39635	0.13520	0.01251	0.09251	79.95442	10.80998	8.43518	91.18415
27	7.98806	0.12519	0.01145	0.09145	87.35077	10.93516	8.63627	94.43901
28	8.62711	0.11591	0.01049	0.09049	95.33883	11.05108	8.82888	97.56868
29	9.31727	0.10733	0.00962	0.08962	103.96594	11.15841	9.01328	100.57385
30	10.06266	0.09938	0.00883	0.08883	113.28321	11.25778	9.18971	103.45579
31	10.86767	0.09202	0.00811	0.08811	123.34587	11.34980	9.35843	106.21627
32	11.73708	0.08520	0.00745	0.08745	134.21354	11.43500	9.51967	108.85747
33	12.67605	0.07889	0.00685	0.08685	145.95062	11.51389	9.67370	111.38192
34	13.69013	0.07305	0.00630	0.08630	158.62667	11.58693	9.82075	113.79242
35	14.78534	0.06763	0.00580	0.08580	172.31680	11.65457	9.96107	116.09199
40	21.72452	0.04603	0.00386	0.08386	259.05652	11.92461	10.56992	126.04220
45	31.92045	0.03133	0.00259	0.08259	386.50562	12.10840	11.04465	133.73309
50	46.90161	0.02132	0.00174	0.08174	573.77016	12.23348	11.41071	139.59279
55	68.91386	0.01451	0.00118	0.08118	848.92320	12.31861	11.69015	144.00645
60	101.25706	0.00988	0.00080	0.08080	1253.21330	12.37655	11.90154	147.30001
65	148.77985	0.00672	0.00054	0.08054	1847.24808	12.41598	12.06016	149.73870
70	218.60641	0.00457	0.00037	0.08037	2720.08007	12.44282	12.17832	151.53262
75	321.20453	0.00311	0.00025	0.08025	4002.55662	12.46108	12.26577	152.84485
80	471.95483	0.00212	0.00017	0.08017	5886.93543	12.47351	12.33013	153.80008
85	693.45649	0.00144	0.00012	0.08012	8655.70611	12.48197	12.37725	154.49250
90	1018.91509	0.00098	0.00008	0.08008	12723.93862	12.48773	12.41158	154.99254
95	1497.12055	0.00067	0.00005	0.08005	18701.50686	12.49165	12.43650	155.35244
100	2199.76126	0.00045	0.00004	0.08004	27484.51570	12.49432	12.45452	155.61073

Interest rate 9%	Single Payment		Uniform Payment Series				Arithmetic Gradient	
	Compound Amount	Present Worth	Sinking Fund	Capital Recovery	Compound Amount	Present Worth	Gradient Uniform Series	Gradient Present Worth
n	(F/P)	(P/F)	(A/F)	(A/P)	(F/A)	(P/A)	(A/G)	(P/G)
1	1.09000	0.91743	1.00000	1.09000	1.00000	0.91743	0.00000	0.00000
2	1.18810	0.84168	0.47847	0.56847	2.09000	1.75911	0.47847	0.84168
3	1.29503	0.77218	0.30505	0.39505	3.27810	2.53129	0.94262	2.38605
4	1.41158	0.70843	0.21867	0.30867	4.57313	3.23972	1.39250	4.51132
5	1.53862	0.64993	0.16709	0.25709	5.98471	3.88965	1.82820	7.11105
6	1.67710	0.59627	0.13292	0.22292	7.52333	4.48592	2.24979	10.09238
7	1.82804	0.54703	0.10869	0.19869	9.20043	5.03295	2.65740	13.37459
8	1.99256	0.50187	0.09067	0.18067	11.02847	5.53482	3.05117	16.88765
9	2.17189	0.46043	0.07680	0.16680	13.02104	5.99525	3.43123	20.57108
10	2.36736	0.42241	0.06582	0.15582	15.19293	6.41766	3.79777	24.37277
11	2.58043	0.38753	0.05695	0.14695	17.56029	6.80519	4.15096	28.24810
12	2.81266	0.35553	0.04965	0.13965	20.14072	7.16073	4.49102	32.15898
13	3.06580	0.32618	0.04357	0.13357	22.95338	7.48690	4.81816	36.07313
14	3.34173	0.29925	0.03843	0.12843	26.01919	7.78615	5.13262	39.96333
15	3.64248	0.27454	0.03406	0.12406	29.36092	8.06069	5.43463	43.80686
16	3.97031	0.25187	0.03030	0.12030	33.00340	8.31256	5.72446	47.58491
17	4.32763	0.23107	0.02705	0.11705	36.97370	8.54363	6.00238	51.28208
18	4.71712	0.21199	0.02421	0.11421	41.30134	8.75563	6.26865	54.88598
19	5.14166	0.19449	0.02173	0.11173	46.01846	8.95011	6.52358	58.38679
20	5.60441	0.17843	0.01955	0.10955	51.16012	9.12855	6.76745	61.77698
21	6.10881	0.16370	0.01762	0.10762	56.76453	9.29224	7.00056	65.05094
22	6.65860	0.15018	0.01590	0.10590	62.87334	9.44243	7.22322	68.20475
23	7.25787	0.13778	0.01438	0.10438	69.53194	9.58021	7.43574	71.23594
24	7.91108	0.12640	0.01302	0.10302	76.78981	9.70661	7.63843	74.14326
25	8.62308	0.11597	0.01181	0.10181	84.70090	9.82258	7.83160	76.92649
26	9.39916	0.10639	0.01072	0.10072	93.32398	9.92897	8.01556	79.58630
27	10.24508	0.09761	0.00973	0.09973	102.72313	10.02658	8.19064	82.12410
28	11.16714	0.08955	0.00885	0.09885	112.96822	10.11613	8.35714	84.54191
29	12.17218	0.08215	0.00806	0.09806	124.13536	10.19828	8.51538	86.84224
30	13.26768	0.07537	0.00734	0.09734	136.30754	10.27365	8.66566	89.02800
31	14.46177	0.06915	0.00669	0.09669	149.57522	10.34280	8.80829	91.10243
32	15.76333	0.06344	0.00610	0.09610	164.03699	10.40624	8.94358	93.06902
33	17.18203	0.05820	0.00556	0.09556	179.80032	10.46444	9.07181	94.93144
34	18.72841	0.05339	0.00508	0.09508	196.98234	10.51784	9.19329	96.69346
35	20.41397	0.04899	0.00464	0.09464	215.71075	10.56682	9.30829	98.35899
40	31.40942	0.03184	0.00296	0.09296	337.88245	10.75736	9.79573	105.37619
45	48.32729	0.02069	0.00190	0.09190	525.85873	10.88120	10.16029	110.55607
50	74.35752	0.01345	0.00123	0.09123	815.08356	10.96168	10.42952	114.32507
55	114.40826	0.00874	0.00079	0.09079	1260.09180	11.01399	10.62614	117.03621
60	176.03129	0.00568	0.00051	0.09051	1944.79213	11.04799	10.76832	118.96825
65	270.84596	0.00369	0.00033	0.09033	2998.28847	11.07009	10.87023	120.33443
70	416.73009	0.00240	0.00022	0.09022	4619.22318	11.08445	10.94273	121.29416
75	641.19089	0.00156	0.00014	0.09014	7113.23215	11.09378	10.99396	121.96458
80	986.55167	0.00101	0.00009	0.09009	10950.57409	11.09985	11.02994	122.43064
85	1517.93203	0.00066	0.00006	0.09006	16854.80033	11.10379	11.05508	122.75327
90	2335.52658	0.00043	0.00004	0.09004	25939.18425	11.10635	11.07256	122.97576
95	3593.49715	0.00028	0.00003	0.09003	39916.63496	11.10802	11.08467	123.12869
100	5529.04079	0.00018	0.00002	0.09002	61422.67546	11.10910	11.09302	123.23350

Interest rate 10%	Single Payment		Uniform Payment Series				Arithmetic Gradient	
	Compound Amount	Present Worth	Sinking Fund	Capital Recovery	Compound Amount	Present Worth	Gradient Uniform Series	Gradient Present Worth
n	(F/P)	(P/F)	(A/F)	(A/P)	(F/A)	(P/A)	(A/G)	(P/G)
1	1.10000	0.90909	1.00000	1.10000	1.00000	0.90909	0.00000	0.00000
2	1.21000	0.82645	0.47619	0.57619	2.10000	1.73554	0.47619	0.82645
3	1.33100	0.75131	0.30211	0.40211	3.31000	2.48685	0.93656	2.32908
4	1.46410	0.68301	0.21547	0.31547	4.64100	3.16987	1.38117	4.37812
5	1.61051	0.62092	0.16380	0.26380	6.10510	3.79079	1.81013	6.86180
6	1.77156	0.56447	0.12961	0.22961	7.71561	4.35526	2.22356	9.68417
7	1.94872	0.51316	0.10541	0.20541	9.48717	4.86842	2.62162	12.76312
8	2.14359	0.46651	0.08744	0.18744	11.43589	5.33493	3.00448	16.02867
9	2.35795	0.42410	0.07364	0.17364	13.57948	5.75902	3.37235	19.42145
10	2.59374	0.38554	0.06275	0.16275	15.93742	6.14457	3.72546	22.89134
11	2.85312	0.35049	0.05396	0.15396	18.53117	6.49506	4.06405	26.39628
12	3.13843	0.31863	0.04676	0.14676	21.38428	6.81369	4.38840	29.90122
13	3.45227	0.28966	0.04078	0.14078	24.52271	7.10336	4.69879	33.37719
14	3.79750	0.26333	0.03575	0.13575	27.97498	7.36669	4.99553	36.80050
15	4.17725	0.23939	0.03147	0.13147	31.77248	7.60608	5.27893	40.15199
16	4.59497	0.21763	0.02782	0.12782	35.94973	7.82371	5.54934	43.41642
17	5.05447	0.19784	0.02466	0.12466	40.54470	8.02155	5.80710	46.58194
18	5.55992	0.17986	0.02193	0.12193	45.59917	8.20141	6.05256	49.63954
19	6.11591	0.16351	0.01955	0.11955	51.15909	8.36492	6.28610	52.58268
20	6.72750	0.14864	0.01746	0.11746	57.27500	8.51356	6.50808	55.40691
21	7.40025	0.13513	0.01562	0.11562	64.00250	8.64869	6.71888	58.10952
22	8.14027	0.12285	0.01401	0.11401	71.40275	8.77154	6.91889	60.68929
23	8.95430	0.11168	0.01257	0.11257	79.54302	8.88322	7.10848	63.14621
24	9.84973	0.10153	0.01130	0.11130	88.49733	8.98474	7.28805	65.48130
25	10.83471	0.09230	0.01017	0.11017	98.34706	9.07704	7.45798	67.69640
26	11.91818	0.08391	0.00916	0.10916	109.18177	9.16095	7.61865	69.79404
27	13.10999	0.07628	0.00826	0.10826	121.09994	9.23722	7.77044	71.77726
28	14.42099	0.06934	0.00745	0.10745	134.20994	9.30657	7.91372	73.64953
29	15.86309	0.06304	0.00673	0.10673	148.63093	9.36961	8.04886	75.41463
30	17.44940	0.05731	0.00608	0.10608	164.49402	9.42691	8.17623	77.07658
31	19.19434	0.05210	0.00550	0.10550	181.94342	9.47901	8.29617	78.63954
32	21.11378	0.04736	0.00497	0.10497	201.13777	9.52638	8.40905	80.10777
33	23.22515	0.04306	0.00450	0.10450	222.25154	9.56943	8.51520	81.48559
34	25.54767	0.03914	0.00407	0.10407	245.47670	9.60857	8.61494	82.77729
35	28.10244	0.03558	0.00369	0.10369	271.02437	9.64416	8.70860	83.98715
40	45.25926	0.02209	0.00226	0.10226	442.59256	9.77905	9.09623	88.95254
45	72.89048	0.01372	0.00139	0.10139	718.90484	9.86281	9.37405	92.45443
50	117.39085	0.00852	0.00086	0.10086	1163.90853	9.91481	9.57041	94.88887
55	189.05914	0.00529	0.00053	0.10053	1880.59142	9.94711	9.70754	96.56192
60	304.48164	0.00328	0.00033	0.10033	3034.81640	9.96716	9.80229	97.70101
65	490.37073	0.00204	0.00020	0.10020	4893.70725	9.97961	9.86718	98.47054
70	789.74696	0.00127	0.00013	0.10013	7887.46957	9.98734	9.91125	98.98702
75	1271.89537	0.00079	0.00008	0.10008	12708.95371	9.99214	9.94099	99.33171
80	2048.40021	0.00049	0.00005	0.10005	20474.00215	9.99512	9.96093	99.56063
85	3298.96903	0.00030	0.00003	0.10003	32979.69030	9.99697	9.97423	99.71203
90	5313.02261	0.00019	0.00002	0.10002	53120.22612	9.99812	9.98306	99.81178
95	8556.67605	0.00012	0.00001	0.10001	85556.76047	9.99883	9.98890	99.87729
100	13780.61234	0.00007	0.00001	0.10001	137796.12340	9.99927	9.99274	99.92018

Interest rate 12%	Single Payment		Uniform Payment Series				Arithmetic Gradient	
	Compound Amount	Present Worth	Sinking Fund	Capital Recovery	Compound Amount	Present Worth	Gradient Uniform Series	Gradient Present Worth
n	(F/P)	(P/F)	(A/F)	(A/P)	(F/A)	(P/A)	(A/G)	(P/G)
1	1.12000	0.89286	1.00000	1.12000	1.00000	0.89286	0.00000	0.00000
2	1.25440	0.79719	0.47170	0.59170	2.12000	1.69005	0.47170	0.79719
3	1.40493	0.71178	0.29635	0.41635	3.37440	2.40183	0.92461	2.22075
4	1.57352	0.63552	0.20923	0.32923	4.77933	3.03735	1.35885	4.12731
5	1.76234	0.56743	0.15741	0.27741	6.35285	3.60478	1.77459	6.39702
6	1.97382	0.50663	0.12323	0.24323	8.11519	4.11141	2.17205	8.93017
7	2.21068	0.45235	0.09912	0.21912	10.08901	4.56376	2.55147	11.64427
8	2.47596	0.40388	0.08130	0.20130	12.29969	4.96764	2.91314	14.47145
9	2.77308	0.36061	0.06768	0.18768	14.77566	5.32825	3.25742	17.35633
10	3.10585	0.32197	0.05698	0.17698	17.54874	5.65022	3.58465	20.25409
11	3.47855	0.28748	0.04842	0.16842	20.65458	5.93770	3.89525	23.12885
12	3.89598	0.25668	0.04144	0.16144	24.13313	6.19437	4.18965	25.95228
13	4.36349	0.22917	0.03568	0.15568	28.02911	6.42355	4.46830	28.70237
14	4.88711	0.20462	0.03087	0.15087	32.39260	6.62817	4.73169	31.36242
15	5.47357	0.18270	0.02682	0.14682	37.27971	6.81086	4.98030	33.92017
16	6.13039	0.16312	0.02339	0.14339	42.75328	6.97399	5.21466	36.36700
17	6.86604	0.14564	0.02046	0.14046	48.88367	7.11963	5.43530	38.69731
18	7.68997	0.13004	0.01794	0.13794	55.74971	7.24967	5.64274	40.90798
19	8.61276	0.11611	0.01576	0.13576	63.43968	7.36578	5.83752	42.99790
20	9.64629	0.10367	0.01388	0.13388	72.05244	7.46944	6.02020	44.96757
21	10.80385	0.09256	0.01224	0.13224	81.69874	7.56200	6.19132	46.81876
22	12.10031	0.08264	0.01081	0.13081	92.50258	7.64465	6.35141	48.55425
23	13.55235	0.07379	0.00956	0.12956	104.60289	7.71843	6.50101	50.17759
24	15.17863	0.06588	0.00846	0.12846	118.15524	7.78432	6.64064	51.69288
25	17.00006	0.05882	0.00750	0.12750	133.33387	7.84314	6.77084	53.10464
26	19.04007	0.05252	0.00665	0.12665	150.33393	7.89566	6.89210	54.41766
27	21.32488	0.04689	0.00590	0.12590	169.37401	7.94255	7.00491	55.63689
28	23.88387	0.04187	0.00524	0.12524	190.69889	7.98442	7.10976	56.76736
29	26.74993	0.03738	0.00466	0.12466	214.58275	8.02181	7.20712	57.81409
30	29.95992	0.03338	0.00414	0.12414	241.33268	8.05518	7.29742	58.78205
31	33.55511	0.02980	0.00369	0.12369	271.29261	8.08499	7.38110	59.67610
32	37.58173	0.02661	0.00328	0.12328	304.84772	8.11159	7.45858	60.50097
33	42.09153	0.02376	0.00292	0.12292	342.42945	8.13535	7.53025	61.26122
34	47.14252	0.02121	0.00260	0.12260	384.52098	8.15656	7.59649	61.96123
35	52.79962	0.01894	0.00232	0.12232	431.66350	8.17550	7.65765	62.60517
40	93.05097	0.01075	0.00130	0.12130	767.09142	8.24378	7.89879	65.11587
45	163.98760	0.00610	0.00074	0.12074	1358.23003	8.28252	8.05724	66.73421
50	289.00219	0.00346	0.00042	0.12042	2400.01825	8.30450	8.15972	67.76241
55	509.32061	0.00196	0.00024	0.12024	4236.00505	8.31697	8.22513	68.40821
60	897.59693	0.00111	0.00013	0.12013	7471.64111	8.32405	8.26641	68.81003
65	1581.87249	0.00063	0.00008	0.12008	13173.93742	8.32807	8.29222	69.05812
70	2787.79983	0.00036	0.00004	0.12004	23223.33190	8.33034	8.30821	69.21029
75	4913.05584	0.00020	0.00002	0.12002	40933.79867	8.33164	8.31806	69.30310
80	8658.48310	0.00012	0.00001	0.12001	72145.69250	8.33237	8.32409	69.35943
85	15259.20568	0.00007	0.00001	0.12001	127151.71400	8.33279	8.32776	69.39347
90	26891.93422	0.00004	0.00000	0.12000	224091.11853	8.33302	8.32999	69.41397
95	47392.77662	0.00002	0.00000	0.12000	394931.47186	8.33316	8.33133	69.42627
100	83522.26573	0.00001	0.00000	0.12000	696010.54772	8.33323	8.33214	69.43364

Interest rate 15%	Single Payment		Uniform Payment Series				Arithmetic Gradient	
	Compound Amount	Present Worth	Sinking Fund	Capital Recovery	Compound Amount	Present Worth	Gradient Uniform Series	Gradient Present Worth
n	(F/P)	(P/F)	(A/F)	(A/P)	(F/A)	(P/A)	(A/G)	(P/G)
1	1.15000	0.86957	1.00000	1.15000	1.00000	0.86957	0.00000	0.00000
2	1.32250	0.75614	0.46512	0.61512	2.15000	1.62571	0.46512	0.75614
3	1.52088	0.65752	0.28798	0.43798	3.47250	2.28323	0.90713	2.07118
4	1.74901	0.57175	0.20027	0.35027	4.99338	2.85498	1.32626	3.78644
5	2.01136	0.49718	0.14832	0.29832	6.74238	3.35216	1.72281	5.77514
6	2.31306	0.43233	0.11424	0.26424	8.75374	3.78448	2.09719	7.93678
7	2.66002	0.37594	0.09036	0.24036	11.06680	4.16042	2.44985	10.19240
8	3.05902	0.32690	0.07285	0.22285	13.72682	4.48732	2.78133	12.48072
9	3.51788	0.28426	0.05957	0.20957	16.78584	4.77158	3.09223	14.75481
10	4.04556	0.24718	0.04925	0.19925	20.30372	5.01877	3.38320	16.97948
11	4.65239	0.21494	0.04107	0.19107	24.34928	5.23371	3.65494	19.12891
12	5.35025	0.18691	0.03448	0.18448	29.00167	5.42062	3.90820	21.18489
13	6.15279	0.16253	0.02911	0.17911	34.35192	5.58315	4.14376	23.13522
14	7.07571	0.14133	0.02469	0.17469	40.50471	5.72448	4.36241	24.97250
15	8.13706	0.12289	0.02102	0.17102	47.58041	5.84737	4.56496	26.69302
16	9.35762	0.10686	0.01795	0.16795	55.71747	5.95423	4.75225	28.29599
17	10.76126	0.09293	0.01537	0.16537	65.07509	6.04716	4.92509	29.78280
18	12.37545	0.08081	0.01319	0.16319	75.83636	6.12797	5.08431	31.15649
19	14.23177	0.07027	0.01134	0.16134	88.21181	6.19823	5.23073	32.42127
20	16.36654	0.06110	0.00976	0.15976	102.44358	6.25933	5.36514	33.58217
21	18.82152	0.05313	0.00842	0.15842	118.81012	6.31246	5.48832	34.64479
22	21.64475	0.04620	0.00727	0.15727	137.63164	6.35866	5.60102	35.61500
23	24.89146	0.04017	0.00628	0.15628	159.27638	6.39884	5.70398	36.49884
24	28.62518	0.03493	0.00543	0.15543	184.16784	6.43377	5.79789	37.30232
25	32.91895	0.03038	0.00470	0.15470	212.79302	6.46415	5.88343	38.03139
26	37.85680	0.02642	0.00407	0.15407	245.71197	6.49056	5.96123	38.69177
27	43.53531	0.02297	0.00353	0.15353	283.56877	6.51353	6.03190	39.28899
28	50.06561	0.01997	0.00306	0.15306	327.10408	6.53351	6.09600	39.82828
29	57.57545	0.01737	0.00265	0.15265	377.16969	6.55088	6.15408	40.31460
30	66.21177	0.01510	0.00230	0.15230	434.74515	6.56598	6.20663	40.75259
31	76.14354	0.01313	0.00200	0.15200	500.95692	6.57911	6.25412	41.14658
32	87.56507	0.01142	0.00173	0.15173	577.10046	6.59053	6.29700	41.50060
33	100.69983	0.00993	0.00150	0.15150	664.66552	6.60046	6.33567	41.81838
34	115.80480	0.00864	0.00131	0.15131	765.36535	6.60910	6.37051	42.10334
35	133.17552	0.00751	0.00113	0.15113	881.17016	6.61661	6.40187	42.35864
40	267.86355	0.00373	0.00056	0.15056	1779.09031	6.64178	6.51678	43.28299
45	538.76927	0.00186	0.00028	0.15028	3585.12846	6.65429	6.58299	43.80513
50	1083.65744	0.00092	0.00014	0.15014	7217.71628	6.66051	6.62048	44.09583
55	2179.62218	0.00046	0.00007	0.15007	14524.14789	6.66361	6.64142	44.25583
60	4383.99875	0.00023	0.00003	0.15003	29219.99164	6.66515	6.65298	44.34307
65	8817.78739	0.00011	0.00002	0.15002	58778.58258	6.66591	6.65929	44.39026
70	17735.72004	0.00006	0.00001	0.15001	118231.46693	6.66629	6.66272	44.41563
75	35672.86798	0.00003	0.00000	0.15000	237812.45317	6.66648	6.66456	44.42918
80	71750.87940	0.00001	0.00000	0.15000	478332.52934	6.66657	6.66555	44.43639
85	144316.64699	0.00001	0.00000	0.15000	962104.31329	6.66662	6.66608	44.44021
90	290272.32521	0.00000	0.00000	0.15000	1935142.16804	6.66664	6.66636	44.44222
95	583841.32764	0.00000	0.00000	0.15000	3892268.85091	6.66666	6.66650	44.44328
100	1174313.45070	0.00000	0.00000	0.15000	7828749.67134	6.66666	6.66658	44.44384

Interest rate 18%	Single Payment		Uniform Payment Series				Arithmetic Gradient	
	Compound Amount	Present Worth	Sinking Fund	Capital Recovery	Compound Amount	Present Worth	Gradient Uniform Series	Gradient Present Worth
n	(F/P)	(P/F)	(A/F)	(A/P)	(F/A)	(P/A)	(A/G)	(P/G)
1	1.18000	0.84746	1.00000	1.18000	1.00000	0.84746	0.00000	0.00000
2	1.39240	0.71818	0.45872	0.63872	2.18000	1.56564	0.45872	0.71818
3	1.64303	0.60863	0.27992	0.45992	3.57240	2.17427	0.89016	1.93545
4	1.93878	0.51579	0.19174	0.37174	5.21543	2.69006	1.29470	3.48281
5	2.28776	0.43711	0.13978	0.31978	7.15421	3.12717	1.67284	5.23125
6	2.69955	0.37043	0.10591	0.28591	9.44197	3.49760	2.02522	7.08341
7	3.18547	0.31393	0.08236	0.26236	12.14152	3.81153	2.35259	8.96696
8	3.75886	0.26604	0.06524	0.24524	15.32700	4.07757	2.65581	10.82922
9	4.43545	0.22546	0.05239	0.23239	19.08585	4.30302	2.93581	12.63287
10	5.23384	0.19106	0.04251	0.22251	23.52131	4.49409	3.19363	14.35245
11	6.17593	0.16192	0.03478	0.21478	28.75514	4.65601	3.43033	15.97164
12	7.28759	0.13722	0.02863	0.20863	34.93107	4.79322	3.64703	17.48106
13	8.59936	0.11629	0.02369	0.20369	42.21866	4.90951	3.84489	18.87651
14	10.14724	0.09855	0.01968	0.19968	50.81802	5.00806	4.02504	20.15765
15	11.97375	0.08352	0.01640	0.19640	60.96527	5.09158	4.18866	21.32687
16	14.12902	0.07078	0.01371	0.19371	72.93901	5.16235	4.33688	22.38852
17	16.67225	0.05998	0.01149	0.19149	87.06804	5.22233	4.47084	23.34820
18	19.67325	0.05083	0.00964	0.18964	103.74028	5.27316	4.59161	24.21231
19	23.21444	0.04308	0.00810	0.18810	123.41353	5.31624	4.70026	24.98769
20	27.39303	0.03651	0.00682	0.18682	146.62797	5.35275	4.79778	25.68130
21	32.32378	0.03094	0.00575	0.18575	174.02100	5.38368	4.88514	26.30004
22	38.14206	0.02622	0.00485	0.18485	206.34479	5.40990	4.96324	26.85061
23	45.00763	0.02222	0.00409	0.18409	244.48685	5.43212	5.03292	27.33942
24	53.10901	0.01883	0.00345	0.18345	289.49448	5.45095	5.09498	27.77249
25	62.66863	0.01596	0.00292	0.18292	342.60349	5.46691	5.15016	28.15546
26	73.94898	0.01352	0.00247	0.18247	405.27211	5.48043	5.19914	28.49353
27	87.25980	0.01146	0.00209	0.18209	479.22109	5.49189	5.24255	28.79149
28	102.96656	0.00971	0.00177	0.18177	566.48089	5.50160	5.28096	29.05371
29	121.50054	0.00823	0.00149	0.18149	669.44745	5.50983	5.31489	29.28416
30	143.37064	0.00697	0.00126	0.18126	790.94799	5.51681	5.34484	29.48643
31	169.17735	0.00591	0.00107	0.18107	934.31863	5.52272	5.37123	29.66376
32	199.62928	0.00501	0.00091	0.18091	1103.49598	5.52773	5.39445	29.81905
33	235.56255	0.00425	0.00077	0.18077	1303.12526	5.53197	5.41487	29.95490
34	277.96381	0.00360	0.00065	0.18065	1538.68781	5.53557	5.43280	30.07362
35	327.99729	0.00305	0.00055	0.18055	1816.65161	5.53862	5.44852	30.17728
40	750.37834	0.00133	0.00024	0.18024	4163.21303	5.54815	5.50218	30.52692
45	1716.68388	0.00058	0.00010	0.18010	9531.57711	5.55232	5.52933	30.70059
50	3927.35686	0.00025	0.00005	0.18005	21813.09367	5.55414	5.54282	30.78561
55	8984.84112	0.00011	0.00002	0.18002	49910.22844	5.55494	5.54943	30.82675
60	20555.13997	0.00005	0.00001	0.18001	114189.66648	5.55529	5.55264	30.84648
65	47025.18090	0.00002	0.00000	0.18000	261245.44944	5.55544	5.55417	30.85586
70	107582.22237	0.00001	0.00000	0.18000	597673.45760	5.55550	5.55490	30.86030
75	246122.06372	0.00000	0.00000	0.18000	1367339.24287	5.55553	5.55525	30.86238
80	563067.66039	0.00000	0.00000	0.18000	3128148.11325	5.55555	5.55541	30.86335
85	1288162.40765	0.00000	0.00000	0.18000	7156452.26472	5.55555	5.55549	30.86381
90	2947003.54012	0.00000	0.00000	0.18000	16372236.33400	5.55555	5.55553	30.86402
95	6742030.20823	0.00000	0.00000	0.18000	37455717.82349	5.55555	5.55554	30.86411
100	15424131.90545	0.00000	0.00000	0.18000	85689616.14141	5.55556	5.55555	30.86416

Interest rate **20%**	**Single Payment**		**Uniform Payment Series**				**Arithmetic Gradient**	
	Compound Amount	**Present Worth**	**Sinking Fund**	**Capital Recovery**	**Compound Amount**	**Present Worth**	**Gradient Uniform Series**	**Gradient Present Worth**
n	(F/P)	(P/F)	(A/F)	(A/P)	(F/A)	(P/A)	(A/G)	(P/G)
1	1.20000	0.83333	1.00000	1.20000	1.00000	0.83333	0.00000	0.00000
2	1.44000	0.69444	0.45455	0.65455	2.20000	1.52778	395.45455	0.69444
3	1.72800	0.57870	0.27473	0.47473	3.64000	2.10648	395.87912	1.85185
4	2.07360	0.48225	0.18629	0.38629	5.36800	2.58873	396.27422	3.29861
5	2.48832	0.40188	0.13438	0.33438	7.44160	2.99061	396.64051	4.90612
6	2.98598	0.33490	0.10071	0.30071	9.92992	3.32551	396.97883	6.58061
7	3.58318	0.27908	0.07742	0.27742	12.91590	3.60459	397.29016	8.25510
8	4.29982	0.23257	0.06061	0.26061	16.49908	3.83716	397.57562	9.88308
9	5.15978	0.19381	0.04808	0.24808	20.79890	4.03097	397.83642	11.43353
10	6.19174	0.16151	0.03852	0.23852	25.95868	4.19247	398.07386	12.88708
11	7.43008	0.13459	0.03110	0.23110	32.15042	4.32706	398.28929	14.23296
12	8.91610	0.11216	0.02526	0.22526	39.58050	4.43922	398.48410	15.46668
13	10.69932	0.09346	0.02062	0.22062	48.49660	4.53268	398.65970	16.58825
14	12.83918	0.07789	0.01689	0.21689	59.19592	4.61057	398.81749	17.60078
15	15.40702	0.06491	0.01388	0.21388	72.03511	4.67547	398.95884	18.50945
16	18.48843	0.05409	0.01144	0.21144	87.44213	4.72956	399.08511	19.32077
17	22.18611	0.04507	0.00944	0.20944	105.93056	4.77463	399.19759	20.04194
18	26.62333	0.03756	0.00781	0.20781	128.11667	4.81219	399.29752	20.68048
19	31.94800	0.03130	0.00646	0.20646	154.74000	4.84350	399.38607	21.24390
20	38.33760	0.02608	0.00536	0.20536	186.68800	4.86958	399.46435	21.73949
21	46.00512	0.02174	0.00444	0.20444	225.02560	4.89132	399.53339	22.17423
22	55.20614	0.01811	0.00369	0.20369	271.03072	4.90943	399.59414	22.55462
23	66.24737	0.01509	0.00307	0.20307	326.23686	4.92453	399.64750	22.88671
24	79.49685	0.01258	0.00255	0.20255	392.48424	4.93710	399.69426	23.17603
25	95.39622	0.01048	0.00212	0.20212	471.98108	4.94759	399.73516	23.42761
26	114.47546	0.00874	0.00176	0.20176	567.37730	4.95632	399.77088	23.64600
27	137.37055	0.00728	0.00147	0.20147	681.85276	4.96360	399.80201	23.83527
28	164.84466	0.00607	0.00122	0.20122	819.22331	4.96967	399.82911	23.99906
29	197.81359	0.00506	0.00102	0.20102	984.06797	4.97472	399.85265	24.14061
30	237.37631	0.00421	0.00085	0.20085	1181.88157	4.97894	399.87308	24.26277
31	284.85158	0.00351	0.00070	0.20070	1419.25788	4.98245	399.89079	24.36809
32	341.82189	0.00293	0.00059	0.20059	1704.10946	4.98537	399.90611	24.45878
33	410.18627	0.00244	0.00049	0.20049	2045.93135	4.98781	399.91935	24.53680
34	492.22352	0.00203	0.00041	0.20041	2456.11762	4.98984	399.93079	24.60384
35	590.66823	0.00169	0.00034	0.20034	2948.34115	4.99154	399.94064	24.66140
40	1469.77157	0.00068	0.00014	0.20014	7343.85784	4.99660	399.97277	24.84691
45	3657.26199	0.00027	0.00005	0.20005	18281.30994	4.99863	399.98769	24.93164
50	9100.43815	0.00011	0.00002	0.20002	45497.19075	4.99945	399.99451	24.96978
55	22644.80226	0.00004	0.00001	0.20001	113219.01129	4.99978	399.99757	24.98675
60	56347.51435	0.00002	0.00000	0.20000	281732.57177	4.99991	399.99894	24.99423
65	140210.64692	0.00001	0.00000	0.20000	701048.23458	4.99996	399.99954	24.99750
70	348888.95693	0.00000	0.00000	0.20000	1744439.78466	4.99999	399.99980	24.99893
75	868147.36931	0.00000	0.00000	0.20000	4340731.84657	4.99999	399.99991	24.99954
80	2160228.46201	0.00000	0.00000	0.20000	10801137.31005	5.00000	399.99996	24.99980
85	5375339.68659	0.00000	0.00000	0.20000	26876693.43295	5.00000	399.99998	24.99992
90	13375565.24893	0.00000	0.00000	0.20000	66877821.24467	5.00000	399.99999	24.99996
95	33282686.52023	0.00000	0.00000	0.20000	166413427.60114	5.00000	400.00000	24.99998
100	82817974.52201	0.00000	0.00000	0.20000	414089867.61007	5.00000	400.00000	24.99999

Interest rate 25%	Single Payment		Uniform Payment Series				Arithmetic Gradient	
	Compound Amount	Present Worth	Sinking Fund	Capital Recovery	Compound Amount	Present Worth	Gradient Uniform Series	Gradient Present Worth
n	(F/P)	(P/F)	(A/F)	(A/P)	(F/A)	(P/A)	(A/G)	(P/G)
1	1.25000	0.80000	1.00000	1.25000	1.00000	0.80000	0.00000	0.00000
2	1.56250	0.64000	0.44444	0.69444	2.25000	1.44000	0.44444	0.64000
3	1.95313	0.51200	0.26230	0.51230	3.81250	1.95200	0.85246	1.66400
4	2.44141	0.40960	0.17344	0.42344	5.76563	2.36160	1.22493	2.89280
5	3.05176	0.32768	0.12185	0.37185	8.20703	2.68928	1.56307	4.20352
6	3.81470	0.26214	0.08882	0.33882	11.25879	2.95142	1.86833	5.51424
7	4.76837	0.20972	0.06634	0.31634	15.07349	3.16114	2.14243	6.77253
8	5.96046	0.16777	0.05040	0.30040	19.84186	3.32891	2.38725	7.94694
9	7.45058	0.13422	0.03876	0.28876	25.80232	3.46313	2.60478	9.02068
10	9.31323	0.10737	0.03007	0.28007	33.25290	3.57050	2.79710	9.98705
11	11.64153	0.08590	0.02349	0.27349	42.56613	3.65640	2.96631	10.84604
12	14.55192	0.06872	0.01845	0.26845	54.20766	3.72512	3.11452	11.60195
13	18.18989	0.05498	0.01454	0.26454	68.75958	3.78010	3.24374	12.26166
14	22.73737	0.04398	0.01150	0.26150	86.94947	3.82408	3.35595	12.83341
15	28.42171	0.03518	0.00912	0.25912	109.68684	3.85926	3.45299	13.32599
16	35.52714	0.02815	0.00724	0.25724	138.10855	3.88741	3.53660	13.74820
17	44.40892	0.02252	0.00576	0.25576	173.63568	3.90993	3.60838	14.10849
18	55.51115	0.01801	0.00459	0.25459	218.04460	3.92794	3.66979	14.41473
19	69.38894	0.01441	0.00366	0.25366	273.55576	3.94235	3.72218	14.67414
20	86.73617	0.01153	0.00292	0.25292	342.94470	3.95388	3.76673	14.89320
21	108.42022	0.00922	0.00233	0.25233	429.68087	3.96311	3.80451	15.07766
22	135.52527	0.00738	0.00186	0.25186	538.10109	3.97049	3.83646	15.23262
23	169.40659	0.00590	0.00148	0.25148	673.62636	3.97639	3.86343	15.36248
24	211.75824	0.00472	0.00119	0.25119	843.03295	3.98111	3.88613	15.47109
25	264.69780	0.00378	0.00095	0.25095	1054.79118	3.98489	3.90519	15.56176
26	330.87225	0.00302	0.00076	0.25076	1319.48898	3.98791	3.92118	15.63732
27	413.59031	0.00242	0.00061	0.25061	1650.36123	3.99033	3.93456	15.70019
28	516.98788	0.00193	0.00048	0.25048	2063.95153	3.99226	3.94574	15.75241
29	646.23485	0.00155	0.00039	0.25039	2580.93941	3.99381	3.95506	15.79574
30	807.79357	0.00124	0.00031	0.25031	3227.17427	3.99505	3.96282	15.83164
31	1009.74196	0.00099	0.00025	0.25025	4034.96783	3.99604	3.96927	15.86135
32	1262.17745	0.00079	0.00020	0.25020	5044.70979	3.99683	3.97463	15.88591
33	1577.72181	0.00063	0.00016	0.25016	6306.88724	3.99746	3.97907	15.90619
34	1972.15226	0.00051	0.00013	0.25013	7884.60905	3.99797	3.98275	15.92293
35	2465.19033	0.00041	0.00010	0.25010	9856.76132	3.99838	3.98580	15.93672
40	7523.16385	0.00013	0.00003	0.25003	30088.65538	3.99947	3.99468	15.97661
45	22958.87404	0.00004	0.00001	0.25001	91831.49616	3.99983	3.99804	15.99146
50	70064.92322	0.00001	0.00000	0.25000	280255.69286	3.99994	3.99929	15.99692
55	213821.17681	0.00000	0.00000	0.25000	855280.70723	3.99998	3.99974	15.99890
60	652530.44680	0.00000	0.00000	0.25000	2610117.78720	3.99999	3.99991	15.99961
65	1991364.88892	0.00000	0.00000	0.25000	7965455.55566	4.00000	3.99997	15.99986
70	6077163.35729	0.00000	0.00000	0.25000	24308649.42915	4.00000	3.99999	15.99995
75	18546030.75344	0.00000	0.00000	0.25000	74184119.01375	4.00000	4.00000	15.99998
80	56597994.24267	0.00000	0.00000	0.25000	226391972.97067	4.00000	4.00000	15.99999
85	172723371.10189	0.00000	0.00000	0.25000	690893480.40756	4.00000	4.00000	16.00000
90	527109897.16153	0.00000	0.00000	0.25000	2108439584.64611	4.00000	4.00000	16.00000
95	1608611746.70876	0.00000	0.00000	0.25000	6434446982.83504	4.00000	4.00000	16.00000
100	4909093465.29773	0.00000	0.00000	0.25000	19636373857.19090	4.00000	4.00000	16.00000

Interest rate **30%**	**Single Payment**		**Uniform Payment Series**				**Arithmetic Gradient**	
	Compound Amount	**Present Worth**	**Sinking Fund**	**Capital Recovery**	**Compound Amount**	**Present Worth**	**Gradient Uniform Series**	**Gradient Present Worth**
n	**(F/P)**	**(P/F)**	**(A/F)**	**(A/P)**	**(F/A)**	**(P/A)**	**(A/G)**	**(P/G)**
1	1.30000	0.76923	1.00000	1.30000	1.00000	0.76923	0.00000	0.00000
2	1.69000	0.59172	0.43478	0.73478	2.30000	1.36095	0.43478	0.59172
3	2.19700	0.45517	0.25063	0.55063	3.99000	1.81611	0.82707	1.50205
4	2.85610	0.35013	0.16163	0.46163	6.18700	2.16624	1.17828	2.55243
5	3.71293	0.26933	0.11058	0.41058	9.04310	2.43557	1.49031	3.62975
6	4.82681	0.20718	0.07839	0.37839	12.75603	2.64275	1.76545	4.66563
7	6.27485	0.15937	0.05687	0.35687	17.58284	2.80211	2.00628	5.62183
8	8.15731	0.12259	0.04192	0.34192	23.85769	2.92470	2.21559	6.47995
9	10.60450	0.09430	0.03124	0.33124	32.01500	3.01900	2.39627	7.23435
10	13.78585	0.07254	0.02346	0.32346	42.61950	3.09154	2.55122	7.88719
11	17.92160	0.05580	0.01773	0.31773	56.40535	3.14734	2.68328	8.44518
12	23.29809	0.04292	0.01345	0.31345	74.32695	3.19026	2.79517	8.91732
13	30.28751	0.03302	0.01024	0.31024	97.62504	3.22328	2.88946	9.31352
14	39.37376	0.02540	0.00782	0.30782	127.91255	3.24867	2.96850	9.64369
15	51.18589	0.01954	0.00598	0.30598	167.28631	3.26821	3.03444	9.91721
16	66.54166	0.01503	0.00458	0.30458	218.47220	3.28324	3.08921	10.14263
17	86.50416	0.01156	0.00351	0.30351	285.01386	3.29480	3.13451	10.32759
18	112.45541	0.00889	0.00269	0.30269	371.51802	3.30369	3.17183	10.47876
19	146.19203	0.00684	0.00207	0.30207	483.97343	3.31053	3.20247	10.60189
20	190.04964	0.00526	0.00159	0.30159	630.16546	3.31579	3.22754	10.70186
21	247.06453	0.00405	0.00122	0.30122	820.21510	3.31984	3.24799	10.78281
22	321.18389	0.00311	0.00094	0.30094	1067.27963	3.32296	3.26462	10.84819
23	417.53905	0.00239	0.00072	0.30072	1388.46351	3.32535	3.27812	10.90088
24	542.80077	0.00184	0.00055	0.30055	1806.00257	3.32719	3.28904	10.94326
25	705.64100	0.00142	0.00043	0.30043	2348.80334	3.32861	3.29785	10.97727
26	917.33330	0.00109	0.00033	0.30033	3054.44434	3.32970	3.30496	11.00452
27	1192.53329	0.00084	0.00025	0.30025	3971.77764	3.33054	3.31067	11.02632
28	1550.29328	0.00065	0.00019	0.30019	5164.31093	3.33118	3.31526	11.04374
29	2015.38126	0.00050	0.00015	0.30015	6714.60421	3.33168	3.31894	11.05763
30	2619.99564	0.00038	0.00011	0.30011	8729.98548	3.33206	3.32188	11.06870
31	3405.99434	0.00029	0.00009	0.30009	11349.98112	3.33235	3.32423	11.07751
32	4427.79264	0.00023	0.00007	0.30007	14755.97546	3.33258	3.32610	11.08451
33	5756.13043	0.00017	0.00005	0.30005	19183.76810	3.33275	3.32760	11.09007
34	7482.96956	0.00013	0.00004	0.30004	24939.89853	3.33289	3.32879	11.09448
35	9727.86043	0.00010	0.00003	0.30003	32422.86808	3.33299	3.32974	11.09798
40	36118.86481	0.00003	0.00001	0.30001	120392.88269	3.33324	3.33223	11.10711
45	134106.81671	0.00001	0.00000	0.30000	447019.38904	3.33331	3.33300	11.10991
50	497929.22298	0.00000	0.00000	0.30000	1659760.74326	3.33333	3.33323	11.11075
55	1848776.34988	0.00000	0.00000	0.30000	6162584.49959	3.33333	3.33330	11.11101
60	6864377.17274	0.00000	0.00000	0.30000	22881253.90915	3.33333	3.33332	11.11108
65	25486951.93600	0.00000	0.00000	0.30000	84956503.12000	3.33333	3.33333	11.11110
70	94631268.45173	0.00000	0.00000	0.30000	315437558.17243	3.33333	3.33333	11.11111
75	351359275.57248	0.00000	0.00000	0.30000	1171197581.90826	3.33333	3.33333	11.11111
80	1304572395.05132	0.00000	0.00000	0.30000	4348574646.83773	3.33333	3.33333	11.11111
85	4843785982.75790	0.00000	0.00000	0.30000	16145953272.52630	3.33333	3.33333	11.11111
90	17984638288.96130	0.00000	0.00000	0.30000	59948794293.20430	3.33333	3.33333	11.11111
95	66775703042.23300	0.00000	0.00000	0.30000	222585676804.11000	3.33333	3.33333	11.11111
100	247933511096.59800	0.00000	0.00000	0.30000	826445036985.32800	3.33333	3.33333	11.11111

Interest rate	Single Payment		Uniform Payment Series				Arithmetic Gradient	
35%	Compound Amount	Present Worth	Sinking Fund	Capital Recovery	Compound Amount	Present Worth	Gradient Uniform Series	Gradient Present Worth
n	(F/P)	(P/F)	(A/F)	(A/P)	(F/A)	(P/A)	(A/G)	(P/G)
1	1.35000	0.74074	1.00000	1.35000	1.00000	0.74074	0.00000	0.00000
2	1.82250	0.54870	0.42553	0.77553	2.35000	1.28944	0.42553	0.54870
3	2.46038	0.40644	0.23966	0.58966	4.17250	1.69588	0.80288	1.36158
4	3.32151	0.30107	0.15076	0.50076	6.63288	1.99695	1.13412	2.26479
5	4.48403	0.22301	0.10046	0.45046	9.95438	2.21996	1.42202	3.15684
6	6.05345	0.16520	0.06926	0.41926	14.43841	2.38516	1.66983	3.98282
7	8.17215	0.12237	0.04880	0.39880	20.49186	2.50752	1.88115	4.71702
8	11.03240	0.09064	0.03489	0.38489	28.66401	2.59817	2.05973	5.35151
9	14.89375	0.06714	0.02519	0.37519	39.69641	2.66531	2.20937	5.88865
10	20.10656	0.04974	0.01832	0.36832	54.59016	2.71504	2.33376	6.33626
11	27.14385	0.03684	0.01339	0.36339	74.69672	2.75188	2.43639	6.70467
12	36.64420	0.02729	0.00982	0.35982	101.84057	2.77917	2.52048	7.00486
13	49.46967	0.02021	0.00722	0.35722	138.48476	2.79939	2.58893	7.24743
14	66.78405	0.01497	0.00532	0.35532	187.95443	2.81436	2.64433	7.44209
15	90.15847	0.01109	0.00393	0.35393	254.73848	2.82545	2.68890	7.59737
16	121.71393	0.00822	0.00290	0.35290	344.89695	2.83367	2.72460	7.72061
17	164.31381	0.00609	0.00214	0.35214	466.61088	2.83975	2.75305	7.81798
18	221.82364	0.00451	0.00158	0.35158	630.92469	2.84426	2.77563	7.89462
19	299.46192	0.00334	0.00117	0.35117	852.74834	2.84760	2.79348	7.95473
20	404.27359	0.00247	0.00087	0.35087	1152.21025	2.85008	2.80755	8.00173
21	545.76935	0.00183	0.00064	0.35064	1556.48384	2.85191	2.81859	8.03837
22	736.78862	0.00136	0.00048	0.35048	2102.25319	2.85327	2.82724	8.06687
23	994.66463	0.00101	0.00035	0.35035	2839.04180	2.85427	2.83400	8.08899
24	1342.79725	0.00074	0.00026	0.35026	3833.70643	2.85502	2.83926	8.10612
25	1812.77629	0.00055	0.00019	0.35019	5176.50369	2.85557	2.84334	8.11936
26	2447.24799	0.00041	0.00014	0.35014	6989.27998	2.85598	2.84651	8.12957
27	3303.78479	0.00030	0.00011	0.35011	9436.52797	2.85628	2.84897	8.13744
28	4460.10947	0.00022	0.00008	0.35008	12740.31276	2.85650	2.85086	8.14350
29	6021.14778	0.00017	0.00006	0.35006	17200.42223	2.85667	2.85233	8.14815
30	8128.54950	0.00012	0.00004	0.35004	23221.57000	2.85679	2.85345	8.15172
31	10973.54183	0.00009	0.00003	0.35003	31350.11951	2.85688	2.85432	8.15445
32	14814.28147	0.00007	0.00002	0.35002	42323.66133	2.85695	2.85498	8.15654
33	19999.27998	0.00005	0.00002	0.35002	57137.94280	2.85700	2.85549	8.15814
34	26999.02797	0.00004	0.00001	0.35001	77137.22278	2.85704	2.85588	8.15936
35	36448.68776	0.00003	0.00001	0.35001	104136.25075	2.85706	2.85618	8.16030
40	163437.13468	0.00001	0.00000	0.35000	466960.38480	2.85713	2.85690	8.16252
45	732857.57684	0.00000	0.00000	0.35000	2093875.93382	2.85714	2.85708	8.16308
50	3286157.87946	0.00000	0.00000	0.35000	9389019.65559	2.85714	2.85713	8.16322
55	14735241.81239	0.00000	0.00000	0.35000	42100688.03540	2.85714	2.85714	8.16325
60	66073316.99641	0.00000	0.00000	0.35000	188780902.84689	2.85714	2.85714	8.16326
65	296274962.73844	0.00000	0.00000	0.35000	846499890.68126	2.85714	2.85714	8.16326
70	1328506839.61324	0.00000	0.00000	0.35000	3795733824.60925	2.85714	2.85714	8.16327
75	5957069090.77322	0.00000	0.00000	0.35000	17020197399.35200	2.85714	2.85714	8.16327
80	26711696992.52480	0.00000	0.00000	0.35000	76319134261.49950	2.85714	2.85714	8.16327
85	119776142486.85000	0.00000	0.00000	0.35000	342217549959.57000	2.85714	2.85714	8.16327
90	537080227925.79800	0.00000	0.00000	0.35000	1534514936927.99000	2.85714	2.85714	8.16327
95	2408285700639.40000	0.00000	0.00000	0.35000	6880816287538.29000	2.85714	2.85714	8.16327
100	10798833608720.20000	0.00000	0.00000	0.35000	30853810310626.30000	2.85714	2.85714	8.16327

Interest rate 40%	Single Payment		Uniform Payment Series				Arithmetic Gradient	
	Compound Amount	Present Worth	Sinking Fund	Capital Recovery	Compound Amount	Present Worth	Gradient Uniform Series	Gradient Present Worth
n	(F/P)	(P/F)	(A/F)	(A/P)	(F/A)	(P/A)	(A/G)	(P/G)
1	1.40000	0.71429	1.00000	1.40000	1.00000	0.71429	0.00000	0.00000
2	1.96000	0.51020	0.41667	0.81667	2.40000	1.22449	0.41667	0.51020
3	2.74400	0.36443	0.22936	0.62936	4.36000	1.58892	0.77982	1.23907
4	3.84160	0.26031	0.14077	0.54077	7.10400	1.84923	1.09234	2.01999
5	5.37824	0.18593	0.09136	0.49136	10.94560	2.03516	1.35799	2.76373
6	7.52954	0.13281	0.06126	0.46126	16.32384	2.16797	1.58110	3.42778
7	10.54135	0.09486	0.04192	0.44192	23.85338	2.26284	1.76635	3.99697
8	14.75789	0.06776	0.02907	0.42907	34.39473	2.33060	1.91852	4.47129
9	20.66105	0.04840	0.02034	0.42034	49.15262	2.37900	2.04224	4.85849
10	28.92547	0.03457	0.01432	0.41432	69.81366	2.41357	2.14190	5.16964
11	40.49565	0.02469	0.01013	0.41013	98.73913	2.43826	2.22149	5.41658
12	56.69391	0.01764	0.00718	0.40718	139.23478	2.45590	2.28454	5.61060
13	79.37148	0.01260	0.00510	0.40510	195.92869	2.46850	2.33412	5.76179
14	111.12007	0.00900	0.00363	0.40363	275.30017	2.47750	2.37287	5.87878
15	155.56810	0.00643	0.00259	0.40259	386.42024	2.48393	2.40296	5.96877
16	217.79533	0.00459	0.00185	0.40185	541.98833	2.48852	2.42620	6.03764
17	304.91347	0.00328	0.00132	0.40132	759.78367	2.49180	2.44406	6.09012
18	426.87885	0.00234	0.00094	0.40094	1064.69714	2.49414	2.45773	6.12994
19	597.63040	0.00167	0.00067	0.40067	1491.57599	2.49582	2.46815	6.16006
20	836.68255	0.00120	0.00048	0.40048	2089.20639	2.49701	2.47607	6.18277
21	1171.35558	0.00085	0.00034	0.40034	2925.88894	2.49787	2.48206	6.19984
22	1639.89781	0.00061	0.00024	0.40024	4097.24452	2.49848	2.48658	6.21265
23	2295.85693	0.00044	0.00017	0.40017	5737.14232	2.49891	2.48998	6.22223
24	3214.19970	0.00031	0.00012	0.40012	8032.99925	2.49922	2.49253	6.22939
25	4499.87958	0.00022	0.00009	0.40009	11247.19895	2.49944	2.49444	6.23472
26	6299.83141	0.00016	0.00006	0.40006	15747.07853	2.49960	2.49587	6.23869
27	8819.76398	0.00011	0.00005	0.40005	22046.90994	2.49972	2.49694	6.24164
28	12347.66957	0.00008	0.00003	0.40003	30866.67392	2.49980	2.49773	6.24382
29	17286.73740	0.00006	0.00002	0.40002	43214.34349	2.49986	2.49832	6.24544
30	24201.43236	0.00004	0.00002	0.40002	60501.08089	2.49990	2.49876	6.24664
31	33882.00530	0.00003	0.00001	0.40001	84702.51324	2.49993	2.49909	6.24753
32	47434.80742	0.00002	0.00001	0.40001	118584.51854	2.49995	2.49933	6.24818
33	66408.73038	0.00002	0.00001	0.40001	166019.32596	2.49996	2.49950	6.24866
34	92972.22254	0.00001	0.00000	0.40000	232428.05634	2.49997	2.49963	6.24902
35	130161.11155	0.00001	0.00000	0.40000	325400.27888	2.49998	2.49973	6.24928
40	700037.69659	0.00000	0.00000	0.40000	1750091.74148	2.50000	2.49994	6.24985
45	3764970.74131	0.00000	0.00000	0.40000	9412424.35328	2.50000	2.49999	6.24997
50	20248916.23976	0.00000	0.00000	0.40000	50622288.09941	2.50000	2.50000	6.24999
55	108903531.27735	0.00000	0.00000	0.40000	272258825.69338	2.50000	2.50000	6.25000
60	585709328.05709	0.00000	0.00000	0.40000	1464273317.64274	2.50000	2.50000	6.25000
65	3150085336.52978	0.00000	0.00000	0.40000	7875213338.82446	2.50000	2.50000	6.25000
70	16941914960.33790	0.00000	0.00000	0.40000	42354787398.34490	2.50000	2.50000	6.25000
75	91117684716.28790	0.00000	0.00000	0.40000	227794211788.22000	2.50000	2.50000	6.25000
80	490052776648.52800	0.00000	0.00000	0.40000	1225131941618.82000	2.50000	2.50000	6.25000
85	2635621445482.18000	0.00000	0.00000	0.40000	6589053613702.95000	2.50000	2.50000	6.25000
90	14175004682950.10000	0.00000	0.00000	0.40000	35437511707372.70000	2.50000	2.50000	6.25000
95	76236577186029.30000	0.00000	0.00000	0.40000	190591442965071.00000	2.50000	2.50000	6.25000
100	410018608884990.00000	0.00000	0.00000	0.40000	1025046522212470.00000	2.50000	2.50000	6.25000

Interest rate 45%	Single Payment		Uniform Payment Series				Arithmetic Gradient	
	Compound Amount	Present Worth	Sinking Fund	Capital Recovery	Compound Amount	Present Worth	Gradient Uniform Series	Gradient Present Worth
n	(F/P)	(P/F)	(A/F)	(A/P)	(F/A)	(P/A)	(A/G)	(P/G)
1	1.45000	0.68966	1.00000	1.45000	1.00000	0.68966	0.00000	0.00000
2	2.10250	0.47562	0.40816	0.85816	2.45000	1.16528	0.40816	0.47562
3	3.04863	0.32802	0.21966	0.66966	4.55250	1.49330	0.75783	1.13166
4	4.42051	0.22622	0.13156	0.58156	7.60113	1.71951	1.05280	1.81031
5	6.40973	0.15601	0.08318	0.53318	12.02163	1.87553	1.29796	2.43436
6	9.29411	0.10759	0.05426	0.50426	18.43137	1.98312	1.49882	2.97234
7	13.47647	0.07420	0.03607	0.48607	27.72548	2.05733	1.66117	3.41756
8	19.54088	0.05117	0.02427	0.47427	41.20195	2.10850	1.79074	3.77578
9	28.33427	0.03529	0.01646	0.46646	60.74282	2.14379	1.89297	4.05813
10	41.08469	0.02434	0.01123	0.46123	89.07709	2.16813	1.97275	4.27719
11	59.57280	0.01679	0.00768	0.45768	130.16178	2.18492	2.03442	4.44505
12	86.38056	0.01158	0.00527	0.45527	189.73458	2.19650	2.08168	4.57239
13	125.25182	0.00798	0.00362	0.45362	276.11515	2.20448	2.11760	4.66820
14	181.61513	0.00551	0.00249	0.45249	401.36696	2.20999	2.14471	4.73978
15	263.34194	0.00380	0.00172	0.45172	582.98209	2.21378	2.16504	4.79294
16	381.84582	0.00262	0.00118	0.45118	846.32403	2.21640	2.18021	4.83222
17	553.67643	0.00181	0.00081	0.45081	1228.16985	2.21821	2.19146	4.86112
18	802.83083	0.00125	0.00056	0.45056	1781.84628	2.21945	2.19977	4.88230
19	1164.10470	0.00086	0.00039	0.45039	2584.67711	2.22031	2.20589	4.89776
20	1687.95181	0.00059	0.00027	0.45027	3748.78181	2.22091	2.21037	4.90902
21	2447.53013	0.00041	0.00018	0.45018	5436.73362	2.22131	2.21364	4.91719
22	3548.91869	0.00028	0.00013	0.45013	7884.26375	2.22160	2.21602	4.92310
23	5145.93210	0.00019	0.00009	0.45009	11433.18244	2.22179	2.21775	4.92738
24	7461.60154	0.00013	0.00006	0.45006	16579.11454	2.22192	2.21901	4.93046
25	10819.32224	0.00009	0.00004	0.45004	24040.71609	2.22202	2.21991	4.93268
26	15688.01725	0.00006	0.00003	0.45003	34860.03833	2.22208	2.22056	4.93427
27	22747.62501	0.00004	0.00002	0.45002	50548.05557	2.22212	2.22104	4.93542
28	32984.05626	0.00003	0.00001	0.45001	73295.68058	2.22215	2.22137	4.93624
29	47826.88158	0.00002	0.00001	0.45001	106279.73684	2.22218	2.22162	4.93682
30	69348.97829	0.00001	0.00001	0.45001	154106.61842	2.22219	2.22179	4.93724
31	100556.01852	0.00001	0.00000	0.45000	223455.59671	2.22220	2.22191	4.93754
32	145806.22685	0.00001	0.00000	0.45000	324011.61523	2.22221	2.22200	4.93775
33	211419.02894	0.00000	0.00000	0.45000	469817.84209	2.22221	2.22207	4.93790
34	306557.59196	0.00000	0.00000	0.45000	681236.87103	2.22221	2.22211	4.93801
35	444508.50835	0.00000	0.00000	0.45000	987794.46299	2.22222	2.22214	4.93809
40	2849181.32701	0.00000	0.00000	0.45000	6331511.83780	2.22222	2.22221	4.93824
45	18262494.60198	0.00000	0.00000	0.45000	40583319.11552	2.22222	2.22222	4.93827
50	117057733.71655	0.00000	0.00000	0.45000	260128294.92567	2.22222	2.22222	4.93827
55	750308943.08204	0.00000	0.00000	0.45000	1667353204.62675	2.22222	2.22222	4.93827
60	4809280789.87130	0.00000	0.00000	0.45000	10687290641.93620	2.22222	2.22222	4.93827
65	30826210894.96500	0.00000	0.00000	0.45000	68502690875.47770	2.22222	2.22222	4.93827
70	197587813991.26600	0.00000	0.00000	0.45000	439084031089.47900	2.22222	2.22222	4.93827
75	1266485341674.73000	0.00000	0.00000	0.45000	2814411870386.06000	2.22222	2.22222	4.93827
80	8117834234189.46000	0.00000	0.00000	0.45000	18039631631529.90000	2.22222	2.22222	4.93827
85	52033158604612.80000	0.00000	0.00000	0.45000	115629241343582.00000	2.22222	2.22222	4.93827
90	333518709087452.00000	0.00000	0.00000	0.45000	741152686861001.00000	2.22222	2.22222	4.93827
95	2137766230118870.00000	0.00000	0.00000	0.45000	4750591622486370.00000	2.22222	2.22222	4.93827
100	13702513022855100.00000	0.00000	0.00000	0.45000	30450028939678000.00000	2.22222	2.22222	4.93827

Interest rate 50%	Single Payment		Uniform Payment Series				Arithmetic Gradient	
	Compound Amount	Present Worth	Sinking Fund	Capital Recovery	Compound Amount	Present Worth	Gradient Uniform Series	Gradient Present Worth
n	(F/P)	(P/F)	(A/F)	(A/P)	(F/A)	(P/A)	(A/G)	(P/G)
1	1.50000	0.66667	1.00000	1.50000	1.00000	0.66667	0.00000	0.00000
2	2.25000	0.44444	0.40000	0.90000	2.50000	1.11111	0.40000	0.44444
3	3.37500	0.29630	0.21053	0.71053	4.75000	1.40741	0.73684	1.03704
4	5.06250	0.19753	0.12308	0.62308	8.12500	1.60494	1.01538	1.62963
5	7.59375	0.13169	0.07583	0.57583	13.18750	1.73663	1.24171	2.15638
6	11.39063	0.08779	0.04812	0.54812	20.78125	1.82442	1.42256	2.59534
7	17.08594	0.05853	0.03108	0.53108	32.17188	1.88294	1.56484	2.94650
8	25.62891	0.03902	0.02030	0.52030	49.25781	1.92196	1.67518	3.21963
9	38.44336	0.02601	0.01335	0.51335	74.88672	1.94798	1.75964	3.42773
10	57.66504	0.01734	0.00882	0.50882	113.33008	1.96532	1.82352	3.58380
11	86.49756	0.01156	0.00585	0.50585	170.99512	1.97688	1.87134	3.69941
12	129.74634	0.00771	0.00388	0.50388	257.49268	1.98459	1.90679	3.78419
13	194.61951	0.00514	0.00258	0.50258	387.23901	1.98972	1.93286	3.84585
14	291.92926	0.00343	0.00172	0.50172	581.85852	1.99315	1.95188	3.89038
15	437.89389	0.00228	0.00114	0.50114	873.78778	1.99543	1.96567	3.92236
16	656.84084	0.00152	0.00076	0.50076	1311.68167	1.99696	1.97560	3.94519
17	985.26125	0.00101	0.00051	0.50051	1968.52251	1.99797	1.98273	3.96143
18	1477.89188	0.00068	0.00034	0.50034	2953.78376	1.99865	1.98781	3.97293
19	2216.83782	0.00045	0.00023	0.50023	4431.67564	1.99910	1.99143	3.98105
20	3325.25673	0.00030	0.00015	0.50015	6648.51346	1.99940	1.99398	3.98677
21	4987.88510	0.00020	0.00010	0.50010	9973.77019	1.99960	1.99579	3.99078
22	7481.82764	0.00013	0.00007	0.50007	14961.65529	1.99973	1.99706	3.99358
23	11222.74146	0.00009	0.00004	0.50004	22443.48293	1.99982	1.99795	3.99554
24	16834.11220	0.00006	0.00003	0.50003	33666.22439	1.99988	1.99857	3.99691
25	25251.16829	0.00004	0.00002	0.50002	50500.33659	1.99992	1.99901	3.99786
26	37876.75244	0.00003	0.00001	0.50001	75751.50488	1.99995	1.99931	3.99852
27	56815.12866	0.00002	0.00001	0.50001	113628.25732	1.99996	1.99952	3.99898
28	85222.69299	0.00001	0.00001	0.50001	170443.38598	1.99998	1.99967	3.99930
29	127834.03949	0.00001	0.00000	0.50000	255666.07898	1.99998	1.99977	3.99951
30	191751.05923	0.00001	0.00000	0.50000	383500.11847	1.99999	1.99984	3.99967
31	287626.58885	0.00000	0.00000	0.50000	575251.17770	1.99999	1.99989	3.99977
32	431439.88327	0.00000	0.00000	0.50000	862877.76655	2.00000	1.99993	3.99984
33	647159.82491	0.00000	0.00000	0.50000	1294317.64982	2.00000	1.99995	3.99989
34	970739.73737	0.00000	0.00000	0.50000	1941477.47473	2.00000	1.99996	3.99993
35	1456109.60605	0.00000	0.00000	0.50000	2912217.21210	2.00000	1.99998	3.99995
40	11057332.32094	0.00000	0.00000	0.50000	22114662.64188	2.00000	2.00000	3.99999
45	83966617.31214	0.00000	0.00000	0.50000	167933232.62428	2.00000	2.00000	4.00000
50	637621500.21405	0.00000	0.00000	0.50000	1275242998.42810	2.00000	2.00000	4.00000
55	4841938267.25044	0.00000	0.00000	0.50000	9683876532.50088	2.00000	2.00000	4.00000
60	36768468716.93300	0.00000	0.00000	0.50000	73536937431.86600	2.00000	2.00000	4.00000
65	279210559319.21000	0.00000	0.00000	0.50000	558421118636.42000	2.00000	2.00000	4.00000
70	2120255184830.25000	0.00000	0.00000	0.50000	4240510369658.50000	2.00000	2.00000	4.00000
75	16100687809804.70000	0.00000	0.00000	0.50000	32201375619607.40000	2.00000	2.00000	4.00000
80	122264598055705.00000	0.00000	0.00000	0.50000	244529196111407.00000	2.00000	2.00000	4.00000
85	928446791485507.00000	0.00000	0.00000	0.50000	1856893582971010.00000	2.00000	2.00000	4.00000
90	7050392822843070.00000	0.00000	0.00000	0.50000	14100785645686100.00000	2.00000	2.00000	4.00000
95	53538920498464600.00000	0.00000	0.00000	0.50000	107077840996929000.00000	2.00000	2.00000	4.00000
100	406561177535215000.00000	0.00000	0.00000	0.50000	813122355070431000.00000	2.00000	2.00000	4.00000

Interest rate 60%	Single Payment		Uniform Payment Series				Arithmetic Gradient	
	Compound Amount	Present Worth	Sinking Fund	Capital Recovery	Compound Amount	Present Worth	Gradient Uniform Series	Gradient Present Worth
n	(F/P)	(P/F)	(A/F)	(A/P)	(F/A)	(P/A)	(A/G)	(P/G)
1	1.60000	0.62500	1.00000	1.60000	1.00000	0.62500	0.00000	0.00000
2	2.56000	0.39063	0.38462	0.98462	2.60000	1.01563	0.38462	0.39063
3	4.09600	0.24414	0.19380	0.79380	5.16000	1.25977	0.69767	0.87891
4	6.55360	0.15259	0.10804	0.70804	9.25600	1.41235	0.94641	1.33667
5	10.48576	0.09537	0.06325	0.66325	15.80960	1.50772	1.13956	1.71814
6	16.77722	0.05960	0.03803	0.63803	26.29536	1.56733	1.28637	2.01616
7	26.84355	0.03725	0.02322	0.62322	43.07258	1.60458	1.39581	2.23968
8	42.94967	0.02328	0.01430	0.61430	69.91612	1.62786	1.47596	2.40266
9	68.71948	0.01455	0.00886	0.60886	112.86579	1.64241	1.53377	2.51908
10	109.95116	0.00909	0.00551	0.60551	181.58527	1.65151	1.57488	2.60093
11	175.92186	0.00568	0.00343	0.60343	291.53643	1.65719	1.60378	2.65778
12	281.47498	0.00355	0.00214	0.60214	467.45829	1.66075	1.62388	2.69685
13	450.35996	0.00222	0.00134	0.60134	748.93327	1.66297	1.63774	2.72350
14	720.57594	0.00139	0.00083	0.60083	1199.29323	1.66435	1.64721	2.74154
15	1152.92150	0.00087	0.00052	0.60052	1919.86917	1.66522	1.65364	2.75368
16	1844.67441	0.00054	0.00033	0.60033	3072.79068	1.66576	1.65799	2.76182
17	2951.47905	0.00034	0.00020	0.60020	4917.46509	1.66610	1.66090	2.76724
18	4722.36648	0.00021	0.00013	0.60013	7868.94414	1.66631	1.66285	2.77084
19	7555.78637	0.00013	0.00008	0.60008	12591.31062	1.66645	1.66415	2.77322
20	12089.25820	0.00008	0.00005	0.60005	20147.09699	1.66653	1.66501	2.77479
21	19342.81311	0.00005	0.00003	0.60003	32236.35519	1.66658	1.66558	2.77582
22	30948.50098	0.00003	0.00002	0.60002	51579.16830	1.66661	1.66596	2.77650
23	49517.60157	0.00002	0.00001	0.60001	82527.66929	1.66663	1.66620	2.77695
24	79228.16251	0.00001	0.00001	0.60001	132045.27086	1.66665	1.66636	2.77724
25	126765.06002	0.00001	0.00000	0.60000	211273.43337	1.66665	1.66647	2.77743
26	202824.09604	0.00000	0.00000	0.60000	338038.49339	1.66666	1.66654	2.77755
27	324518.55366	0.00000	0.00000	0.60000	540862.58943	1.66666	1.66658	2.77763
28	519229.68585	0.00000	0.00000	0.60000	865381.14309	1.66666	1.66661	2.77768
29	830767.49737	0.00000	0.00000	0.60000	1384610.82894	1.66666	1.66663	2.77772
30	1329227.99578	0.00000	0.00000	0.60000	2215378.32631	1.66667	1.66664	2.77774

rn	Compound Amount Factor F/P	Present Worth Factor P/F
0.01	1.0100502	0.9900498
0.02	1.0202013	0.9801987
0.03	1.0304545	0.9704455
0.04	1.0408108	0.9607894
0.05	1.0512711	0.9512294
0.06	1.0618365	0.9417645
0.07	1.0725082	0.9323938
0.08	1.0832871	0.9231163
0.09	1.0941743	0.9139312
0.1	1.1051709	0.9048374
0.11	1.1162781	0.8958341
0.12	1.1274969	0.8869204
0.13	1.1388284	0.8780954
0.14	1.1502738	0.8693582
0.15	1.1618342	0.860708
0.16	1.1735109	0.8521438
0.17	1.1853049	0.8436648
0.18	1.1972174	0.8352702
0.19	1.2092496	0.8269591
0.2	1.2214028	0.8187308
0.21	1.2336781	0.8105842
0.22	1.2460767	0.8025188
0.23	1.2586	0.7945336
0.24	1.2712492	0.7866279
0.25	1.2840254	0.7788008
0.26	1.2969301	0.7710516
0.27	1.3099645	0.7633795
0.28	1.3231298	0.7557837
0.29	1.3364275	0.7482636
0.3	1.3498588	0.7408182
0.31	1.3634251	0.733447
0.32	1.3771278	0.726149
0.33	1.3909681	0.7189237
0.34	1.4049476	0.7117703
0.35	1.4190675	0.7046881
0.36	1.4333294	0.6976763
0.37	1.4477346	0.6907343
0.38	1.4622846	0.6838614
0.39	1.4769808	0.6770569
0.4	1.4918247	0.67032
0.41	1.5068178	0.6636503
0.42	1.5219616	0.6570468
0.43	1.5372575	0.6505091
0.44	1.5527072	0.6440364
0.45	1.5683122	0.6376282
0.46	1.584074	0.6312836
0.47	1.5999942	0.6250023
0.48	1.6160744	0.6187834
0.49	1.6323162	0.6126264
0.5	1.6487213	0.6065307

rn	Compound Amount Factor F/P	Present Worth Factor P/F
0.51	1.6652912	0.6004956
0.52	1.6820276	0.5945205
0.53	1.6989323	0.588605
0.54	1.7160069	0.5827483
0.55	1.733253	0.5769498
0.56	1.7506725	0.5712091
0.57	1.7682671	0.5655254
0.58	1.7860384	0.5598984
0.59	1.8039884	0.5543273
0.6	1.8221188	0.5488116
0.61	1.8404314	0.5433509
0.62	1.858928	0.5379444
0.63	1.8776106	0.5325918
0.64	1.8964809	0.5272924
0.65	1.9155408	0.5220458
0.66	1.9347923	0.5168513
0.67	1.9542373	0.5117086
0.68	1.9738777	0.506617
0.69	1.9937155	0.5015761
0.7	2.0137527	0.4965853
0.71	2.0339913	0.4916442
0.72	2.0544332	0.4867523
0.73	2.0750806	0.481909
0.74	2.0959355	0.4771139
0.75	2.117	0.4723666
0.76	2.1382762	0.4676664
0.77	2.1597663	0.4630131
0.78	2.1814723	0.458406
0.79	2.2033964	0.4538448
0.8	2.2255409	0.449329
0.81	2.247908	0.4448581
0.82	2.2704998	0.4404317
0.83	2.2933187	0.4360493
0.84	2.316367	0.4317105
0.85	2.3396469	0.4274149
0.86	2.3631607	0.4231621
0.87	2.3869109	0.4189515
0.88	2.4108997	0.4147829
0.89	2.4351297	0.4106558
0.9	2.4596031	0.4065697
0.91	2.4843225	0.4025242
0.92	2.5092904	0.398519
0.93	2.5345092	0.3945537
0.94	2.5599814	0.3906278
0.95	2.5857097	0.386741
0.96	2.6116965	0.3828929
0.97	2.6379445	0.379083
0.98	2.6644562	0.3753111
0.99	2.6912345	0.3715767
1	2.7182818	0.3678794

References

1. Adithan, M and Pabla, B. 1989. *Production Engineering, Estimating and Costing,* Konark Publishers, New Delhi, pp. 149–177.
2. Allen, J. 2006. An Investigation into the Comparative Costs of Additive Manufacture vs. Machine from Solid for Aero Engine parts, DTIC Document.
3. Boothroyd, G and Knight, W. 1989. *Fundamentals of Machining and Machine Tools,* Marcel Dekker, pp. 145–204; 399–465.
4. Cary, H. 1989. *Modern Welding Technology,* Prentice Hall, Upper Saddle River, NJ p. 616.
5. Creese, R, Adithan, M and Pabla, B. 1992. *Estimating and Costing for the Metal Manufacturing Industries,* Marcel Dekker Inc., New York.
6. *The Procedure Handbook of Arc Welding,* 1973. The Lincoln Electric Company, Cleveland, Ohio.
7. Hopkinson, N and Dickens, P. 2003. Analysis of Rapid Manufacturing-using layer manufacturing processes for production, *Proceedings of the Institution of Mechanical Engineers, Part C: Journal of Mechanical Engineering Science,* Vol: 217, No: 1, pp. 31–39.
8. Hubka, V. 1982. *Bewertein und Entscheiden beim Konstruieren",* Konstruktionsmethoden-Reading, heurista WDK-8, pp. 164.
9. Knight, W and Poli, C. 1958. A Systematic Approach to Forging Design, *Machine Design,* pp. 94–99.
10. Mahmoud, MAM and Pugh, S. 1979. The Costing of turned components at the design stage, *Information for Designers. Fourth International Symposium,* University of Southampton, pp. 37–42.
11. Morrow, W, Qi, H, Kim, I, Mazumder, J and Skerlos, S. 2007. Environmental Aspects of Laser based and Conventional Tool and Die Manufacturing, *Journal of Cleaner Production,* Vol: 15, No: 10, pp. 932–943.
12. Ostwald, P. 1982. *American Machinist Manufacturing Cost Estimating Guide,* McGraw-Hill, New York, pp. 382.
13. Peters, MS and Timmerhaus, KD. 1980. *Plant Design and Economics for Chemical Engineers,* Third Edition, McGraw-Hill, New York.
14. Ruffo, M and Hague, R. 2007. Cost estimation for Rapid Manufacturing—Simultaneous Production of Mixed Components using Laser Sintering, *Proceedings of the Institution of Mechanical Engineers, Part B: Journal of Engineering Manufacture,* Vol: 221, No: 11, pp. 1585–1591.
15. Ruffo, M, Tuck, C and hague, R. 2006. Cost Estimation for Rapid Manufacturing-Laser Sintering Production for Low to Medium Volumes, *Proceedings of the Institution of Mechanical Engineers, Part B: Journal of Engineering Manufacture,* Vol: 220, No: 9, pp. 1417–1427.
16. Schey, J. 1987. *Introduction to Manufacturing Processes,* McGraw-Hill, New York, pp. 505.
17. Shaw, M. 1984. *Metal Cutting Principles,* Oxford University Press, England, New York p. 474.
18. Wierda, LF. 1988. *Detailed Cost Estimation Concept and Database Structure for DIDACOE,* Report: IM K 177, Delft University of Technology, The Netherlands, p. 122.

Index

D

E

G

H

L

M

U

V

W

Y